D0421513

SHEFFIELD HALLAM UNIVERSITY
LEARNING CENTRE
WITHDRAWN FROM STOCK

Introduction to

Engineering Mechanics

A Continuum Approach

Introduction to

Engineering Mechanics

A Continuum Approach

Jenn Stroud Rossmann
Lafayette College
Easton, Pennsylvania, USA

Clive L. Dym
Harvey Mudd College
Claremont, California, USA

CRC Press
Taylor & Francis Group
Boca Raton London New York

CRC Press is an imprint of the
Taylor & Francis Group, an **informa** business

CRC Press
Taylor & Francis Group
6000 Broken Sound Parkway NW, Suite 300
Boca Raton, FL 33487-2742

Printed in the United States of America on acid-free paper
10 9 8 7 6 5 4 3 2 1

International Standard Book Number-13: 978-1-4200-6271-7 (Hardcover)

Library of Congress Cataloging-in-Publication Data

Rossman, Jenn Stroud.
Introduction to engineering mechanics: A continuum approach / Jenn Stroud Rossman, Clive L. Dym.
p. cm.
Includes bibliographical references and index.
ISBN 978-1-4200-6271-7 (alk. paper)
1. Mechanics, Applied. I. Dym, Clive L. II. Title.

TA350.B348 1986
620.1--dc22 2008033432

Visit the Taylor & Francis Web site at
http://www.taylorandfrancis.com

and the CRC Press Web site at
http://www.crcpress.com

Contents

If science teaches us anything, it's to accept our failures, as well as our successes, with quiet dignity and grace.

Gene Wilder, *Young Frankenstein,* **1974**

Preface

This book is intended to provide a unified introduction to solid and fluid mechanics and to convey the underlying principles of continuum mechanics to undergraduates. We assume that students using this book have taken courses in calculus, physics, and vector analysis. By demonstrating both the connections and the distinctions between solid and fluid mechanics, this book will prepare students for further study in either field or in fields such as bioengineering that blur traditional disciplinary boundaries.

The use of a continuum approach to make connections between solid and fluid mechanics is a perspective typically provided only to advanced undergraduates and graduate students. This book *introduces* the concepts of stress and strain in the continuum context, showing the relationships between solid and fluid behavior and the mathematics that describe them. It is an introductory textbook in strength of materials and in fluid mechanics and also includes the mathematical connective tissue between these fields. We have decided to begin with the *a-ha!* of continuum mechanics rather than requiring students to wait for it.

This approach was first developed at Harvey Mudd College (HMC) for a sophomore-level course called "Continuum Mechanics." The broad, unspecialized engineering program at HMC requires that curriculum planners ask themselves, "What specific knowledge is essential for an engineer who may practice, or continue study, in one of a wide variety of fields?" This course was our answer to the question, what *engineering mechanics* knowledge is essential?

An engineer of any type, we felt, should have an understanding of how materials respond to loading: how solids deform and incur stress; how fluids flow. We conceived of a spectrum of material behavior, with the idealizations of Hookean solids and Newtonian fluids at the extremes. Most modern engineering materials—biological materials, for example—lie between these two extremes, and we believe that students who are aware of the entire spectrum from their first introduction to engineering mechanics will be well prepared to understand this complex middle ground of nonlinearity and viscoelasticity.

Our integrated introduction to the mechanics of solids and fluids has evolved. As initially taught by CLD, the HMC course emphasized the underlying principles from a mathematical, applied mechanics viewpoint. This focus on the structure of elasticity problems made it difficult for students to relate formulation to applications. In subsequent offerings, JSR chose to embed continuum concepts and mathematics into introductory problems, and to build gradually to the strain and stress tensors. We now establish a "continuum checklist"—compatibility [deformation], constitutive law, and equilibrium—that we return to repeatedly. This checklist provides a framework for a wide variety of problems in solid and fluid mechanics.

We make the necessary definitions and present the template for our continuum approach in Chapter 1. In Chapter 2, we introduce strain and stress in one dimension, develop a constitutive law, and apply these concepts to the simple case of an axially loaded bar. In Chapter 3, we extend these concepts to higher dimensions, introducing Poisson's ratio and the strain and stress tensors. In Chapters 4–7 we apply our continuum sense of solid mechanics to problems including torsion, pressure vessels, beams, and columns. In Chapter 8, we make connections between solid and fluid mechanics, introducing properties of fluids and the strain *rate* tensor. Chapter 9 addresses fluid statics. Applications in fluid mechanics are considered in Chapters 10 and 11. We develop the governing equations in both control volume and differential forms. In Chapter 12, we see that the equations for solid *dynamics* strongly resemble those we've used to study fluid dynamics. Throughout, we emphasize real-world design applications. We maintain a continuum "big picture" approach, tempered with worked examples, problems, and a set of case studies.

The six case studies included in this book illustrate important applications of the concepts. In some cases, students' developing understanding of solid and fluid mechanics will help them understand "what went wrong" in famous failures; in others, students will see how the textbook theories can be extended and applied in other fields such as bioengineering. The essence of continuum mechanics, the internal response of materials to external loading, is often obscured by the complex mathematics of its formulation. By building gradually from one-dimensional to two- and three-dimensional formulations and by including these illustrative real-world case studies, we hope to help students develop physical intuition for solid and fluid behavior.

We've written this book for our students, and we hope that reading it is very much like sitting in our classes. We have tried to keep the tone conversational and have included many asides that describe the historical context for the ideas we describe and hints at how some concepts may become even more useful later on.

We are grateful to the students who have helped us refine our approach. We are deeply appreciative of our colleague and friend Lori Bassman (HMC)—of her sense of pure joy in structural mechanics and her ability to communicate that joy. Lori has been a sounding board, contributor of elegant (and fun) homework problems, and defender of the integrity of "second moment of area" despite the authors' stubbornly abiding affection for "moment of inertia." We also thank Joseph A. King (HMC), Harry E. Williams (HMC), Josh Smith (Lafayette), James Ferri (Lafayette), Diane Windham Shaw (Lafayette), Brian Storey (Olin), Borjana Mikic (Smith), and Drew Guswa (Smith). We thank Michael Slaughter and Jonathan Plant, our editors at Taylor & Francis/CRC, and their staff.

We want to convey our warmest gratitude to our families. First are Toby, Leda, and Cleo Rossmann. Thanks especially to Toby, for his direct and indirect support of this project. And then there's Joan Dym, Jordana, and Miriam, and Matt and Ryan and spouses and partners, and a growing number of grandchildren. We are grateful for their support, love, and patience.

About the Authors

Jenn Stroud Rossmann is assistant professor of mechanical engineering at Lafayette College. She earned her B.S. and Ph.D. degrees from the University of California, Berkeley. Her current research includes the study of blood flow in vessels affected by atherosclerosis and aneurysms. She has a strong commitment to teaching engineering methods and literacy to non-engineers and has developed several courses and workshops for liberal arts majors.

Clive L. Dym is the Fletcher Jones Professor of Engineering Design at Harvey Mudd College. He earned his B.S. from Cooper Union and his Ph.D. from Stanford University. His primary interests are in engineering design and structural mechanics. He is the author of eleven books and has edited nine others; his two most recent books are *Engineering Design: A Project-Based Introduction,* 3rd ed. (with Patrick Little, and with Elizabeth J. Orwin and R. Erik Spjut, John Wiley, 2008) and *Principles of Mathematical Modeling,* 2nd ed. (Academic Press, 2004). Among his awards are the Fred Merryfield Design Award (American Society for Engineering Education [ASEE], 2002) and the Joel and Ruth Spira Outstanding Design Educator Award (American Society of Mechanical Engineers [ASME], 2004). Dr. Dym is a fellow of the ASCE, ASME, and ASEE.

1

Introduction

This textbook, *Introduction to Engineering Mechanics: A Continuum Approach,* is intended to demonstrate the connections between solid and fluid mechanics, and the larger mathematical concepts shared by both fields, while introducing the fundamentals of both solid and fluid engineering mechanics.

Mechanics is the study of the motion or equilibrium of matter and the forces that cause such motion or equilibrium. The reader is likely already familiar with the sort of "billiard ball" mechanics formulated in physics courses—for example, when two such billiard balls collide, applying Newton's second law will help us learn the velocities of both balls after the collision. *Engineering mechanics* mandates that we also consider how the impact will affect the balls: Will they deform or even crack? How many such collisions can they sustain? How does the material chosen for their construction affect both these answers? What design decisions will optimize the strength, cost, or other properties of the balls? Taking a *continuum* approach to engineering mechanics means, essentially, that we will consider what's going on inside the billiard balls and will quantify the *internal response* to external loading.

This book provides an introduction to the mechanics of both solids and fluids and emphasizes both distinctions and connections between these fields. We will see that the material behaviors of ideal solids and fluids are at the far ends of a *spectrum* of material behavior and that many materials of interest to modern engineers—particularly biomaterials—lie between these two extremes, combining elements of both "solid" and "fluid" behavior.

Our objectives are to learn how to formulate problems in mechanics and how to reduce vague questions and ideas into precise mathematical statements. The floor of a building may be strong enough to support us, our furniture, and even the occasional fatiguing dance party, but if not designed carefully, the floor may deflect considerably and sag. By learning how to predict the effects of forces, stresses, and strains, we will become better designers and better engineers.

1.1 A Motivating Example: Remodeling an Underwater Structure

Underwater rigs like that shown in Figure 1.1 are commonly used by the petroleum industry to harvest offshore oil. Over the life of a structure, many sea creatures and plants attach themselves to the supports. When wells have dried up, the underwater structures can be removed in manageable segments and towed to shore. However, this process results in the loss of both the reef dwellers attached to the platform's trusses and the larger fish who feed there. Corporations often abandon their rigs rather than incurring the financial and environmental expense of removal. An engineering firm would like to make use of a decommissioned rig by remodeling it as an artificial reef, providing a hospitable sea habitat. This firm must find ways to strengthen the supports and to affix the reef components to sustain sea life.

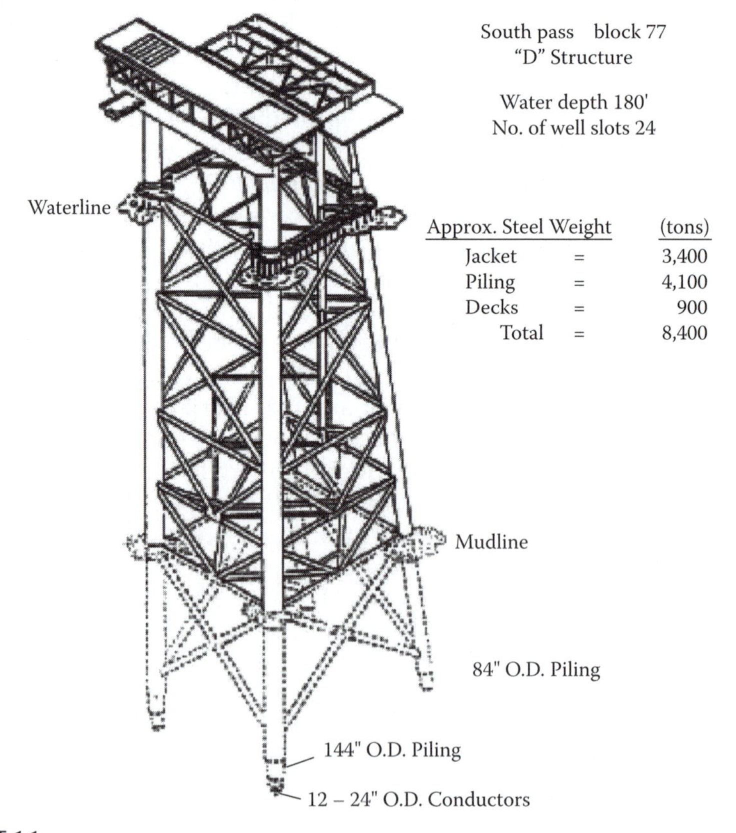

FIGURE 1.1
Mud-slide-type platform. (From the Committee on Techniques for Removing Fixed Offshore Structures and the Marine Board Commission on Engineering and Technical Systems, National Research Council, An Assessment of Techniques for Removing Offshore Structures, Washington, DC: National Academy Press, 1996. With permission.)

The rig support structure was initially designed to support the drilling platform above the water level. As the oil drill itself was mobile, the structure was built so that it could remain balanced, without listing, under this dynamic loading. In its new life as the support for an artificial reef, this structure must continue to withstand the weight of the platform and the changing loads of wind and sea currents, and it must also support the additional loading of concrete "reef balls" and other reef-mimicking assemblies (Figure 1.2), as well as the weight of the reef dwellers.

FIGURE 1.2
Concrete reef ball. (Courtesy of the Reef Ball Foundation, Athens, GA.)

To remodel the underwater rig, a team of engineers must dive below the water surface to attach the necessary reef balls and other attachments. The reef balls themselves may be lowered using a crane. A conceptualization of this is shown in Figure 1.3.

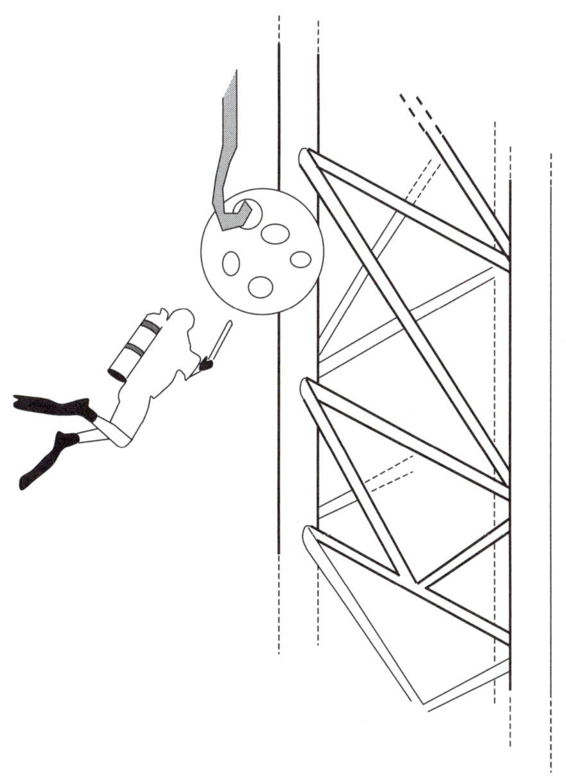

FIGURE 1.3
Rendering of scuba diver at work remodeling underwater rig structure.

Among the factors that must be considered in the redesign process is the structural performance of the modified structure, its ability to withstand the required loading. An additional challenge to the engineering firm is the undersea location of the structure. What materials should be chosen so that the structure remains sound? How should the additional supports and reef assemblies be added? What precautions must engineers and fabricators take when they work underwater? What effects will the exposure to the ocean environment have on their structure, equipment, and bodies? We address many of these issues in this book. Throughout, we return to this problem to demonstrate the utility of various theoretical results, and we rely on first principles that look familiar.

1.2 Newton's Laws: The First Principles of Mechanics

Newton's laws provide us with the *first principles* that, along with conservation equations, guide the work we do in continuum mechanics. Many of the equations we use in problem solving are directly descended from these elegant statements. These laws were formulated by Sir Isaac Newton (1642–1727), based on his own experimental work and on the observations of others, including Galileo Galilei (1564–1642). Newton's laws are expressed as follows:

> Newton's first law: A body remains at rest or moves in a straight line with constant velocity if there is no unbalanced force acting on it.
>
> Newton's second law: The time rate of change of momentum of a body is equal to (and in the same direction as) the resultant of the forces acting on it:
>
> $$\sum \underline{F} = \frac{d}{dt}(m\ \underline{V}). \tag{1.1}$$
>
> When the mass of the body of interest is constant, this has the form
>
> $$\Sigma\, \underline{F} = m\ \underline{a}, \tag{1.2}$$
>
> and when $\underline{a} = \underline{0}$, this means that we have
>
> $$\Sigma\, F = 0. \tag{1.3}$$
>
> (This last class of problems is often called "statics.")
>
> Newton's third law: To every action there is an equal and opposite reaction. That is, the forces of action and reaction between interacting bodies are equal in magnitude and exactly opposite in direction.

Forces always occur, according to Newton's third law, in pairs of equal and opposite forces. The downward force exerted on the desk by your pencil is accompanied by an upward force of equal magnitude exerted on your pencil by the desk.

1.3 Equilibrium

We have alluded to the concept of equilibrium (also known as *static* equilibrium) in our discussion of Newton's second law. To be in equilibrium, a three-dimensional object must satisfy six equations. In Cartesian coordinates, these are as follows:

$$\sum F_x = 0$$
$$\sum F_y = 0$$
$$\sum F_z = 0 \tag{1.4a}$$

$$\sum M_x = 0$$
$$\sum M_y = 0$$
$$\sum M_z = 0 \tag{1.4b}$$

These equations can be written more concisely in vector form as

$$\Sigma \underline{F} = \underline{0} \tag{1.5}$$

$$\Sigma \underline{M} = \underline{0}, \tag{1.6}$$

and represent the statements "the sum of forces equals zero" and "the sum of moments (about some reference axis) equals zero." One advantage of writing these equations in vector form is that we don't have to specify a coordinate system!

For planar (two-dimensional) situations or models, equilibrium requires the satisfaction of only three equations, usually

$$\sum F_x = 0 \tag{1.7a}$$

$$\sum F_y = 0 \tag{1.7b}$$

$$\sum F_z = 0 \tag{1.7c}$$

These equations essentially state that the object is neither translating (in the x or y directions) nor rotating (about the z axis) in the xy plane as a result of applied forces.

It is useful to distinguish between forces that act externally and those that act internally. *External* loads are applied to a structure by, for example, gravity or wind. Reaction forces are also external: They occur at supports and at points where the structure is prevented from moving in response to the external loads. These supports may be surfaces, rollers, hinges; fixed or free. *Internal* forces, on the other hand, *result* from the applied external loads and are what we are concerned with when we study continuum mechanics. These are forces that act within a body as a result of all external forces. Chapter 2 shows how the principle of equilibrium helps us calculate these internal forces.

1.4 Definition of a Continuum

In elementary physics, we concerned ourselves with particles and bodies that behaved like inert billiard balls, bouncing off each other and interacting without deformation or other changes. In continuum mechanics, we consider the effects of deformation, of internal forces within bodies, to get a fuller sense of how bodies react to external forces.

We would like to be able to consider these bodies as whole entities and not have to account for each individual particle composing each body. It would be much more convenient for us to treat the properties (e.g., density, momentum, forces) of such bodies as continuous functions. We may do this if the body in question is a *continuum*.

We may treat a body as a continuum if the ensemble of particles making up the body acts like a continuum. We can then consider the average or "bulk" properties of the body and can neglect the details of the individual particle dynamics. Acting like a continuum means that no matter how small a chunk of the body we consider, the chunk will have the same properties (e.g., density) as the bulk material.

Mathematically, we define a continuum as a continuous distribution of matter in space and time. For a mass m_n contained in a small volume of space, V_n, surrounding a point P, as in Figure 1.4, we can define a mass density ρ:

$$\rho(P) = \lim_{\substack{n\to\infty \\ V_n\to 0}} \frac{m_n}{V_n}. \tag{1.8}$$

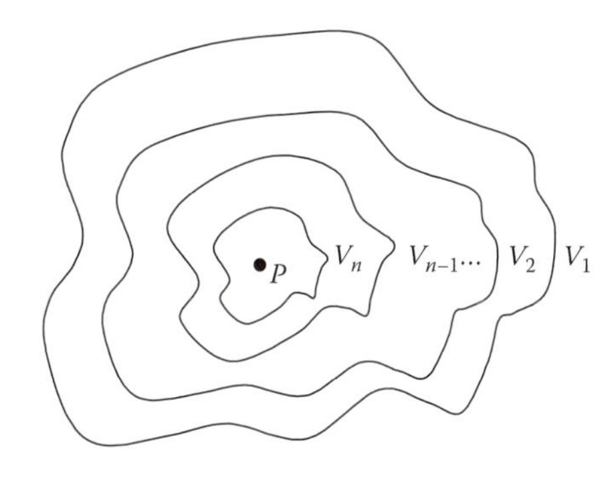

FIGURE 1.4
Volumes V_i surrounding point *P*.

So, a material continuum is a material for which density (of mass, momentum, or energy) exists in a mathematical sense. We are able to define its properties as continuous functions and to neglect what's happening on the microscopic, molecular level in favor of the macroscopic, bulk behaviors.

Note that if V_n truly goes to zero, gases and liquids will not satisfy this equation: Density will be undefined. (If the volume goes to zero, it will not have a chance to enclose any atoms—so naturally, the density will be undefined!) Yet we still think of these materials as continua. So physically, our definition of a continuum is a material for which

$$\left| \rho - \frac{m_n}{V_n} \right| < \varepsilon \text{ as } n \to \infty. \tag{1.9}$$

Here, ϵ represents a *very* small number approaching zero, indicating that the mathematical definition of density approaches a usable value, ρ.

Sometimes it is easier to get a grasp on what is not a continuum than on what is. Almost all solids satisfy the definition handily. Solids are generally much denser than fluids. For fluids, it can be harder to pin down a "density" once gas molecules get sparse. Interstellar space, for example, where the objects of interest (e.g., planets, asteroids) are not much farther apart than the molecules of the interstellar medium, is surely stretching the limits of the definition of a continuum. Fortunately, another test for continuity is available. It's especially applicable to fluids.

A given material may be called a *continuum* if the Knudsen[1] number, *Kn,* is less than about 0.1. The Knudsen number is defined as

$$Kn = \frac{\lambda}{L}, \tag{1.10}$$

where L is a problem-specific characteristic length, such as a diameter or width, and λ is the material's "mean free path," or average distance between particle collisions, obtainable from

$$\lambda = 0.225 \frac{m}{\rho d^2}, \tag{1.11}$$

where m is the mass of a molecule, ρ is its density, and d is the diameter of a molecule. For example, for air $m = 4.8 \times 10^{-26}$ kg, $d = 3.7 \times 10^{-10}$ m, and at atmospheric conditions λ is approximately 6×10^{-6} cm; at an altitude of 100 km it is 10 cm, and at 160 km it's 5000 cm. So at higher altitudes, the continuum assumption is unacceptable and the molecular dynamics must be considered in the governing equations.

The ease with which we can define density, and continuity, is not the only difference between solids and fluids:

> A *solid* is a three-dimensional continuum that supports both tensile and shear forces and stresses. The atoms making up a solid have a fixed spatial arrangement—often a crystal lattice structure—in which atoms are able to vibrate and spin and their electrons can fly and dance around but the microstructure is fixed. Because of this, although it's possible to distort or destroy the shape taken by a solid, it is generally said that a solid object retains its own shape. For solids, we will be able to relate *stresses* and *strains* by a *constitutive law.*

> A *fluid* may be a liquid or a gas. A fluid, it's been said, is something that flows: Liquids assume the shape of their containers, and gases expand to fill any container. This is because the atoms comprising a fluid are not spatially constrained like those of a solid. More formally, a fluid is a three-dimensional continuum that (a) cannot support tensile forces or stresses, and (b) deforms continuously under the smallest shearing forces or stresses. For fluids, we will be able to relate stresses and strain rates by a constitutive law.

We note that the distinction between solid and fluid behavior is not always clear-cut; there are classes of materials whose behavior situates them in a sort of middle ground. We explore this middle ground further in Case Study 5. The existence of this middle ground provides us with more motivation to understand the broad field of continuum mechanics and the connections between solid and fluid behavior.

In this text we are interested in how Newton's laws apply to continua. Some of the relevant consequences of Newton's laws, which we discuss in more detail later, are as follows:

- Momentum is always conserved, in both solids and fluids. *Equilibrium* equations (see Section 1.3) are the mathematical expressions of the conservation of momentum.
- Equilibrium must apply both to entire bodies and to sections of, or particles within, those bodies. This is one of the reasons why free-body diagrams (FBDs) are so valuable: They illustrate the equilibrium of a section of a larger body or system. This is also why we use *control volumes* to analyze fluid flows.
- Mass is conserved.
- Area is a vector, having both magnitude (size) and direction, which is defined by a unit vector normal to the area and directed outward from the free body or volume of interest.
- Forces produce changes in shape and geometry, which are characterized in terms of *strains* for solids and *strain rates* for fluids.

In the real world, material objects are subjected to *body* forces (e.g., gravitational and electromagnetic forces), which do not require direct contact, and *surface* forces (e.g., atmospheric pressure, wind and rain, burdens to be carried), which do. We want to know how the material in the body reacts to external forces. To do this, we will need to (1) characterize the deformation of a continuous material, (2) define the *internal* loading, (3) relate this to the body's deformation, and (4) make sure that the body is in equilibrium. This is what continuum mechanics is all about.

1.5 Mathematical Basics: Scalars and Vectors

The familiar distinction between scalars and vectors is that a vector, unlike a scalar, has direction as well as magnitude. Examples of scalar quantities are time, volume, density, speed, energy, and mass. Velocity, acceleration, force, and momentum are vectors and contain the extra directional information. We typically denote vectors with a bold font or an underline. This book underlines all vectors.

A vector $\underline{V}$ may be expressed mathematically by multiplying its magnitude, V, by a unit vector $\underline{n}$ (note: $|\underline{n}| = 1$, and $\underline{n}$'s direction coincides with $\underline{V}$):

$$\underline{V} = V\underline{n}. \tag{1.12}$$

We may also write a vector $\underline{V}$ in terms of its *components* along the primary directions, whether these are the Cartesian (x, y, z) directions or cylindrical (r, θ, z) or another set. In Cartesian coordinates this is simply written as

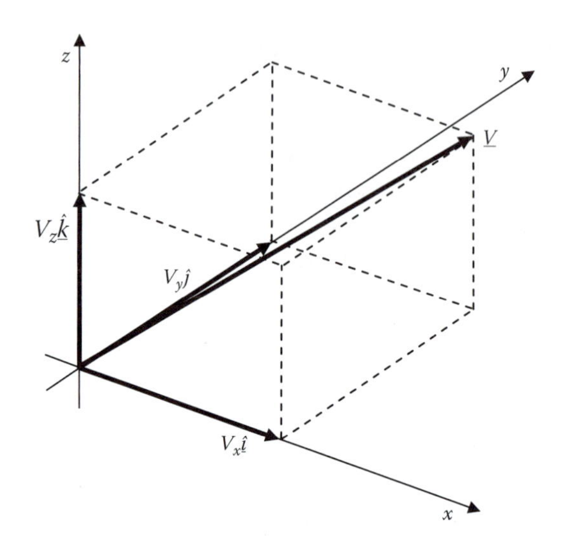

FIGURE 1.5
Decomposition of vector $\underline{V}$ in x, y, z coordinates.

$$\underline{V} = V_x\hat{i} + V_y\hat{j} + V_z\hat{k} \tag{1.13}$$

based on a situation like that shown in Figure 1.5. In general, in coordinates (x_1, x_2, x_3) with unit vectors $\hat{e}_1, \hat{e}_2, \hat{e}_3$, we will be able to write any vector $\underline{V}$ as

$$\underline{V} = V_1\hat{e}_1 + V_2\hat{e}_2 + V_3\hat{e}_3 \tag{1.14}$$

or as (V_1, V_2, V_3)—what we called a column vector in linear algebra.[2] We remember that the magnitude of $\underline{V}$ can be obtained:

$$V = |\,\underline{V}\,| = \sqrt{(V_1^2 + V_2^2 + V_3^2)}, \tag{1.15}$$

so $\underline{V} = \underline{0}$ if, and only if, $V_1 = V_2 = V_3 = 0$.

The calculated dot and cross products are also of interest. Remember that the result of taking a dot product is a *scalar* and that the result of a cross product is a *vector*. Briefly,

$$\underline{u} \bullet \underline{v} = |\,\underline{u}\,|\;|\,\underline{v}\,| \cos\theta, \tag{1.16}$$

where θ is the angle between vectors $\underline{u}$ and $\underline{v}$, and $0 \le \theta \le \pi$. Physically, the scalar or *dot product* can be thought of as the magnitude of $\underline{u}$ times the component of $\underline{v}$ along $\underline{u}$. In terms of components,

$$\underline{u} = u_1 \hat{e}_1 + u_2 \hat{e}_2 + u_3 \hat{e}_3 , \tag{1.17}$$

$$\underline{v} = v_1 \hat{e}_1 + v_2 \hat{e}_2 + v_3 \hat{e}_3 , \tag{1.18}$$

$$\underline{u} \bullet \underline{v} = u_1 v_1 + u_2 v_2 + u_3 v_3 . \tag{1.19}$$

Also, the cross product results in a vector that is perpendicular to both $\underline{u}$ and $\underline{v}$:

$$\underline{u} \times \underline{v} = \underline{w}, \tag{1.20}$$

where

$$| \underline{w} | = | \underline{u} | \; | \underline{v} | \sin \theta, \tag{1.21}$$

and

$$\underline{u} \times \underline{v} = (u_2 v_3 - u_3 v_2) \; \hat{e}_1 + (u_3 v_1 - u_1 v_3) \; \hat{e}_2 + (u_1 v_2 - u_2 v_1) \; \hat{e}_3 . \tag{1.22}$$

We notice that this has the form of a determinant:

$$\underline{u} \times \underline{v} = \begin{vmatrix} \hat{e}_1 & \hat{e}_2 & \hat{e}_3 \\ u_1 & u_2 & u_3 \\ v_1 & v_2 & v_3 \end{vmatrix} . \tag{1.23}$$

When we work with vectors, we may find ourselves getting stuck carrying around a set of variables, x_1, x_2, … x_n. This can become unwieldy, and so we may use a shortcut known as *index notation*. Using this shortcut, we write x_i, $i = 1, 2, \ldots n$, and call i the *index*. If, for example, we are working with the equation

$$a_1 x_1 + a_2 x_2 + a_3 x_3 = p, \tag{1.24}$$

we may write this as

$$\sum_{i=1}^{3} a_i x_i = p \tag{1.25}$$

and may further simplify life by writing

$$a_i x_i = p. \tag{1.26}$$

This substantially more efficient shortcut is known as the *summation convention*: The repetition of the index represents summation with respect to that index over its range. Using index notation and the summation convention, we could rewrite the definition of dot product (1.19) as

$$\underline{u} \cdot \underline{v} = u_i v_i \ . \tag{1.27}$$

We understand scalars to contain the least possible amount of information—only a magnitude—while a vector contains more information and can be manipulated in more ways. The curious student may be wondering whether there is any type of variable that can contain more information than a vector. That provocative question is answered in Chapter 3.

1.6 Problem Solving

Any reader of your solution to a given problem should be able to follow the reasoning behind it. To test yourself you may find a stranger on the street and ask whether your logic is clear, or you may simply make sure that you have included each of the following steps:

1. State what is given: The speed of major league fastball and distance from pitcher's mound to home plate, 60 feet 6 inches are given.
2. State what is sought: Find the time a batter has to react to an incoming pitch.
3. Draw relevant sketches or pictures: In particular, isolate the body (or relevant control volume) to see the forces involved, by means of a free-body diagram.
4. Identify the governing principles (e.g., Newton's second law).
5. Calculations: Keep in symbolic form (e.g., $v = d/t$).
6. Check the physical dimensions of your answer: Will answer have dimensions of time? If it looks like it will be a length, go back.
7. Complete calculations: Substitute in numbers; wait as long as possible before plugging in numbers. This gives you time to do a dimensional check and to think about whether the dependencies you've found make sense (should the answer depend on the pitcher's wingspan?) and allows you to reuse the model for similar problems that may arise.
8. State answers and conclusions.

In the worked example problems that follow each chapter in this textbook, these steps are followed.

1.7 Examples

Example 1.1

A force $\underline{F}$ with magnitude 100 N passes through the points (1, 2, 1) and (3, –2, 2) (pointing toward (3, –2, 2)) where coordinates are in meters. Determine the following:

(a) The magnitudes of the x, y, and z scalar components of $\underline{F}$
(b) The moment of $\underline{F}$ about the origin
(c) The moment of $\underline{F}$ about the point (2, 0.3, 1)

Given: Force vector.

Find: Components of vector and moment of vector about two points.

Assume: No assumptions are necessary.

Solution

We can obtain a solution using either a holistic "vector approach" or a piece-by-piece "component approach." We will demonstrate both approaches.

Vector Approach

(a) The force can be written as $\underline{F} = F\,\underline{n}$ where $\underline{n}$ is the unit vector in the direction of the force:

$$\underline{n} = \frac{2\hat{\underline{i}} - 4\hat{\underline{j}} + 1\hat{\underline{k}}}{\sqrt{2^2 + (-4)^2 + 1^2}} = 0.436\hat{\underline{i}} - 0.873\hat{\underline{j}} + 0.218\hat{\underline{k}}$$

$$\underline{F} = 100\,\underline{n} = 43.6\,\hat{\underline{i}} - 87.3\,\hat{\underline{j}} + 21.8\,\hat{\underline{k}}\ \text{N}$$

so the scalar components of $\underline{F}$ are $Fx = 43.6$ N, $Fy = -87.3$ N, and $Fz = 21.8$ N.

(b) The moment of F about the origin is found using $\underline{Mo} = \underline{r} \times \underline{F}$, where $\underline{r}$ is a vector from the origin to any point on the line of action of $\underline{F}$.

Using $\underline{r} = 1\,\hat{\underline{i}} + 2\,\hat{\underline{j}} + 1\,\hat{\underline{k}}$, $\underline{r} \times \underline{F}$ may be written as a determinant:

$$\underline{M}_o = \begin{vmatrix} \hat{\underline{i}} & \hat{\underline{j}} & \hat{\underline{k}} \\ r_x & r_y & r_z \\ F_x & F_y & F_z \end{vmatrix} = \begin{vmatrix} \hat{\underline{i}} & \hat{\underline{j}} & \hat{\underline{k}} \\ 1 & 2 & 1 \\ 43.6 & -87.3 & 21.8 \end{vmatrix}$$

$$= [2\,(21.8) - 1\,(-87.3)]\,\hat{\underline{i}} - [1\,(21.8) - 1\,(43.6)]\,\hat{\underline{j}} + [1\,(-87.3) - 2\,(43.6)]\,\hat{\underline{k}}$$

$$\underline{M}_o = 130.9\,\hat{\underline{i}} + 21.8\,\hat{\underline{j}} - 174.5\,\hat{\underline{k}} \text{ N·m.}$$

(c) A vector $\underline{r}$ is needed from the point P (2, 0.3, 1) to any point on the line of action of $\underline{F}$. We see that $\underline{r} = -1\,\hat{\underline{i}} + 1.7\,\hat{\underline{j}} + 0\,\hat{\underline{k}}$ is such a vector (goes to the point (1, 2, 1)). Then $\underline{M}_p = \underline{r} \times \underline{F}$:

$$\underline{M}_p = \begin{vmatrix} \hat{\underline{i}} & \hat{\underline{j}} & \hat{\underline{k}} \\ r_x & r_y & r_z \\ F_x & F_y & F_z \end{vmatrix} = \begin{vmatrix} \hat{\underline{i}} & \hat{\underline{j}} & \hat{\underline{k}} \\ -1 & 1.7 & 0 \\ 43.6 & -87.3 & 21.8 \end{vmatrix}$$

$$= [1.7\,(21.8) - 0]\,\hat{\underline{i}} - [-1\,(21.8) - 0]\,\hat{\underline{j}} + [-1\,(-87.3) - 1.7\,(43.6)]\,\hat{\underline{k}}$$

$$\underline{M}_p = 37.1\,\hat{\underline{i}} + 21.8\,\hat{\underline{j}} + 13.2\,\hat{\underline{k}} \text{ N·m.}$$

Scalar (Components) Approach

(a) The length of the segment from (1, 2, 1) to (3, –2, 2) is

$$\sqrt{(3-1)^2 + (-2-2)^2 + (2-1)^2} = \sqrt{22 + (-4)2 + 12} = \sqrt{21}$$

v

Direction Cosines	Then
$l = 2/\sqrt{21} = 0.436$	$F_x = 100\ (0.436) = 43.6$ N
$m = -4/\sqrt{21} = -0.873$	$F_y = 100\ (-0.873) = -87.3$ N
$n = 1/\sqrt{21} = 0.218$	$F_z = 100\ (0.218) = 21.8$ N

(b) Remember that we can consider the force $\underline{F}$ to be acting at any point along its line of action. Choosing (1, 2, 1), the moments about the x, y, and z axes through the origin are

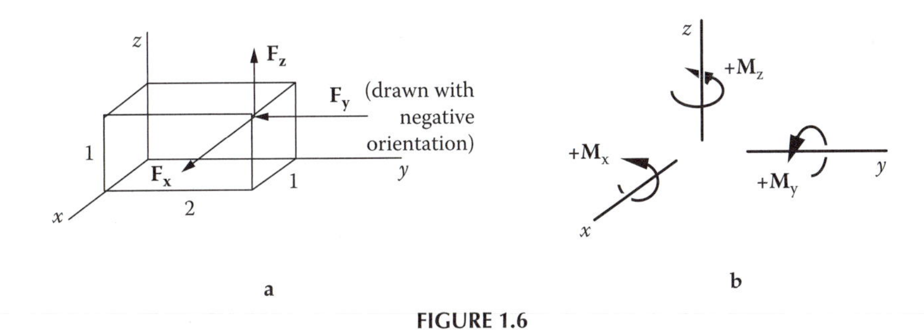

FIGURE 1.6

$$M_{ox} = 1\,(87.3) + 2\,(21.8) = 130.9 \text{ N·m}.$$

(F_x is parallel to the x axis and thus does not have a moment about the x axis.)

$$M_{oy} = 1\,(43.6) - 1\,(21.8) = 21.8 \text{ N·m}$$

$$M_{oz} = -2\,(43.6) - 1\,(87.3) = -174.5 \text{ N·m}.$$

(c) Use the same procedure as part (b). In this case, the distances required are from the point of action of the force (choose (1, 2, 1) as previously) to the point P (2, 0.3, 1):

$$M_{px} = (1 - 1)\,(87.3) + (2 - 0.3)\,(21.8) = 37.1 \text{ N·m},$$

$$M_{py} = (1 - 1)\,(43.6) + (2 - 1)\,(21.8) = 21.8 \text{ N·m},$$

$$M_{pz} = -\,(2 - 0.3)\,(43.6) + (2 - 1)\,(87.3) = 13.2 \text{ N·m}.$$

Example 1.2

A clever sophomore wants to weigh himself but has access only to a scale (A) with capacity limited to 500 N and a small 80 N spring dynamometer

(B). With the rig shown he discovers that when he exerts a pull on the rope so that B registers 76 N, the scale reads 454 N. What are his correct weight and mass?

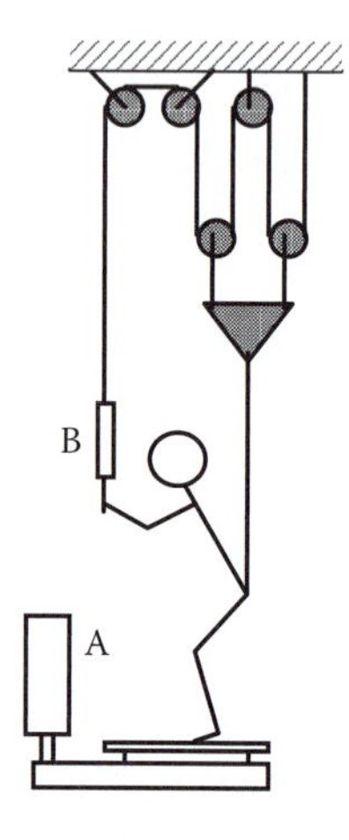

FIGURE 1.7

Given: Geometry of problem, weight indicated on scale A.
Find: True weight and mass of student.
Assume: No assumptions are necessary.

Solution

We assume the tension in the continuous top rope is constant, and we'll neglect the mass of the pulleys. The relevant free-body diagrams are (the circles are the lower pulleys):

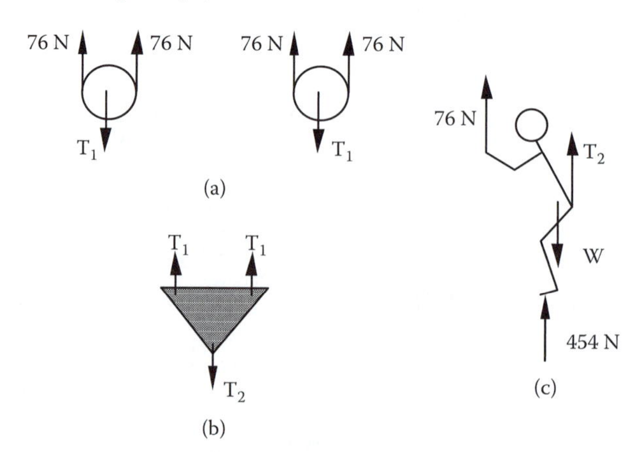

FIGURE 1.8

Next, we ensure that $\Sigma Fy = 0$ holds for each FBD—that is, that each part is in equilibrium.

From diagram (a),

$$T_1 = 76 \text{ N} + 76 \text{ N} = 152 \text{ N}.$$

From diagram (b),

$$T_2 = T_1 + T_1 = 304 \text{ N}.$$

From diagram (c),

$$W = 454 \text{ N} + 76 \text{ N} + T_2 = 834 \text{ N}.$$

So, his mass is

$$\frac{834 \text{ N}}{9.81 \text{ m/s}^2} = 85.0 \text{ kg}.$$

1.8 Problems

1.1 The premixed concrete in a cement truck can be treated as a fluid continuum when it is poured into a mold. Sand flowing from a large bucket can also be considered a fluid. Describe three other examples in which an aggregate of solid objects flows likes a fluid continuum.

1.2 Investigate the reef balls used in creating artificial reef environments. What parameters are most important to the successful maintenance of a stable marine environment?

1.3 Find the angle θ between the two vectors $\underline{F}_1 = 4\hat{i} + 3\hat{k}$ and $\underline{F}_2 = \hat{i} + 7\hat{k}$ using their dot product.

1.4 Find and sketch the cross product $\underline{F}_1 \times \underline{F}_2$, given $\underline{F}_1 = -5\hat{i} + 3\hat{k}$ and $\underline{F}_2 = \hat{i} - 4\hat{k}$.

1.5 Determine the force F and the angle θ required to keep the pulley system shown in static equilibrium.

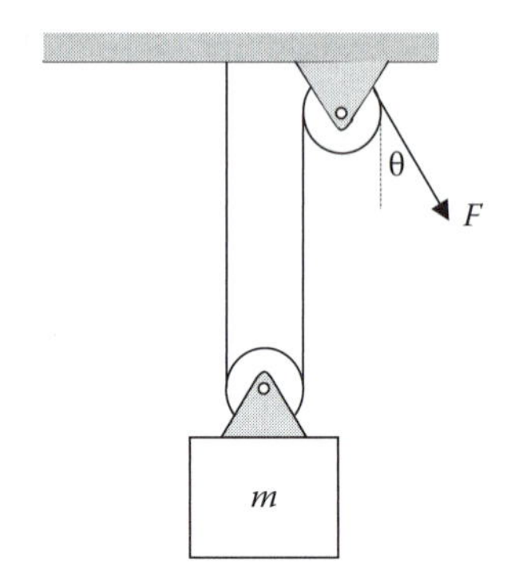

FIGURE 1.9

1.6 A force F acts on a uniform pendulum as shown. Find the reaction forces at the pin connection and the angle θ, letting $F = 100$ N, $d = 1.6$ m, and $W = 300$ N.

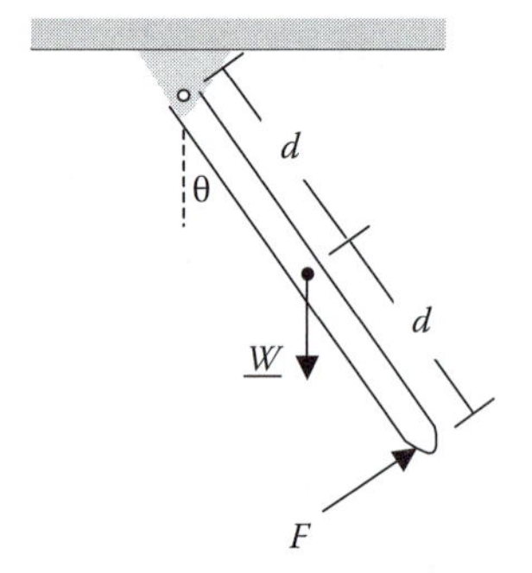

FIGURE 1.10

Notes

1. The Knudsen number is named for Martin Hans Christian Knudsen (1871–1949), professor at the University of Copenhagen and author of *The Kinetic Theory of Gases* (London, 1934). In physical gas dynamics, the Knudsen number defines the extent to which a gas behaves like a collection of independent particles (Kn >>1) or like a viscous fluid (Kn <<1).
2. We have written the column vector of V's components as a row vector to save space.

2

Strain and Stress in One Dimension

In the previous chapter, we stated that in order to study continuum mechanics—that is, to characterize the response of a continuous material to external loading—we must (1) characterize the material's deformation, (2) define its internal loading and (3) relate this to its deformation, and (4) ensure that the body is in equilibrium.[1] We begin this chapter by considering the deformation of a material under loading.

Returning to our example of the remodeling of an underwater oil rig as an artificial reef, we want to examine the trusses of the existing rig. As we have seen (Figure 1.1 and Figure 1.3), the rig is composed of many slender steel members that must withstand the cyclic loading of ocean currents as well as other loads. Each member may be pulled or pushed along its axis, as in Figure 2.1, and by isolating each member we can begin to determine whether the members can withstand this loading.

This raises the question of what it means to "withstand" a load. Is it sufficient for the member to sustain the load without incurring damage or breaking, or is it necessary for it to sustain the load without deforming or bending?

You may have noticed that a standard office table or desk can support far more weight or force than it does when serving as a writing table or computer desk and that some chairs can support the weight of several people without breaking. These are not examples of wasteful or inefficient designs. In fact, these products have been designed for *stiffness* rather than for *strength*. Instead of merely building a chair strong enough to hold the average person, designers have chosen to make the chair stiff enough that its deflections can be limited to some small amount, under a load much larger than it is expected to typically carry. Under normal use, therefore, the chair should not deflect perceptibly. Designing for stiffness means minimizing or limiting deflections and is generally a much more restrictive proposition than designing purely for strength. In this chapter, we discover ways to characterize the stiffness and strength of materials and structures.

To begin to design for stiffness by minimizing deflection, we must understand how to characterize the deformation a loaded body will undergo.

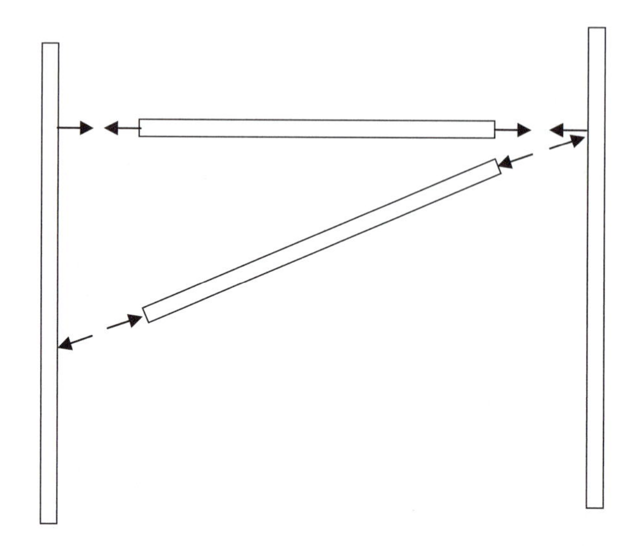

FIGURE 2.1
Isolated members of underwater structure.

2.1 Kinematics: Strain

In continuum mechanics, we want to characterize how bodies respond to the effects of external loading and how these responses are distributed through the bodies. One way a body responds to external loads is by deforming. We develop a way of quantifying its deformation relative to its initial size and shape, and we call this relative deformation *strain*.

2.1.1 Normal Strain

When an axial force is applied to a body, the distance between any two points A and B along the body changes. We call the initial, undeformed length between two points A and B the *gage length* (or *gauge length*). During a tensile experiment such as the one sketched in Figure 2.2, we may measure the change in gage length as a function of applied force. What interests us is how much this gage length changes, relative to its initial value—in other words, the *intensity of deformation*.

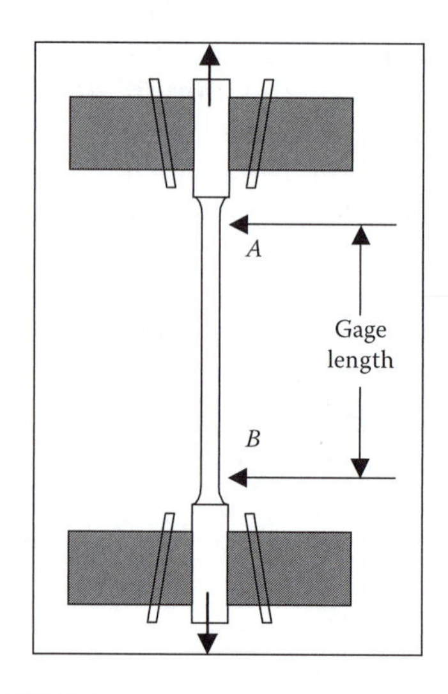

FIGURE 2.2
Tension specimen.

In Figure 2.2, the bar is acted on, or loaded, at its ends by two equal and opposite axial forces. (An axial force is one that coincides with the longitudinal axis of the bar and acts through the centroid of the cross section.) These forces, called *tensile* forces, tend to stretch or elongate the bar. We say that such a bar is in tension.

If L_o is the initial gage length and L is the observed length of the same segment under an applied load, the gage elongation is $\Delta L = L - L_o$. The elongation ε per unit of initial gage length, or "deformation intensity," is then

$$\varepsilon = \frac{L - L_o}{L_o} = \frac{\Delta L}{L_o}. \tag{2.1}$$

This expression for epsilon defines the macroscopic extensional strain.

It is also possible for this apparatus to load a bar with two equal and opposite forces directed toward each other, as in the sketch in Figure 2.3. These forces, called *compressive* forces, tend to shorten or compress the bar. We say that such a bar is in compression. Note that for compressive loading, $\Delta L < 0$, and the normal strain is negative.

Both tensile (tending to elongate) and compressive (tending to shorten) deformations result in *normal* strain, defined as the change in length of our material relative to its initial undeformed length. Normal strain is a dimensionless quantity but is often represented as having dimensions of length/length, in./in., m/m, or mm/mm. Sometimes it is given as a percentage.

In some applications, we use a slightly more careful definition of strain. This is sometimes called the *natural* or *true* strain as distinct from the *engineering* strain defined by equation (2.1). In this true strain definition, a strain increment dε is integrated over the bar:

$$\bar{\varepsilon} = \int_{L_o}^{L} d\varepsilon = \int_{L_o}^{L} \frac{dL}{L} = \ln\left(\frac{L}{L_o}\right) = \ln(1+\varepsilon). \tag{2.2}$$

For very small strains, this natural strain is coincident with the engineering normal strain ε.

In a third definition of strain, we consider that each and every planar section normal to the axis moves a uniform (over the plane) distance along the axis, $u(x)$. An element of the axis that was originally of length dx is thus

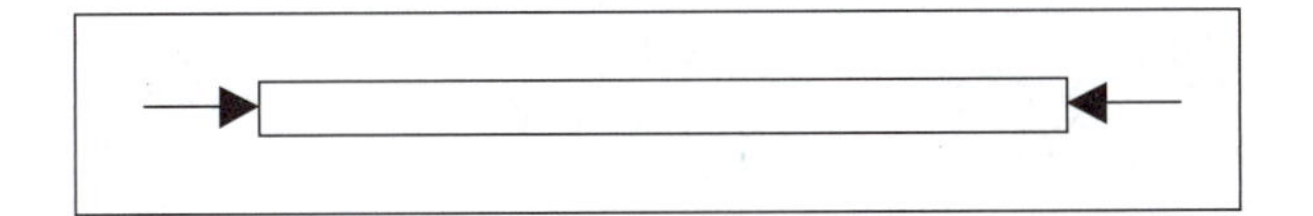

FIGURE 2.3
Bar in compression.

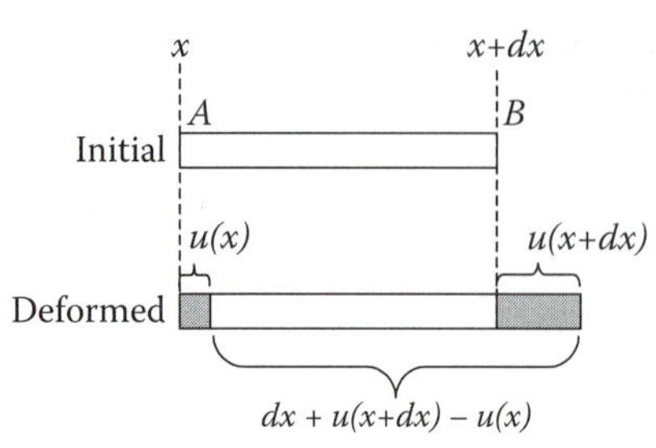

FIGURE 2.4
One-dimensional stretching of a bar.

stretched to a new length, $dx + u(x + dx) - u(x)$. This is illustrated in Figure 2.4.

For this deformation we define strain—in the same spirit as the first definition—as in equation (2.3).

$$\varepsilon = \frac{\text{change in length}}{\text{original length}} = \frac{\left[dx + u(x+dx) - u(x)\right] - dx}{dx}, \tag{2.3}$$

or, retaining only the first-order term in a Taylor series expansion of $u(x + dx)$, we find

$$\varepsilon \cong \frac{[u(x) + u'(x)dx - u(x)]}{dx} = u'(x) = \frac{du}{dx}. \tag{2.4}$$

In Section 2.7, we use equation (2.4) to express equilibrium in terms of the displacement $u(x)$, to illustrate where compatibility is applied, and to obtain a classic result for the extension of an axially loaded bar.

We also note that the quantity $[u(x + dx) - u(x)]$ represents the *relative displacement* of point B (at $x + dx$) with respect to point A (at x). This will provide a useful context for a more general definition of strain that we develop in Section 3.3.

Example

By bending a thin ruler, you are able to deform it into a circular arc. This arc, with a radius of 30 in., encloses an angle of 23° at center, as shown. Find the average normal strain developed in the ruler.

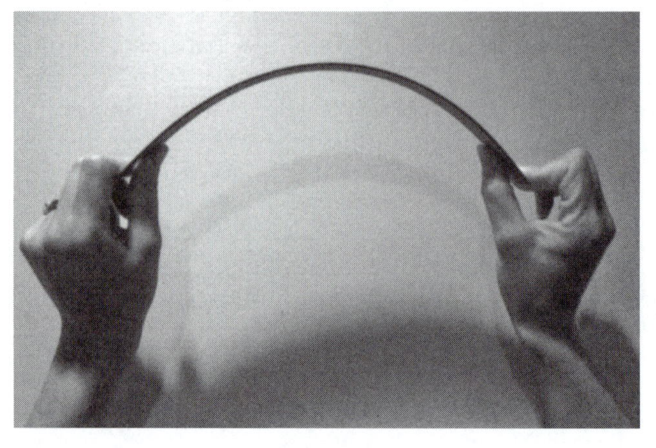

Given the initial length of the ruler, L_o, which we assume to be exactly 12 in., and the characteristics of a circular arc formed when it is deformed under bending, we must find the intensity of deformation, or induced strain. Since we know that strain is a measure of the change in a body's length relative to the original length, we must determine how much the ruler's length of 12 in. changes under this deformation.

Recalling that the arc length of a circular arc is given by the equation

$$\textit{arc length} = r\theta$$

and that in this case, the arc length is the deformed length of the ruler, L, we have

$$L = r\theta = (30 \text{ in.}) \cdot 23° \cdot \frac{2\pi \text{ rad}}{260°}$$

$$= (30 \text{ in.}) \cdot (0.4014 \text{ rad}) = 12.04277 \text{ in.}$$

Normal strain is then calculated

$$\varepsilon = \frac{\text{change in length}}{\text{original length}} = \frac{L - L_o}{L_o} = \frac{0.04277 \text{ in.}}{12 \text{ in.}} = 0.003564 \frac{\text{in.}}{\text{in.}}$$

For convenience, such a small strain might be reported as 3564 micro-inches per inch (μin./in.), or 3564 *microstrain*, or alternatively as a 0.36% strain.

2.1.2 Shear Strain

Bodies may experience both normal and *shear* deformations and, hence, normal and shear strains. When an axial tensile load is applied to a body, it causes a longitudinal tensile deformation: an elongation. Similarly, an axial compressive load will cause a longitudinal compressive deformation: a shortening. When a shear force is applied to a body, it will cause an angular deformation.

To visualize the effect of shear strain, consider a motor mount as shown in Figure 2.5a. The motor mount is composed of a block of elastic material (our "body") with attachments to allow for connection to the base of the motor and the support structure. A force P is applied at the top of the block. This subjects the block/body (of initial height L) to a pair of shear forces, as shown in Figure 2.5b. If we imagine that the block is composed of many thin layers and that each layer will slide slightly with respect to its neighbor, we may visualize how the angular distortion of the block will develop.

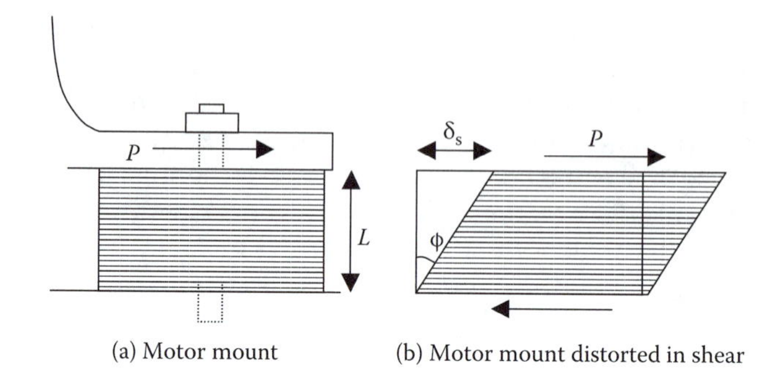

FIGURE 2.5
Shear strain.

As for normal strain, several definitions of shear strain exist. The *engineering* shear deformation incurred is ϕ, the *change* in an initially right angle. This is the formal definition of shear strain: the change in the angle between two initially perpendicular planes. It is measured in radians. However, it is often difficult to take precise measurements of these angular changes, especially for very small deformations. For small deformations, the tangent of the angle ϕ will closely approximate ϕ itself, so that we can approximate the shear strain by

$$\phi \cong \tan\phi = \frac{\delta_s}{L}, \tag{2.5a}$$

so

$$\gamma \cong \frac{\delta_s}{L}. \tag{2.5b}$$

With normal and shear strain defined, we are equipped to address the kinematics of deformation of continuous materials due to loading. We now move on to the second item on our checklist: the internal forces developed in response to external loading.

2.1.3 Measurement of Strain

Until 1930, strain was commonly measured indirectly, using extensometers that measured the displacement ΔL over some initial gage length L to allow strain to be calculated using the equations just discussed. An extensometer system typically included a mechanical or optical lever system. In 1931, the first electrical strain gauge demonstrated that strain could also be measured directly. Most modern strain gauges are resistive electrical meters.

In 1856 Lord Kelvin demonstrated that the resistances of copper and iron wires changed when the wires were stretched, compressed, or otherwise deformed. This concept is at the heart of the electrical strain gauges first implemented by Roy Carlson in 1931 and Edward Simmons in 1938.[2] Advances in materials and fabrication techniques have since refined the design of the resistive strain gauge, whose general construction is shown in Figure 2.6.

When the resistance element (wire grid or metallic foil) is attached to a loaded (and thus deformed) body in such a way that the wire will also be deformed, the measured change in resistance may be calibrated in terms of strain. Important parameters in the design and performance of a strain gauge are (1) the materials used for the wires or foil, and, to a lesser extent, the backing and bonding materials; (2) protection of the gauge; and (3) electrical circuitry, typically involving a Wheatstone bridge. The wires should have a large change in resistance corresponding to the strains expected (sometimes called the wire material's *gauge factor*), a high electrical resistivity, a low temperature sensitivity,[3] and good corrosion resistance, among other factors. Mounting a strain gauge is straightforward (though not always easy) as long as the surface of the body in question is extremely clean and as long as the manufacturer's installation procedures are followed carefully.

2.2 The Method of Sections and Stress

We now want to consider the forces *within* a body that balance the effect of externally applied forces. To do this, we must prepare a free-body diagram (FBD) that shows all the external forces acting on the body at their respective points of application (Figure 2.7a). All of the forces acting on a body, including reactive forces caused by supports and the weight of the body itself (usually not included in a free-body diagram), are considered external forces. This view is valuable but does not allow us to visualize the internal forces

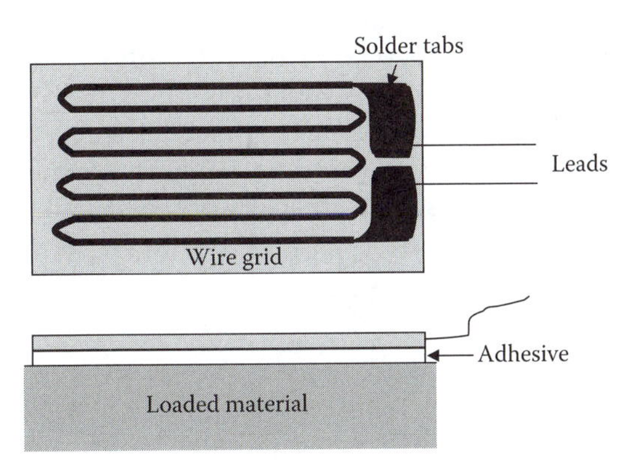

FIGURE 2.6
Construction of a bonded-wire strain gauge.

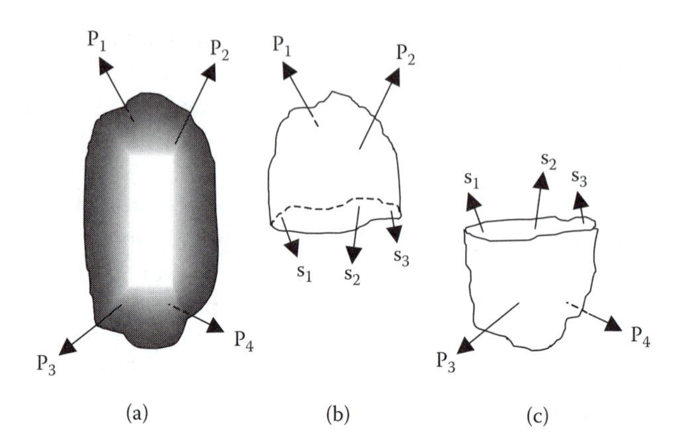

FIGURE 2.7
The method of sections.

we're interested in, so we "slice open" our body (Figure 2.7b and Figure 2.7c). Each sliced section must be in equilibrium, just as the larger body is in equilibrium. The fundamental statement of this is:

> The externally applied forces on one side of an arbitrary cut must be balanced by the internal forces developed at the cut.

The name given to this technique is the *method of sections.*

These internal forces revealed by the method of sections have varying magnitude and direction. They are vectors, and they maintain the externally applied forces in equilibrium. In a solid, these forces determine the solid's resistance to deformations and to external forces.

Physically, these internal forces are what hold the body together: intermolecular forces, or chemical bonds. The application of an external force changes the distance between atoms (i.e., deformation), which changes the forces exerted by these bonds. We could model the internal forces s_i as the resultant of bond forces, but the bookkeeping associated with so many force vectors, and complex atomic arrangements, would be prohibitive. Plus, dealing with continuous materials was supposed to get us off the hook from having to worry about individual atoms, anyway. So, we tend to consider one distributed internal force, and stress is the intensity of that distributed force.

In general, *stress,* represented by sigma, is a force per unit area, or the force's intensity:

$$\sigma = \frac{P}{A}. \tag{2.6}$$

Remember that both the force P and the area A are vectors.[4] The stress depends on the orientations of both P and A, as demonstrated in subsequent chapters. Its units are of force per unit area, generally [N/m^2] or [lb/in.^{2}]. It will be useful for us to resolve the internal force P into its components perpendicular and parallel to the section of interest.

Interestingly, it took a long time for engineers and scientists to conceptualize stress as we now understand it. While this was partly due to the susceptibility of scientific progress to fads and biases, and the tyranny of Isaac Newton as a trendsetter (more on this later), it was also a result of researchers focusing on whole structures and not "looking inside" the body as the method of sections demands. Instead, as J. E. Gordon noted, "All through the eighteenth century and well into the nineteenth, very clever men, such as Leonhard Euler and Thomas Young, performed what must appear to the modern engineer to be the most incredible intellectual contortions"[5] to characterize material behavior without the modern notion of stress.

It was Augustin Cauchy who first conceptualized stress and strain as we now understand them, in 1822: "Cauchy perceived that … the 'stress' in a solid is rather like the 'pressure' in a liquid or a gas. It is a measure of how hard the atoms and molecules which make up the material are being pushed together or pulled apart as a result of external forces" (Gordon, 1988, p. 46).

2.2.1 Normal Stresses

By using the method of sections, we can identify the different types of stress. Consider a straight bar acted on at its ends by two equal and opposite forces, as in Figure 2.8a. Remember that these external forces are called tensile forces. Similarly, the bar in Figure 2.9a is acted on by two equal and opposite forces, directed toward each other; these forces are compressive forces. If we make an imaginary cut through each bar and consider the left-hand segment as a free body, as in Figure 2.8b and Figure 2.9b, we see that for each bar to be in equilibrium, a force P_1, equal and opposite to external force P, must exist. This force P_1 is actually an internal force in the original bar that "resists" the action of force P. Also, we assume that the internal resisting force is uniformly distributed over the cross section of the bar. This force per area (the internal force divided by the cross-sectional area) is what we call stress.

The tensile forces of Figure 2.8 produce internal tensile stresses, and the compressive forces of Figure 2.9 produce internal compressive stresses. By convention, tensile stresses are positive, and compressive are negative. (This sign convention has to do with the outward normal vector of surface A, as is discussed in Chapter 3.) Tensile and compressive stresses are developed in a direction perpendicular (normal) to the surfaces on which they act and,

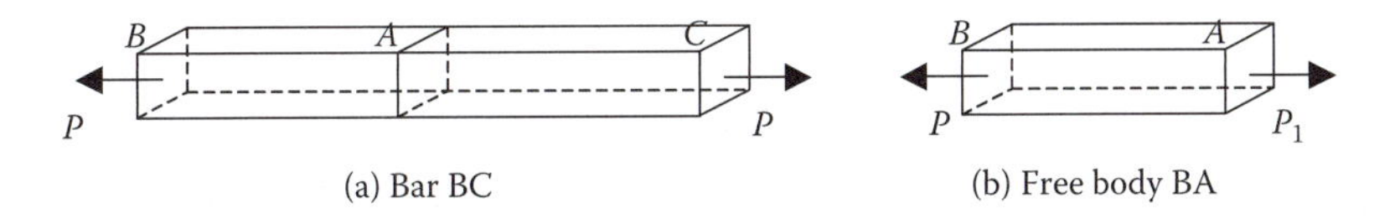

FIGURE 2.8
Bar in tension.

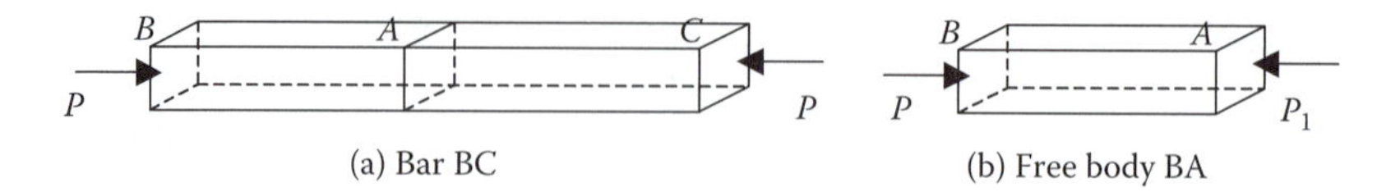

FIGURE 2.9
Bar in compression.

hence, are sometimes called *normal* stresses. We use the Greek letter sigma, σ, to represent normal stress, and we write

$$\sigma \equiv \frac{P}{A}. \tag{2.7}$$

2.2.2 Shear Stresses

Another type of stress, called *shear* stress (sometimes *tangential* stress), is developed in a direction parallel to the surface on which it acts. An example is shown in Figure 2.10. When equal and opposite forces P are applied to two flat plates bonded together by adhesive, the contact (shaded) surface is subjected to a shearing action. In the absence of the adhesive, the two surfaces would slide past one another. The shear force is assumed to be uniformly distributed across the contact area. As a result the shear stress, defined as this shearing force divided by the contact area, is developed. Shear stress can also develop within a single body, when various layers of the material tend to slide with respect to each other.

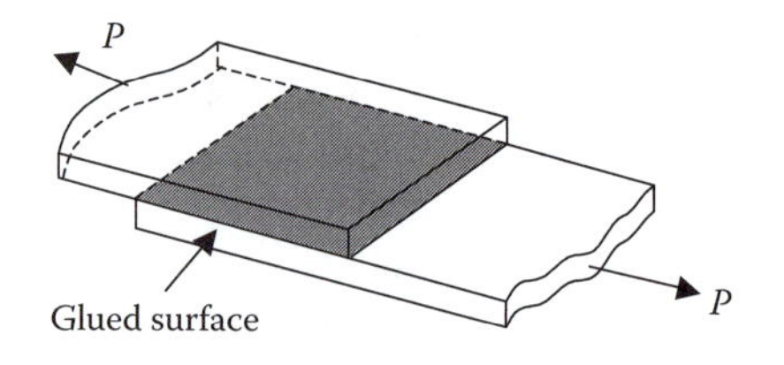

FIGURE 2.10
Shear between two bodies.

Again, stress is the intensity of the internal force and, in this case, is once again P/A, where A is the area of the glued surface; however, for shear stresses, the area A is oriented parallel to the force P, while for normal stresses P is perpendicular to A. (If we more carefully characterize the area A by its outward normal vector, the shear stress is normal to this normal vector, and the normal stress is parallel to it.) We use the Greek letter tau, τ, to represent shear stress:

$$\tau \equiv \frac{P}{A_{\parallel}}. \tag{2.8}$$

We have included a subscript to remind ourselves that the area A in this expression seems to be parallel to the force P. Now that we have defined both strain (kinematics) and stress, we must consider the relationship between them. We do this in Section 2.3.

Example

Let's consider a structure that might be part of an underwater oil rig turned artificial reef. All members of the truss pictured here have a cross-sectional area of 500 mm², and all the bolts and pin connectors have a diameter of 20 mm. Find (a) the axial or normal stresses in members BC and DE, and (b) the shear stress in the bolt at A if it is in double shear.

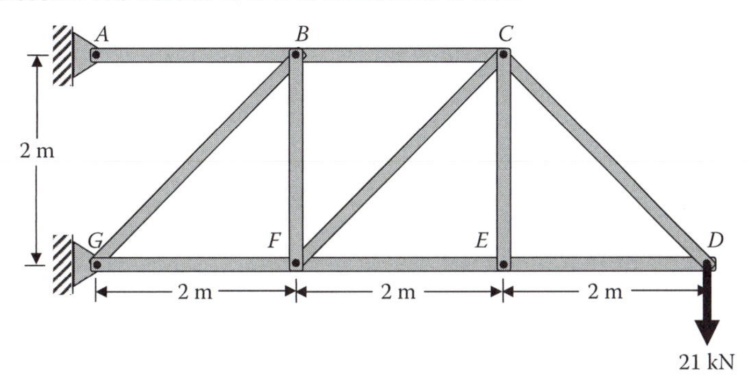

FIGURE 2.11

Given a truss with specified parameters and loading, we must find the requested values of stress. We first examine an FBD of the joint at D:

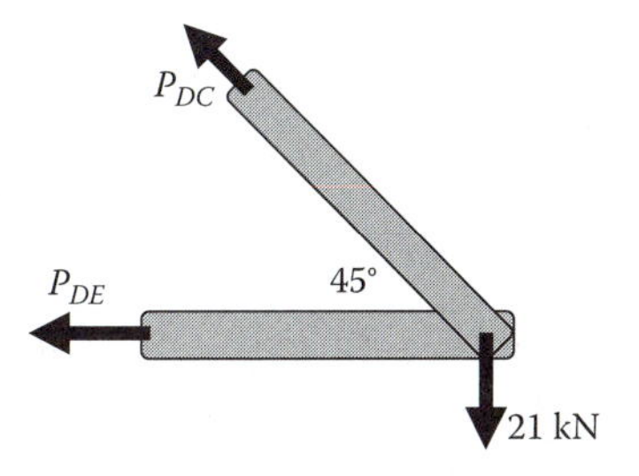

FIGURE 2.12

Note that we have assumed both members DC and DE to be in tension; if we calculate negative values for either internal force, we will know that this assumption was incorrect and that the member is in compression. Since the joint must be in equilibrium we have

$$\Sigma F_y = 0 = P_{DC}\sin 45 - 21\text{ kN} \rightarrow P_{DC} = 29.7\text{ kN}$$

$$\Sigma F_x = 0 = -P_{DE} - N_{DC}\cos 45 \rightarrow P_{DE} = -21\text{ kN}.$$

Using the definition of normal stress we know that

$$\sigma_{DE} = \frac{P_{DE}}{A_{x-\text{sec}}} = -42 \times 10^6 \frac{\text{N}}{\text{m}^2},$$

or

$$\sigma_{DE} = 42 \text{ MPa compressive.}$$

Next, we use the method of sections. We make an imaginary cut between B and C, resulting in an FBD that includes the internal forces in three members of the truss:

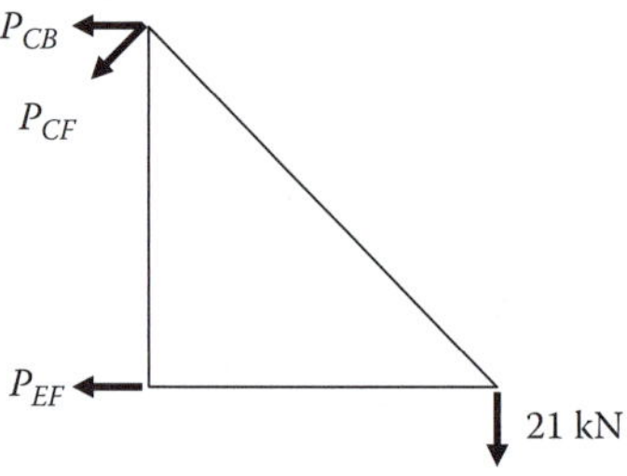

FIGURE 2.13

We apply the third equilibrium equation, summing moments about point F:

$$\Sigma M_F = 0 = P_{CB} \cdot (2 \text{ m}) - 21\text{k N} \cdot (4 \text{ m}) \rightarrow P_{CB} = 42 \text{ kN},$$

so that

$$\sigma_{CB} = \frac{P_{CB}}{A} = 42 \times 10^6 \frac{\text{N}}{\text{m}^2} \quad \sigma_{CB} = 84 \text{ MPa tensile.}$$

We may take this opportunity to check our intuition about this truss. The load, P, is pulling the structure down. Thus, composite member ABC should become longer, and DEFG should become shorter. This would mean that members on the top (like BC) would be in tension and members on the bottom (like DE) in compression. Our results so far are consistent with our physical intuition. This buoys our spirits as we continue to part (b) of the problem, in which we consider the bolt at joint A.

We are told that this bolt is in "double shear." A connection element (bolt or pin) is said to be in "single shear" if one cut between the member and its support is sufficient to break the connection, as shown in Figure 2.14 on the left; "double shear" means that two cuts are needed to break the connection, as on the right. A quick analysis using free-body diagrams of each case should be persuasive evidence that a bolt in double shear experiences half the shear stress of an identically loaded bolt in single shear. This analysis is left as an exercise.

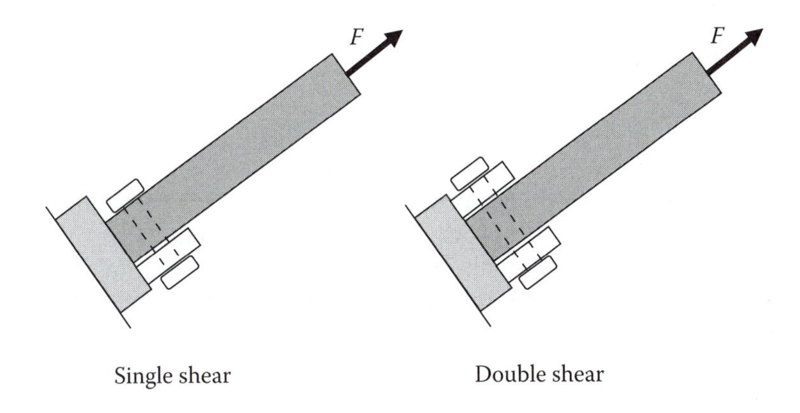

FIGURE 2.14

To find the reaction forces at the supports, we consider an FBD of the entire truss:

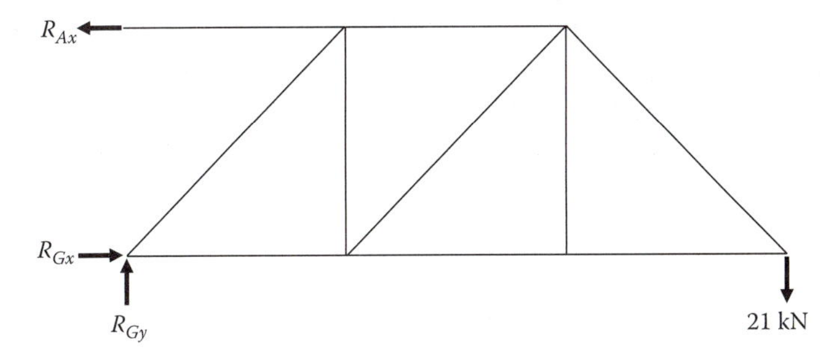

FIGURE 2.15

Summing moments about point G, we have

$$\Sigma M_G = 0 = R_{Ax} \cdot (2\text{ m}) - 21\text{ kN} \cdot (6\text{ m}) \rightarrow R_{Ax} = 63\text{ kN}.$$

So the shear stress in the bolt at A is found:

$$\tau_A = \frac{A_x / 2}{A_{bolt}} = 100 \times 10^6 \frac{\text{N}}{\text{m}^2} = 100\text{ MPa}.$$

2.3 Stress–Strain Relationships

Different materials respond differently to loads. In some materials (e.g., rubber), small loads produce relatively large deformations. Other engineering materials, such as steel, undergo smaller deformations—however, it is still important to consider the effects of such changes. Even very rigid materials, when subjected to a load, will experience a small deformation.

For most engineering materials, a relationship exists between stress and strain. For each increment in stress there is a proportional increase in strain, provided that a certain limit of stress is not exceeded. If the induced stress exceeds the limiting value, the strain will no longer be linearly proportional to the stress. This limiting value is called the *proportional limit*.

Most of the behavior we will consider occurs below the proportional limit, in the regime where stress and strain enjoy a linearly proportional relationship. If we subject a material in this regime to a tensile load P_A, producing a stress σ_A and a strain ε_A, and then subject it to a tensile load P_B, producing stress σ_B and a strain ε_B, and we then plot the stresses and strains, we see a linear relationship between stress and strain, as shown in Figure 2.16.[6]

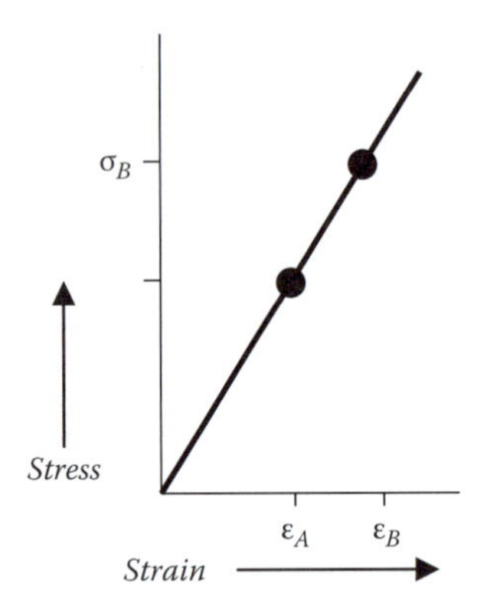

FIGURE 2.16
Linear relationship between stress and strain.

This linear relationship between load and deformation was first stated by Robert Hooke in 1678 and became known as Hooke's law: *Ut tensio, sic vis*. This Latin phrase—in the form of an anagram, *ceiiinosssttuv*—was how Hooke[7] summed up his finding, which he first applied to the extension of a spring. It translates, "As is the extension, so is the force." We have seen his law in this form:

$$F = kx \tag{2.9}$$

and have called k the "spring constant" or "stiffness" of the spring in question. Figure 2.17 shows a representative spring.

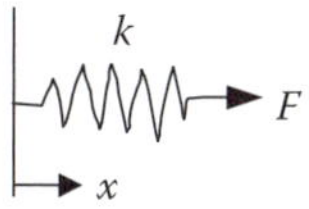

FIGURE 2.17
Linear (Hookean) spring.

The stress–strain diagram is another example of a force (stress being a force per area) being linearly proportional to an extension (strain being extension per initial length). It, too, is Hooke's law: *Ut tensio, sic vis*. It, too, contains a linear constant of proportionality, a *stiffness*.

The ratio of stress to strain, which is also the slope of the line joining these two data points, is constant for loading below the material's proportional limit. This constant is now known as the *modulus of elasticity* or *Young's modulus*, after Thomas Young, who defined it in 1807. (Young's definition was somewhat awkward and ungainly, since Cauchy had yet to clearly define stress. It wasn't until 1826 that Claude Navier defined Young's modulus as we are about to.) The modulus of elasticity for bodies in tension or compression is usually represented by the symbol E and is expressed as

$$E = \frac{stress}{strain} = \frac{\sigma}{\varepsilon}. \tag{2.10}$$

Since strain is a dimensionless quantity (length divided by length), E has the same units as stress: either pounds per square inch (psi) in English units, or N/m^2 or Pascals (Pa) in SI. Table 2.1 shows the values of E for several engineering materials.

Physically, the modulus of elasticity represents the stiffness of a material. A material's stiffness may be defined as the property that enables the material to withstand stress without great strain—in other words, the material's resistance to deformation.

TABLE 2.1
Approximate Design Values (Reflecting Proportional Limits) of Elasticity and Shear Moduli, in Linear Regimes (SI)

Material	Modulus of Elasticity E (MPa)	Modulus of Rigidity G (MPa)
California redwood	7600	
Steel (carbon) ASTM A36	207,000	83,000
Stainless steel	200,000	80,000
Aluminum 6061-T6	70,000	28,000
Glass	48,000–83,000	19,000–35,000
Polycarbonate	2400	800
Concrete	21,500	8970
Bone	1–16,000	4–8000

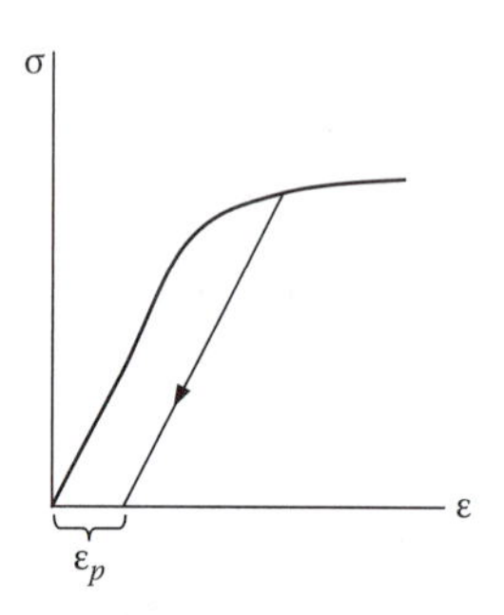

FIGURE 2.18
Plastic deformation incurred when proportional limit is exceeded.

In the Hookean regime, both springs and solid materials are linearly elastic. In the presence of an applied load, stress is linearly related to strain. If an applied load is removed, both stress and strain decrease linearly to zero. However, if a material's proportional limit is exceeded due to an applied load, this is no longer true. In this case, the removal of the applied load causes both stress and strain to decrease linearly, along a line parallel to the linear portion of the stress–strain curve, as shown in Figure 2.18. The strain does not return to zero. By exceeding its proportional limit, the material has undergone a permanent *plastic* deformation. Plastic, as opposed to elastic, deformation represents a permanent set of the material. For most materials, the degree of plastic deformation depends on both the maximum stress value reached and the time elapsed before the load is removed. The stress-dependent portion of plastic deformation is known as *slip*, and the time-dependent part, which can also be influenced by temperature, is known as *creep*.

Shear stress is also proportional to shear strain, as long as the stress is below the proportional limit. The constant of shear proportionality is known as the *shear modulus* or the *modulus of rigidity*. It is represented by G and expressed as

$$G = \frac{\textit{shear stress}}{\textit{shear strain}} = \frac{\tau}{\gamma}. \tag{2.11}$$

Average values of the modulus of rigidity for some common materials are given in Table 2.1. Note that the moduli of elasticity and rigidity differ significantly for each material.

It is interesting to observe the consistency of the ratio of E to G, despite the diversity of materials represented in Table 2.1. In Section 3.1 we reflect further on the relationship between E and G, representing a material's resistance to axial deformation relative to its resistance to shear.

We now have two additional forms of Hooke's law, likenesses of $F = kx$ for one-dimensional loading. We see this likeness clearly by rearranging the two equations:

$$\sigma = E\varepsilon. \tag{2.10}$$

$$\tau = G\gamma. \tag{2.11}$$

In modeling our material body as a linear spring, we are making the assumption of linearity (small deformations, i.e., that we are in the Hookean regime of the material's stress–strain curve). This model incorporates three further assumptions that thus represent limitations—albeit broad ones—on the kinds of materials it can represent. One assumption is that the material is *homogeneous*, by which we mean the material constants (e.g., Young's modulus) do not vary from point to point—that is, are not functions of the coordinates. The second assumption is that the material is *isotropic*, by which we mean that the elastic properties are invariant with respect to any rotation of the coordinate axes. In other words, no matter which axis we look down, we see the same material behavior. The third assumption is that there is no apparent effect of temperature in our simple version of Hooke's law. We incorporate the effects of temperature in Section 2.9.

Each material has its own characteristic stress–strain curve. The extreme values of strain that materials can withstand vary widely, as do the slopes of the Hookean portions of their curves, as shown in Figure 2.19. The terminal point on a stress–strain diagram represents the complete failure (rupture or fracture) of the specimen. Materials that are capable of withstanding large strains without a significant increase in stress (and that may be thought of as "stretchy") are called *ductile* materials. Low-carbon steels, polymers, skin, and rubber are examples of ductile materials. *Brittle* materials, on the other hand, will experience a huge increase in stress from even a small strain and will fail abruptly after a small amount of deformation. Cast iron, glass, ceramics, concrete, and bone are examples of brittle materials. Further discussion of material properties is available in Section 3.6 and Section 3.7.

For the most part, we consider homogeneous, isotropic materials—materials whose behavior does not depend on the direction (e.g., tension or compression) of loading. Many engineering materials such as metals and ceramics may be readily modeled this way; however, some materials, like wood and bone, have different properties in different directions. Wood is strongest against loading along its grain and is much easier to break with loads applied across the grain; compact bone is strongest along its long axis to resist compressive loading. For the time being, we neglect such variations and cling to the assumptions of homogeneity and isotropy.

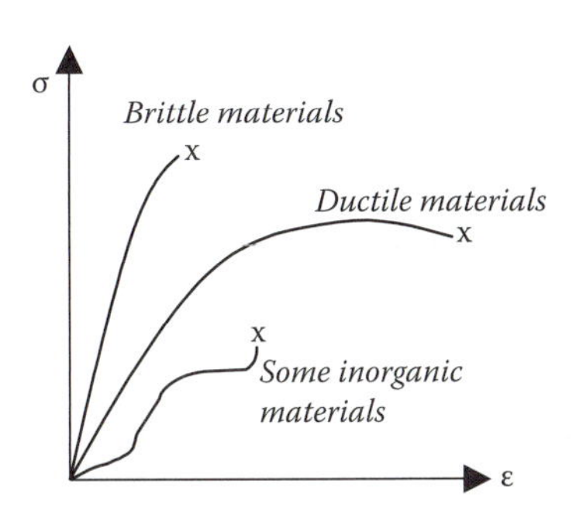

FIGURE 2.19
Schematic of typical stress–strain diagrams. See Sections 3.6 and 3.7 for further discussion of the terms *ductile* and *brittle*.

We recall our checklist of what is needed to apply continuum mechanics to understand the response of a body to external loading: We must (1) characterize the deformation of a continuous material, (2) define the internal loading and (3) relate this to the body's deformation, and (4) make sure that the body is in equilibrium. We have accomplished the first three items on the list and now understand that in doing so we have constructed (1) a kinematic description of deformation, or strain; (2) a definition of stress; and (3) a constitutive law relating stress and strain. The last item on our list, (4) equilibrium, is addressed by the method of sections; we also consider equilibrium more rigorously in the following section.

2.4 Equilibrium

We have used equilibrium and the method of sections to apply Newton's second law on a "macroscopic" basis. Now we will do a "microscopic" equilibrium analysis in terms of the stress resultants at an arbitrary point in the bar, acting on an infinitesimal element of length dx and of volume $d\mathcal{V} = A(x)dx$, as shown in Figure 2.20. Since the point we have chosen is arbitrary, this analysis is valid at every point in the bar—and so for the entire bar.

Summing forces in the x direction on this uniaxially loaded element, we see that the *internal* axial force N balances both the *external* axial load $q(x)$, a *distributed* axial load per unit length of the bar (a force per length, having units of N/m or lb/ft), and an axial *body force*, B_x (a force per volume):

$$\left(N(x) + \frac{dN(x)}{dx}dx\right) - N(x) + q(x)dx + B_x A(x)dx = 0. \tag{2.12}$$

The internally distributed body force allows us to include forces that depend on intrinsic mass or volume, such as gravity or magnetic fields. For example, to consider the weight of a vertical element, we would use $B_x = \rho g$ if x points toward the center of the Earth. Equation (2.12) can then be simpli-

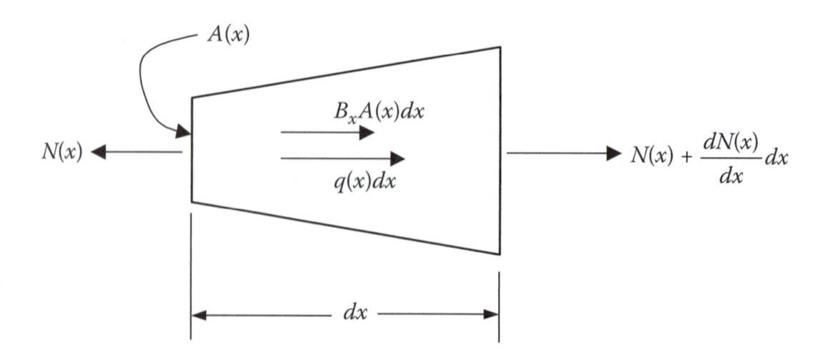

FIGURE 2.20
Equilibrium of an infinitesimal element in one dimension: Internal axial force N balances applied axial load $q(x)$, and body force B_x.

fied, yielding an ordinary differential equation of first order for the axial normal stress resultant:

$$\frac{dN(x)}{dx} + q(x) + B_x A(x) = 0. \tag{2.13}$$

The axial loads are generally "distributed" as concentrated loads P_i located at coordinates x_i, in which case

$$q(x) = \sum_i P_i \delta(x - x_i),$$

and Equation (2.13) takes the form

$$\frac{dN(x)}{dx} + \sum_i P_i \delta(x - x_i) + B_x A(x) = 0. \tag{2.14}$$

Section 2.7 shows that microscopic and macroscopic equilibrium results are in agreement. Our checklist for continuum mechanics analysis is complete:

- ✓ Kinematics (strain)
- ✓ Definition of stress
- ✓ Constitutive law (stress–strain relationship)
- ✓ Equilibrium

Now that we have developed these four items for one-dimensional loading, we will see what they mean for an axially loaded bar like those in our underwater structure.

2.5 Stress in Axially Loaded Bars

Consider a steel ruler—a thin body made of a seemingly compliant material. We know that if we hold such a ruler by one end and push down on the other end (perpendicular to the ruler's broad surface), as in Figure 2.21a, the loaded end will be deflected significantly. In this case of loading, we call the system a cantilever *beam*. On the other hand, if we instead pull on the free end (parallel with the long axis of the ruler), as shown in Figure 2.21b, we would see very little movement. A system with this type of loading is called a *bar*. It is intriguing that the same body can experience such dramatically different behavior due to differences in loading. We hope to be able to postulate and develop models to explain these different behaviors.

Once we remove either load from the ruler—once we stop pushing or pulling—the ruler returns to its original, planar shape. In this way, the ruler behaves like an elastic spring, just as Hooke suggested. In our "beam" and "bar" experiments, the different behavior of the ruler can be explained by its having a different stiffness depending on the loading. Later in this text, we derive the different forms of this stiffness and see in detail that the beam stiffness is much less than the stiffness in the bar mode, which is why we see greater movement or deflection when the ruler acts like a beam.

For now, the important lesson is that the effective stiffness (a measure of how much a body will resist being deflected by a load) of a structural element or mechanical device is dependent on several factors, including the nature of the loading, as well as the element's geometry and the material itself. Since we are interested in how bodies will react to external forces, this stiffness provides us with a way to quantify their reactions.

Let's expand our ruler example of a bar in axial loading (Figure 2.22a). The bar is built in, or attached to a wall, at $x = 0$ and is subjected to a single external (applied) load P at $x = L$. The load P acts along the bar's axis. We know from Newton's second law that to keep the bar in static equilibrium, the attachment point or wall must exert an equal and opposite force P at the left end of the bar.

What we're interested in, of course, is what's happening inside the bar. We can use the method of sections to make an imaginary slice along the bar, exposing a cross section of area A. A free-body diagram will show us that something must be happening on that area to exert a net tensile force P across A. And, if our slice is normal to the bar's axis (as in Figure 2.22b), the exposed area A is also normal to the axis, and we can define the normal stress, σ, acting on that area as we did in Equation (2.7):

$$\sigma \equiv \frac{P}{A}. \tag{2.7}$$

If instead we make our section cut at an angle, θ, the picture will be different (Figure 2.22c). Now, the equilibrating force at the section surface has two

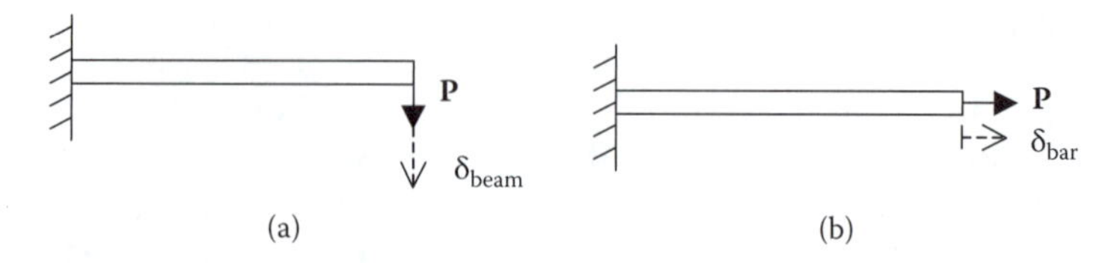

FIGURE 2.21
Illustration of beam and bar modes.

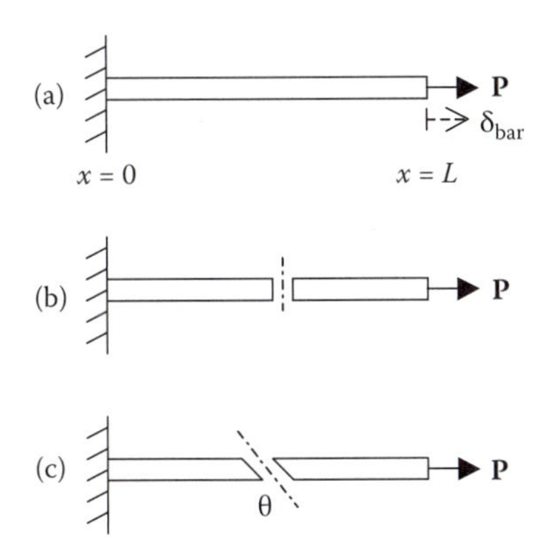

FIGURE 2.22
Stresses on axially loaded bar.

components, as shown in Figure 2.18. The normal force component is $P \cos \theta$, and the shear component (parallel to the section surface) is $P \sin \theta$. (These components may be obtained by summing forces in the x and z directions.)

The area of the inclined cross section is $A/\cos \theta$. From these values we can calculate the normal stress σ_θ and the shear stress τ_θ on this angled section by the two equations:

$$\sigma_\theta = \frac{force}{area} = \frac{P\cos\theta}{A/\cos\theta} = \frac{P}{A}\cos^2\theta, \tag{2.15a}$$

$$\tau_\theta = -\frac{P\sin\theta}{A/\cos\theta} = -\frac{P}{A}\sin\theta\cos\theta. \tag{2.15b}$$

The negative sign in the equation for tau comes about due to the sign convention for shear stresses (the shear force $P \sin \theta$ is in the negative y' direction).

Both normal and shear stresses, we have seen, will vary with the angle θ. Looking at equation (2.15a) and equation (2.15b) for σ_θ, we see that it will reach its maximum value when $\theta = 0°$, that is, when the section is perpendicular to the axis of the bar (as in Figure 2.22b). The corresponding shear stress at $\theta = 0°$ would be zero. Hence we determine the maximum normal stress in an axially loaded bar:

$$\sigma_{max} = \frac{P}{A}. \tag{2.16}$$

A question to think about is: what happens at $\theta = 90°$? Does this make sense?

If we differentiate the equation for shear stress with respect to angle θ and set it equal to zero, we should find the maximum value of $\tau\theta$. We find that $\tau\theta$ has its maximum value when $\tan\theta = \pm 1$, leading us to the conclusion that τ_{max} occurs on planes of either +45° or −45° with the bar axis. If we substitute ±45° into our equation, we find that

$$\left|\tau_{max}\right| = \frac{P}{2A} = \frac{\sigma_{max}}{2}. \tag{2.17}$$

Thus, the maximum shear stress in an axially loaded bar is only half as large as the maximum normal stress.

To consider the stresses on the section formed by a "cut" at the angle $\theta - 90°$, a section perpendicular to the θ section, we can either examine the figure on the right side of Figure 2.23, or substitute $\theta - 90°$ in for θ in the equations we have for σ_θ and τ_θ. Either way, we will find that

$$\sigma_{\theta-90°} = \frac{force}{area} = \frac{P\sin\theta}{A/\sin\theta} = \frac{P}{A}\sin^2\theta, \tag{2.18a}$$

$$\tau_{\theta-90°} = \frac{P\cos\theta}{A/\sin\theta} = \frac{P}{A}\sin\theta\cos\theta. \tag{2.18b}$$

2.6 Deformation of Axially Loaded Bars

We've established expressions for stress, strain, and the modulus of elasticity E. These may now be combined into a convenient expression to directly determine the total deformation δ for an axially loaded bar (Figure 2.22a). We begin with the definition of modulus of elasticity, or Hooke's law, and substitute for stress and strain:

$$E = \frac{\sigma}{\varepsilon} = \frac{P/A}{\delta/L} = \frac{PL}{A\delta}. \tag{2.19}$$

Then, solving for δ, we obtain

$$\delta = \frac{PL}{AE}, \tag{2.20}$$

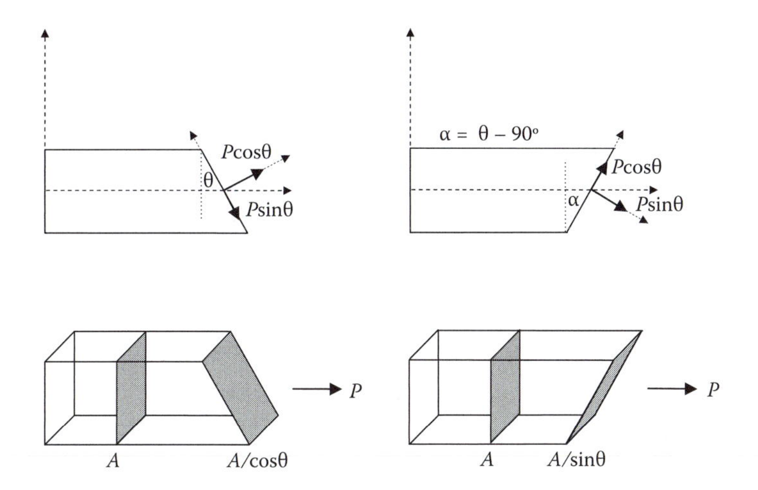

FIGURE 2.23
Sectioning of a bar at angle θ (left) and angle $\theta - 90°$(right) from vertical.

where

- δ = the total axial deformation, with dimensions of Length [(in.), (m, mm)]
- P = the total applied external axial load, with dimensions of Force [(lb, kips), (N)]
- L = the original length of the bar, with dimensions of Length [(in.), (m, mm)]
- A = the cross-sectional area of the member, with dimensions Length^2 [(in.^{2}), (m^2, mm^2)]
- E = the modulus of elasticity with dimensions of Force/Length^2 [(psi, ksi), (Pa, MPa)]

This expression is valid only when the stress in the bar does not exceed the proportional limit. This should make sense, as it is only below this limit that the bar's stress and strain will obey Hooke's law. Also, equation (2.20) assumes that the forces, area, and properties of the bar do not change along its length. For a more complex problem, where quantities vary along the bar's axis (here the x axis), we can obtain a similar relationship that takes such variations into account:

$$\delta = \int_0^L \frac{P(x)}{A(x)E(x)} dx . \tag{2.21}$$

We can cast this relationship in terms of the bar's stiffness, as discussed earlier. If we recall the form Hooke's law took for linear springs, $F = kx$, we can write P as a function of δ using equation (2.20):

$$P = \frac{AE}{L}\delta . \tag{2.22}$$

Comparing equation (2.22) with $F = kx$, we see that the axial deformation δ of this bar due to the axial load P depends on its stiffness, AE/L. Chapter 6 compares this spring stiffness with that for a beam, loaded as in Figure 2.16.

2.7 Equilibrium of an Axially Loaded Bar

Now we want to combine our kinematics (equation 2.4) and constitutive (equation 2.10) and equilibrium (equation 2.14) equations to characterize a uniaxially loaded bar. In principle, this is a system of three equations for three unknowns: the strain ε, the stress σ, and the axial displacement $u(x)$. However, we can simplify the mathematics by eliminating variables and reducing our system to a single differential equation. Since our system of equations includes two first-order differential equations (equilibrium, kinematics) and one algebraic equation (Hooke's law, our constitutive equation), we expect our single equation to be second order. We achieve this result by, first, writing the stress in terms of strain and strain in terms of the displacement, $u(x)$, that is,

$$\sigma = E\varepsilon = E\left(\frac{du(x)}{dx}\right). \tag{2.23}$$

Second, we substitute equation (2.23) into the equilibrium equation (2.14) to find (assuming that the area, the elastic modulus, and the temperature change are all constant—that is, they do not vary with the x coordinate)

$$E\frac{d^2u(x)}{dx^2} + B_x = 0 . \tag{2.24}$$

This is the second-order equation we expected. In the absence of body forces ($B_x = 0$) it is easily integrated, yielding

$$u(x) = C_1 x + C_2 . \tag{2.25}$$

To determine the constants of integration in equation (2.25), we must apply appropriate boundary conditions. As an example, we solve for the diplacement in the bar shown in Figure 2.22a. One boundary condition is clear: The displacement (or movement) of the bar is zero at the left end ($u(0) = 0$) because the bar is attached to the wall and restrained there. At the "free" end, $x = L$, we are pulling with a force P so that we can express this boundary condition in terms of the strain as

$$\frac{du(L)}{dx}=\varepsilon(L)=\frac{\sigma}{E}=\frac{P}{AE}. \tag{2.26}$$

After applying our two boundary conditions, we find the solution (2.25) to be

$$u(x)=u(0)+\frac{Px}{AE}=\frac{Px}{AE}. \tag{2.27}$$

The net extension of an entire bar (or rod) of length L is thus

$$\delta=u(L)-u(0)=\frac{PL}{AE}, \tag{2.28}$$

which is in agreement with equation (2.20) and from which we can recover the expression for the bar stiffness, AE/L.

2.8 Indeterminate Bars

For some structural systems, the equations for static equilibrium expressed in terms of stresses[8] are insufficient for determining reactions. This may be because some of the reactions are superfluous or redundant for maintaining equilibrium. But even a redundant support feels reaction forces—forces we as engineers must calculate. Equilibrium equations may also be insufficient when some internal forces cannot be determined using the equations of statics alone. Both of these situations, called *statical indeterminacy,* may arise in axially loaded bar systems.

We can resolve statical indeterminacy by several methods. In all of the available methods, as in all of our mechanics problems, we must make sure of three things, in no prescribed order:

- Equilibrium conditions for the system must be assured, both locally and globally.
- Geometric compatibility must be satisfied among deformed parts of the body and at boundaries. This has to do with the kinematics of deformation.
- Constitutive relations such as Hooke's law must be obeyed by all materials of the system.

Of the available methods, the two most commonly used are (1) the *force* method, in which we first remove and then restore a redundant reaction;

and (2) the *displacement* method, in which we maintain compatibility of the displacements of adjoining members and at the boundaries and in which solution displacements are obtained from equilibrium equations. The displacement method is the basis for most of the finite element method (FEM) programs that are commonly used to analyze complex structures and is better suited to large systems. Both methods make use of the analogy between Young's modulus E and our old friend the spring constant k. E and k each relate force and displacement in a linear equation: $\sigma = E\varepsilon$ and $F = kx$.

We have just seen that the stiffness of an axially loaded bar may be expressed as $k = AE/L$.

2.8.1 Force (Flexibility) Method

The force method is also sometimes called the force/flexibility method. We will be thinking of our indeterminate bars as elastic members of a system, each bar with a *flexibility* f related to its stiffness k. In fact, f is defined as the reciprocal of k:

$$f = 1/k = \Delta/P,$$

or

$$L/AE.$$

Note that f has physical dimensions of displacement/force, reciprocal dimensions of the stiffness k.

To illustrate the force method, consider the following example. In Figure 2.24a, an axial force P is applied at point B of the varying-diameter bar ABC. This axial load leads to reactions R_1 and R_2 being developed at both ends, and the system deforms to the state seen in Figure 2.24b. The deformations shown are exaggerated.

Since only one nontrivial equation of statics is available ($\Sigma F_x = 0$, with two unknowns R_i), this system is statically indeterminate to the first degree. We will assume positive forces and deflections so that any result with a negative sign will mean that the force or deflection in question is in the opposite direction from that drawn in Figure 2.24b. The force method tells us to "remove" one of the reactions (in the same hypothetical sense that we "slice" bodies open to use the method of sections). We choose to remove the right-hand reaction R_2 first. This permits the system to deform, as in Figure 2.24c.

We see that in Figure 2.24c, the same axial deformation Δ_o occurs at B as at C—in the imagined absence of reaction R_2 (imposed by the right wall), the bar is free to deform in this way. If the flexibility of the narrower elastic bar is f_2, we can use the definition of flexibility to write

$$\Delta_o = f_2 P. \tag{2.29}$$

But this deformation violates the geometric condition that is actually imposed at A: There is, truly, a wall that prevents a deflection of even Δ_o. To comply with geometric compatibility, we must find the deflection Δ_1 that would be caused by R_2 on the unloaded bar, as shown in Figure 2.24d. This deflection is caused by the stretching (if R_2 is in the direction shown; otherwise, the compression) of both constituent bars. Thus,

$$\Delta_2 = \frac{R_2 L_1}{A_1 E_1} + \frac{R_2 L_2}{A_2 E_2} = (f_1 + f_2)\, R_2. \tag{2.30}$$

We may then achieve compatibility by requiring that

$$\Delta_o + \Delta_1 = 0. \tag{2.31}$$

That is, there is no net deformation of the actual bar system. From this expression we find an expression for R_2:

$$R_2 = -\frac{f_2}{f_1 + f_2} P. \tag{2.32}$$

The negative sign here indicates that R_2 acts in the opposite direction from what we'd assumed: The bar is in compression. (The same is true for its deflection, Δ_1. It is negative, reflecting the fact that the bar is being compressed.)

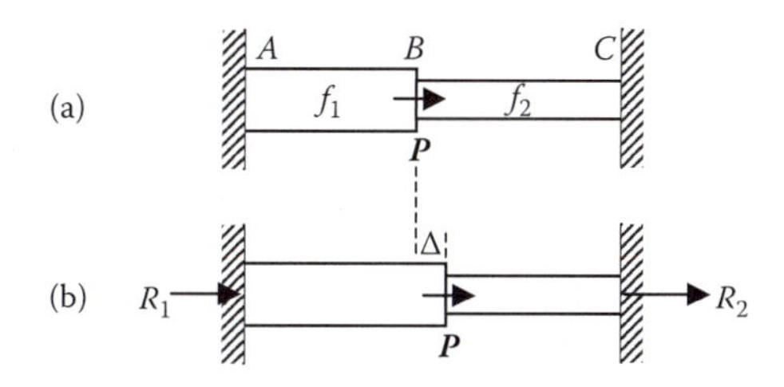

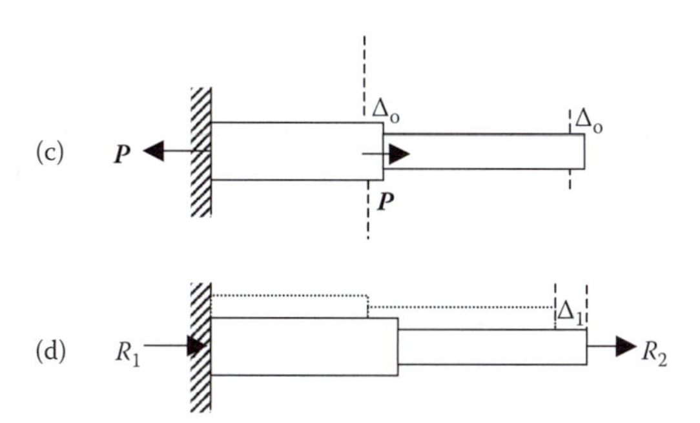

FIGURE 2.24
Decomposition of indeterminate bar by force method.

The idea of the force method is that the complete solution is the sum of the solutions shown in Figure 2.24c and Figure 2.24d; the method is an application of the principle of *superposition.* Our premise is that the resultant stress or strain in a system due to several forces is the algebraic sum of these forces' individual effects. This is only true if each effect is linearly related to the force causing it—that is, if we are in the Hookean range of behavior.

It may be useful to refer to the steps of the force method in problem solving:

1. Determine the number of redundants—that is, the number of forces that cannot be determined from equilibrium alone. The number of forces needed to maintain equilibrium is equal to the number of equations of equilibrium, so any additional forces are "redundant."
2. Choose some of the reactions to be the redundants and remove them from the structure, thus temporarily producing a determinate structure. There is no formal method or set of criteria for making the choice, so convenience, as viewed through the lens of experience, is the guiding principle for choosing redundants.
3. Calculate the displacements at the points from which redundants were removed, as produced by the actual (given) external loading.
4. Calculate the displacements at the points from which redundants were removed but now as produced by the redundants without the given external loading.
5. Sum the two displacements at each point where a redundant has been removed, as calculated in the last two steps—that is, as displacement (step 3) + displacement (step 4). Applying superposition to this linear structure, we see that we must add the actual displacement at that point of the fully loaded, indeterminate structure. We then calculate the values for the redundant forces from these equations. (We are enforcing *compatibility* or *consistency of deformations* when we perform this step.)
6. With the redundants determined in step 5 acting, determine the remaining support reactions of the fully loaded, indeterminate structure by applying equilibrium.

This procedure is very general; in practice, any number of axial loads, bar cross sections, material properties, and thermal effects on the length of a bar system may be included in your analyses. However, for very large systems, application of the force (flexibility) method is very difficult.

2.8.2 Displacement (Stiffness) Method

The displacement method is also known as the stiffness method. We remember that the stiffness of an axially loaded bar may be expressed as

$$k = AE/L.$$

If we are presented with a statically indeterminate elastic axially loaded bar system (like that in Figure 2.25a), we may define the stiffness of each member k_i as

$$K_i = A_i E_i / L_i.$$

An applied force P at point B causes reactions R_1 and R_2. As before, these forces and the displacement Δ at B are considered positive when they act toward the right.

Our objective is to determine the displacement Δ. (Since there is only one unknown Δ to be determined in this example, this problem is said to have one degree of kinematic indeterminacy, or one degree of freedom.) We also hope to find expressions for the reaction forces R_i.

In the problem considered here (Figure 2.25), the displacement Δ at B causes tension in bar AB and compression in bar BC. Because we understand this, we can assume the senses of the reaction forces as shown in Figure 2.25b. So, if k_1 and k_2 are the stiffnesses of the two bars, the respective internal forces are $k_1\Delta$ and $k_2\Delta$. These internal forces and reactions are shown on isolated free bodies at points A, B, and C in Figure 2.25c. These points are called *nodes*, or *nodal points*. The sense of the internal forces is known, since AB is in tension and BC is in compression. Writing an equilibrium equation for the free body at node B, we have

$$-k_1\Delta - k_2\Delta + P = 0, \tag{2.33a}$$

$$\Delta = \frac{P}{k_1 + k_2} b. \tag{2.33b}$$

Equilibrium for free bodies A and C gives us

$$R_1 = k_1\Delta \text{ and } R_2 = k_2\Delta. \tag{2.34}$$

So, synthesizing these three results, we find that

$$R_1 = \frac{k_1}{k_1 + k_2} P \text{ and } R_2 = \frac{k_2}{k_1 + k_2} P \tag{2.35}$$

in the directions indicated in Figure 2.25b, such that AB is in tension and BC in compression.

It may be useful to refer to this sequence of steps for the displacement method:

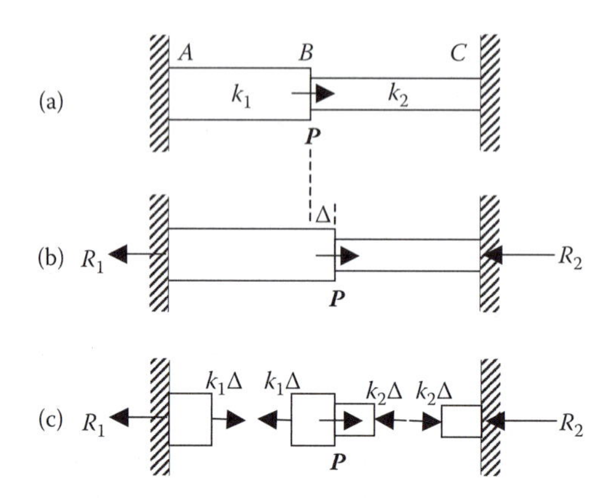

FIGURE 2.25
Displacement method for statically indeterminate bar.

1. Determine the number of redundants—that is, the number of forces that cannot be determined from equilibrium alone.
2. Identify within the structure a number of points equal to the number of redundants, and for each of these points identify a nodal displacement of the structure.
3. Calculate the forces needed to "produce" the nodal displacement, and sum all the forces at the nodes to enforce equilibrium.
4. Eliminate the nodal displacements from the nodal equilibrium equations to calculate the unknown nodal forces.
5. Determine the support reactions of the fully loaded, indeterminate structure by applying equilibrium.

2.9 Thermal Effects

So far, we have considered mechanical stress and externally applied loads as the only sources of strain in materials. With changes of temperature, however, solid bodies expand with increasing temperature and contract with decreasing temperature. These deformations produce *thermal strains.* We define thermal strain ε_T in the following way:

$$\varepsilon_T = \alpha\,(T - T_o) = \alpha\,\Delta T, \tag{2.36}$$

where α is an experimentally determined coefficient of (linear) thermal expansion, and T_o and T are the initial and final temperatures of our material of interest. The thermal expansion coefficient α measures dimensional change per degree of temperature change for a given material. Typical values in SI units of (m/m)/°C, or just $(°C)^{-1}$, range from 9.9×10^{-6} for concrete to 11.7×10^{-6} for carbon steel to 23×10^{-6} for aluminum.

Thermal strain has no directional dependence; equal thermal strains develop in every direction for unconstrained homogeneous isotropic materials. For a body of length L subjected to a temperature change, the extensional deformation δ_T is

$$\delta_T = \alpha\,(\Delta T)\,L, \tag{2.37}$$

where ΔT is allowed to be positive or negative for increasing or decreasing temperature.

If the body in question is free to expand or contract (i.e., the body is not restrained), no stress is induced by these thermal effects. The dimensional change δ_T will simply occur, and the otherwise unloaded bar will continue to be in equilibrium. However, if the body is partially or fully restrained so as to prevent this change δ_T, internal thermal stresses will develop. Thermal stress for a temperature change ΔT is given as

$$\sigma_T = E\,\alpha\,(\Delta T). \tag{2.38}$$

If this body is fully restrained and then cooled, the stress induced is tensile; if the body is fully restrained and then heated, the stress induced is compressive. The stresses and strains due to thermal effects may be combined with the stresses and strains in the same directions by straightforward superposition.

2.10 Saint-Venant's Principle and Stress Concentrations

In applying equations such as $\sigma = P/A$, we have assumed that forces and stresses are distributed uniformly across their surfaces of action. In ideal cases such as the axially loaded bars of the previous sections, this is very nearly the true situation. However, in more realistic scenarios, things are more complex. Fortunately for us, many researchers have performed detailed calculations of stress states, and have learned things from the distributions they found. We may benefit from their conclusions without performing arduous computations ourselves.

An exemplary such result came from the analysis of an elastic block, acted on by concentrated forces at its ends, as in Figure 2.26a. (Of course, in the real world, a truly concentrated force such as this one is not even possible.) The calculated stress distributions at three incremental depths within the bar are shown in Figure 2.26b, Figure 2.26c, and Figure 2.26d. Clearly, these are not uniform distributions across the cross section.

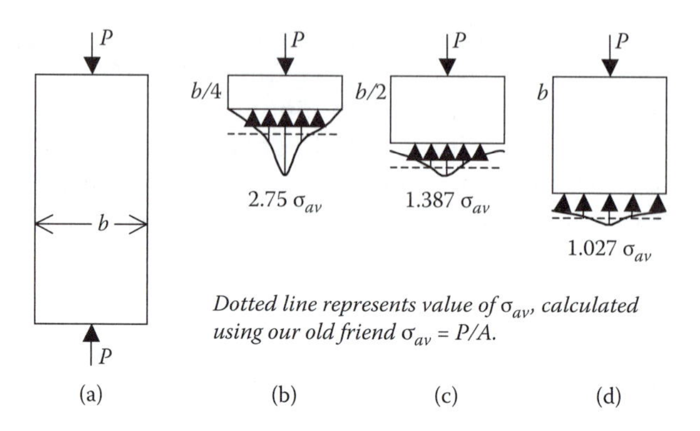

FIGURE 2.26
Stress distribution near concentrated force in plate.

Two important facts may be gleaned from these results. One, the average stresses as calculated by our formulas (stress = force/area) are in agreement with these more carefully obtained numbers. Two, the normal stresses are nearly uniform on a surface whose distance from the applied force is the same as the width of the body. (This is true despite the high spatial variation in stress at surfaces nearer to the force application.) This second point illustrates *Saint-Venant's Principle,* as first stated by the eponymous French elastician in 1855. It means that the manner of force application (point, or evenly distributed, or other) has a significant effect on the stress distribution only in the near vicinity of the force's application. We are applying this principle when we idealize our systems.

Highlighted in Figure 2.26b, Figure 2.26c, and Figure 2.26d are the maximum normal stresses at each cut and their proportionality to the average stress. This maximum stress and its relation to average stress is a function of geometry. In particular, features such as holes and filleted edges cause areas of stress concentration and ruin our idealization of uniform stress distribution. A formula is available for the calculation of maximum normal stress:

$$\sigma_{\max} = K\sigma_{av} = K\frac{P}{A}, \tag{2.39}$$

where K is an experimentally obtained stress concentration factor for the particular geometric feature in question. Figure 2.27 shows stress concentration factors for flat axially loaded members with three types of change in cross section.

Whether the area A used in equation (2.39) is the original area (without a hole) or the reduced area can vary with researcher and data; this naturally affects the value of K. The data in Figure 2.27 are based on the reduced cross section. In cases not covered by the graph in Figure 2.27, another reference (e.g., *Peterson's Stress Concentration Factors,* by Walter Pilkey (1997)) or an online stress concentration calculator may prove useful.

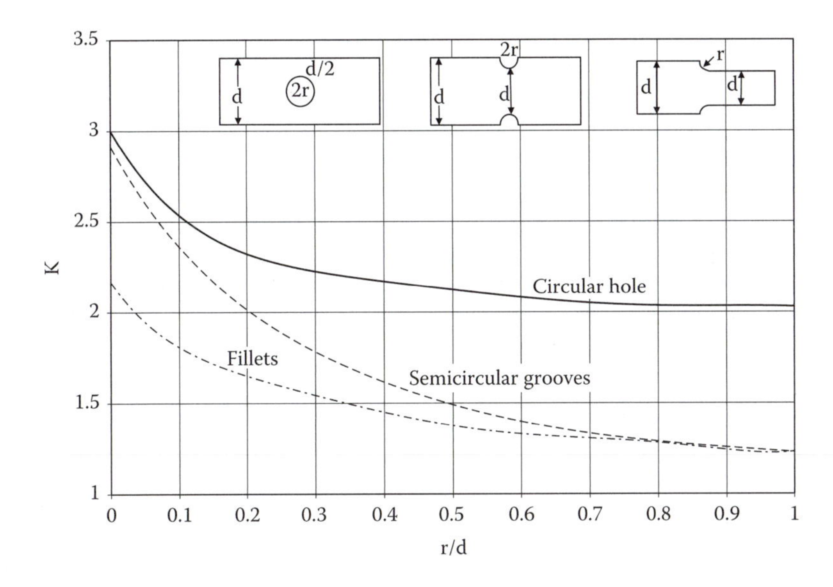

FIGURE 2.27
Stress concentration factors for flat bars. (Adapted from M. M. Frocht, *ASME Journal of Applied Mechanics* 2, A67–A68, 1935.)

In ductile materials, high stress concentration is not necessarily dangerous because these materials can accommodate high stresses through plastic yielding and subsequent stress redistribution. In brittle materials, cracks may occur in areas of high localized stress.

2.11 Strain Energy in One Dimension

Thanks to Robert Hooke, we have recognized that a solid material responds to loading in much the same way as a linear spring, as long as the material remains below its proportional limit. Recall that the linear elastic spring is an energy storage device for which we can calculate the stored energy as

$$U_{spring} = \int_0^x F_s dx = \int_0^x kx dx = \frac{1}{2}kx^2. \tag{2.40}$$

We can also calculate the strain energy stored in a deformed elastic solid. For the elementary one-dimensional Hooke's law, the *strain energy density*, U_0, or strain energy per unit volume (check the dimensions!) can be calculated as the work done by a stress state acting through its corresponding strain:

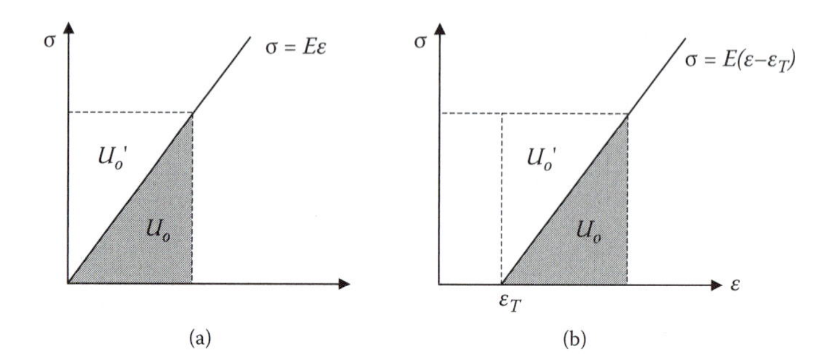

FIGURE 2.28
Stress-strain relationship and energy densities (a) without and (b) with internal stresses.

$$U_0 = \int_0^{\varepsilon} \sigma d\varepsilon = \int_0^{\varepsilon} E\varepsilon d\varepsilon = \frac{1}{2}E\varepsilon^2. \tag{2.41}$$

As with the comparable spring calculation, we recognize U_0 as the area below the stress–strain curve given by Hooke's law, as is shown in Figure 2.28a. The area above the stress–strain curve is the *complementary energy density*:

$$U_o^C = \frac{\sigma^2}{2E}. \tag{2.42}$$

The important point to note here is that while the strain and complementary energy densities are obviously equal, we refer to the strain energy density when the expression is cast in terms of strains or displacements, and we refer to the complementary energy density when the corresponding expression is written in terms of stresses or forces. For the spring the comparable formulas would be

$$U_{spring} = \frac{1}{2}kx^2, \tag{2.43a}$$

$$U_{spring}^C = \frac{F_s^2}{2k}. \tag{2.43b}$$

When thermal stresses are included, the calculation is somewhat less straightforward because of the offset of thermal strain along the strain axis, as shown in Figure 2.28b. Thus, here the strain energy density integration would take the form

$$U_0 = \int_{\varepsilon_T}^{\varepsilon} \sigma d\varepsilon = \int_{\varepsilon_T}^{\varepsilon} E(\varepsilon - \varepsilon_T) d\varepsilon = \frac{1}{2} E(\varepsilon - \varepsilon_T)^2, \tag{2.44}$$

which in expanded form can also be written as

$$U_0 = \frac{1}{2} E\varepsilon^2 - E\varepsilon(\alpha T) + \frac{1}{2} E(\alpha T)^2. \tag{2.45}$$

As a final note, which is easily verified, the strain and complementary energies must always satisfy the requirements (and produce the results) that

$$\frac{\partial U_o(\varepsilon)}{\partial \varepsilon} = E\varepsilon \equiv \sigma, \tag{2.46a}$$

$$\frac{\partial U_o^{\prime C}(\sigma)}{\partial \sigma} = \frac{\sigma}{E} \equiv \varepsilon. \tag{2.46b}$$

2.12 A Road Map for Strength of Materials

For one-dimensional loading, we have addressed the checklist for continuum mechanics, involving (1) kinematics, or description of deformation; (2) a definition of stress; (3) a relationship between stress and strain; and (4) equilibrium. We must next turn our attention to loading in multiple dimensions so that we may model more realistic problems. If we look back at our modeling of stretched or compressed bars, we can discern a pattern of thought that serves as a road map for a more general approach to problems in strength of materials, structural analysis, and elasticity.

Our road map encompasses six major physical elements, beginning with the external loads—the given, applied loads on a solid. These loads or forces are the "drivers" of our analyses because, as engineers, we design structures and machine elements to support, guide, and contain the effects of the external loads. This was illustrated in our analysis of a long, thin bar that was being pulled (or pushed) by an axial force.

The reactions are external forces that support the loaded body and keep it from moving in response the given applied loads. They are determined by requiring the body in its entirety to be in equilibrium under the given externally applied loads. There are many kinds of reactions. We needed only one axially directed support to ensure equilibrium for the stretched (or compressed) bar.

The internal forces $N(x)$ are the force distributions or stress resultants needed to maintain internal equilibrium. Stresses describe the distribution of the internal forces over planar sections drawn through the body's interior. They were defined as point functions of the body's coordinates. So, although it seemed relatively straightforward to define a stress as the quotient $N(x)/A$, where A is the bar's cross-sectional area, we want to extend and generalize this simple definition.

The strains are measures of the deformation of the body that result from the applied forces. There are many definitions of strain, which we reviewed in Section 2.1. The strains are specifically related to the stresses by constitutive laws that describe the properties of the material of which the body is made (cf. Section 2.3). The strains are required to be compatible, by which we mean that their point-by-point variation cannot produce holes in the continuous material of which the body is made, nor can they permit deformation that violates any geometrical constraints relative to the supports that keep the body in place. Simply put, we want our models to reflect "well-behaved" deformation that doesn't produce physically untenable results.

The displacements or the deflections are the (generally) more visible movements of the body. The strains are typically found by differentiating the displacements or deflections with respect to spatial coordinates, as we began to see in Section 2.1.1 and further explore in Chapter 3. The deflections must also be compatible—that is, they must conform with the geometry of the body and its support constraints.

In the language of continuum mechanics, we can now restate our four-item checklist as three major physical considerations that must be applied:

- *Equilibrium* considerations relate external forces, reactions, internal forces, and stresses. That is, we apply Newton's second law to relate external loads to reactions; the method of sections to relate external forces and reactions to internal (resultant) forces; and both Newton's laws and the method of sections to relate internal forces to stresses.
- *Constitutive laws* relate stresses to strains. We invoke constitutive laws to describe the properties of the material of which a body is made.
- *Compatibility* considerations relate strains to displacements or deflections (i.e., kinematics). We pay attention to compatibility both when calculating movements and deflections and when ensuring consistency and continuity with respect to the geometry of the body and its support constraints.

The order in which we apply these criteria, or in which we check off items on our checklist, is not important. The requirement is that our analyses include all of them, no matter the order.

2.13 Examples

Example 2.1

Figure 2.29a shows a diagram of the bones and biceps muscle of a person's arm supporting a mass; Figure 2.29b shows a biomechanical model of the arm, in which the biceps muscle AB is represented by a bar with pin supports. The suspended mass is $m = 2$ kg, and the weight of the forearm is 9 N. If the cross-sectional area of the tendon connecting the biceps to the forearm at A is 28 mm^2, what is the average normal stress in the tendon?

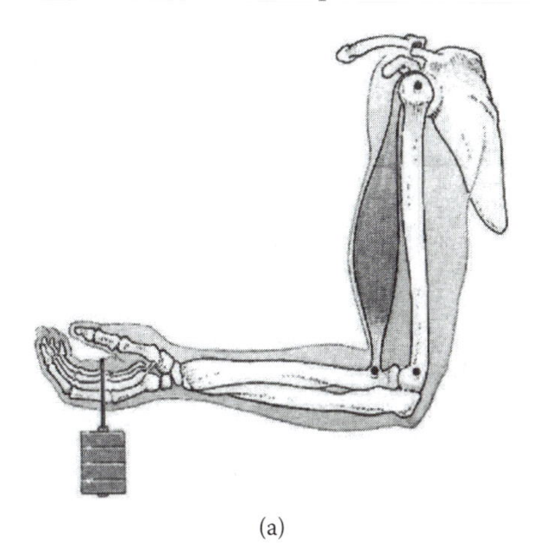

(a)

(b)

FIGURE 2.29

Given: Dimensions of and loading on truss system.

Find: Average normal stress in tendon AB.

Assume: Equilibrium; planar system; neglect weight of muscle and tendon AB.

Solution

We first need to find the internal axial force in AB and then calculate the normal stress by dividing this force by the cross-sectional area. We must construct an FBD of the system (Figure 2.30):

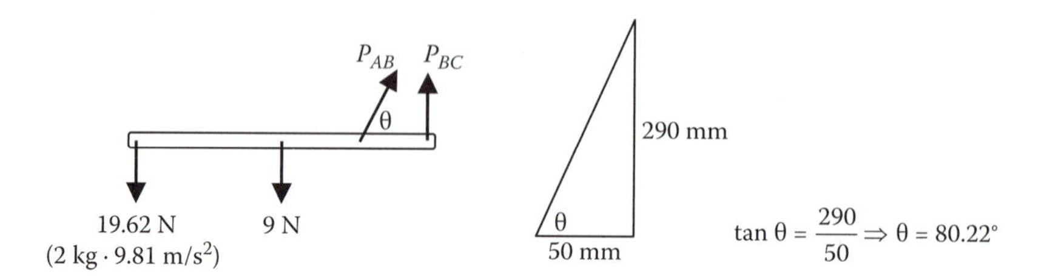

FIGURE 2.30

Equilibrium requires that the sum of moments taken about point C be zero, where a counterclockwise moment is taken to be positive:

$$\Sigma M_C = 0 = 19.62 \text{ N } (0.35 \text{ m}) + 9 \text{ N } (0.15 \text{ m}) - P_{AB} \sin\theta \, (0.05 \text{ m}).$$

Solving for P_{AB},

$$P_{AB} = \frac{8.22 \text{N} \cdot \text{m}}{(0.05\text{m}) \sin\theta} = 166.76 \text{ N}.$$

The average normal stress σ_{AB} is then

$$\sigma_{AB} = \frac{P_{AB}}{A_{AB}} = \frac{166.76\text{N}}{28 \times 10^{-6} m^2} = 5.95 \text{ MPa}.$$

Example 2.2

An infinitesimal rectangle at a point in a reference state of a material becomes the parallelogram shown in a deformed state (Figure 2.31). Determine (a) the extensional strain in the dL_1 direction; (b) the extensional strain in the dL_2 direction; and (c) the shear strain corresponding to the dL_1 and dL_2 directions.

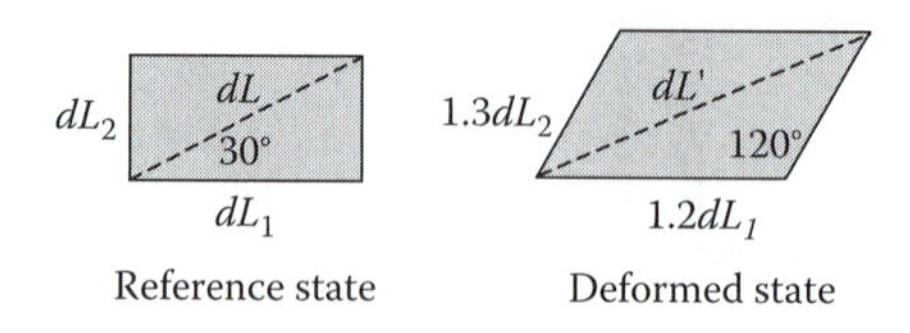

FIGURE 2.31

Given: Reference and deformed geometries of infinitesimal rectangle.

Find: Normal and shear components of strain.

Assume: Strain definitions are adequate; use of "true" strain integral is unnecessary.

Solution

Normal strain in dL_1 direction:

$$\varepsilon_1 = \frac{dL_1' - dL_1}{dL_1} = \frac{1.2dL_1 - dL_1}{dL_1} = 0.2.$$

Normal strain in dL_2 direction:

$$\varepsilon_2 = \frac{dL_2' - dL_2}{dL_2} = \frac{1.3dL_2 - dL_2}{dL_2} = 0.3.$$

Shear strain is the angular deformation, or change in angle between two reference lines. In reference state, the angle between dL_1 and dL_2 is 90°, or $\pi/2$. In the deformed state, the angle between dL_1' and dL_2' is 60°, or $\pi/3$. The shear strain is thus

$$\gamma_{12} = \frac{\pi}{6} = 0.524 \text{ radians.}$$

Note: If we tried to approximate shear strain by the tangent of this angular deformation instead of using the angle itself, we would get

$$\gamma = \frac{1.3dL_2 \sin \pi / 6}{dL_2} = 0.650 \text{ radians.}$$

This is close, but not *that* close, to 0.524 radians. The angular change in this problem is not sufficiently small to justify the use of the tangent in place of the angle itself.

Example 2.3

Three metal balls are suspended by three wires of equal length arranged in sequence as shown in Figure 2.32. The masses of the balls, starting at the top, are 2 kg, 4 kg, and 3 kg, respectively. In the same order, beginning at the top, the wires have diameters 2 mm, 1.5 mm, and 1 mm, respectively. (a) Determine the highest stressed wire, and (b) by changing the location of the balls, optimize the mass locations to achieve a system with minimum stresses.

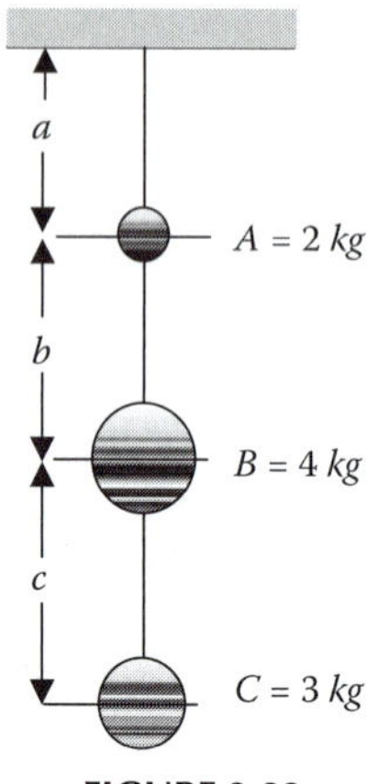

FIGURE 2.32

Given: Dimensions and arrangement of steel balls.

Find: Stresses in each wire; lowest-stress configuration.

Assume: Neglect weights of wires.

Solution

We must find the internal force within each wire and then divide by the wire's cross-sectional area to find the normal stress in each wire. For each wire, the internal force will equal the mass this wire must support times the acceleration of gravity. For example, the top wire, *a*, must support 2 + 4 + 3 kg, so its internal axial force is 88.3 N. We tabulate these calculations:

	P_i (N)	A_i (m^2)	σi (MPa)
Wire a	88.3	3.14×10^{-6}	28.1
Wire b	68.7	1.77×10^{-6}	38.8
Wire c	29.4	0.79×10^{-6}	37.2

The wire subjected to the highest stress is wire *b*.

To achieve a minimum stress system, we recognize that stress is inversely proportional to cross-sectional area. Hence, since $A_a > A_b > A_c$, wire *a* should carry the largest load (which it must), and wire *b* and wire *c* should support as little load as possible. This leads us to the following configuration (Figure 2.33):

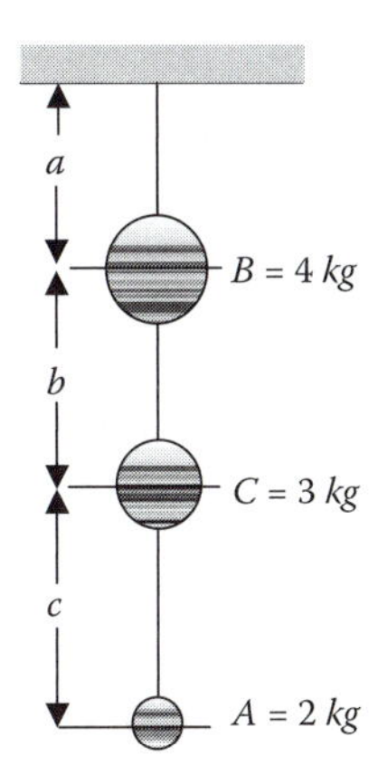

FIGURE 2.33

	σ_i (MPa)
Wire a	28.1
Wire b	27.7
Wire c	24.8

In the configuration of part (a), the total stress in the three-wire system is 104 MPa; in part (b), the total stress is 80.6 MPa.

Example 2.4

A steel bar 10 m long used in a control mechanism must transmit a tensile force of 5 kN without stretching more than 3 mm or exceeding an allowable stress of 150 MN/m^2. What must the diameter of the bar be? State your answer to the nearest millimeter, and use E = 200 GPa.

Given: Dimensions and loading on steel bar.

Find: Required bar diameter to nearest mm.

Assume: Hooke's law applies.

Solution

We will impose both *strength* and *stiffness* constraints on the bar and will see which is the limiting case.

Using the definition of normal stress, we must have

$$\sigma = \frac{P}{A} \leq 150 \frac{\text{MN}}{\text{m}^2},$$

or

$$A \geq \frac{P}{150\ \text{MN/m}^2} = \frac{5000\ \text{N}}{150 \times 10^6 \text{N/m}^2} = 33.33 \times 10^{-6}\ \text{m}^2,$$

that is,

$$A \geq 33.33\ \text{mm}^2.$$

If Hooke's law applies, as we have assumed it does, then

$$\delta = \frac{PL}{AE},$$

and we must have

$$\frac{PL}{AE} \leq 3\ \text{mm}$$

or

$$A \geq \frac{PL}{\delta E} = \frac{5000\ \text{N} \cdot 10\ \text{m}}{(0.003\ \text{m})(200 \times 10^9 \text{N/m}^2)} = 83.33 \times 10^{-6}\ \text{m}^2$$

that is,

$$A \geq 83.33\ \text{mm}^2.$$

We see that stiffness is the limiting case and that we must have a cross-sectional area greater than or equal to 83.33 mm^2 to safely meet our constraint. This is all we need to find the required diameter of the steel bar:

$$\frac{\pi}{4} d^2 \geq 83.33\ \text{mm}^2,$$

so

$$d \geq 10.3\ \text{mm}.$$

So to the nearest millimeter, we must use an 11-mm-diameter bar.

Example 2.5

A solid bar 50 mm in diameter and 2000 mm long consists of a steel and an aluminum section, as shown in Figure 2.34. When axial force P is applied to the system, a strain gauge attached to the aluminum indicates an axial strain of 0.000873 m/m. (a) Determine the magnitude of applied force *P*, and (b) if the system behaves elastically, find the total elongation of the bar.

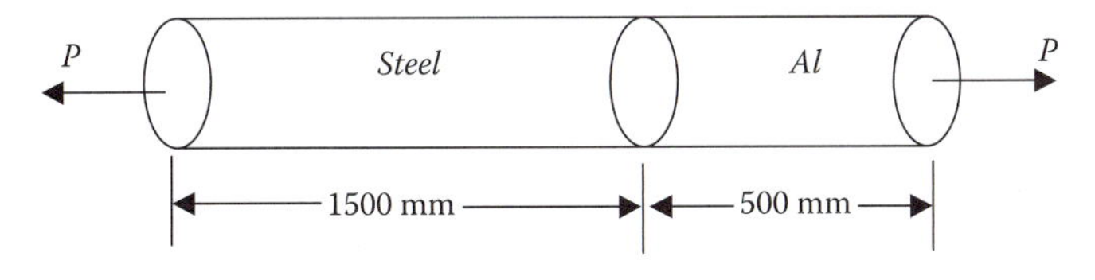

FIGURE 2.34

Given: Dimensions of composite bar and measured normal strain.

Find: Applied force, *P*, and elongation of bar, δ.

Assume: Hooke's law applies.

Solution

The diameter of the bar is 50 mm, so the cross-sectional areas of both parts are equal:

$$A_{St} = A_{Al} = \frac{\pi}{4}\left(0.05\ \text{m}\right)^2 = 0.00196\ \text{m}^2.$$

The elastic moduli for aluminum and steel may be looked up in Table 2.1 or in another reference.

$$E_{Al} \approx 70\ \text{GPa, and } E_{St} \approx 200\ \text{GPa.}$$

If Hooke's law applies, we can relate the strain measured in the aluminum portion to the stress induced by *P* in that portion:

$$\varepsilon_{Al} = 0.000873 = \frac{\sigma_{Al}}{E_{Al}} = \frac{(P/A_{Al})}{E_{Al}},$$

so

$$P = (0.000873)(70 \times 10^9\ \text{Pa})(0.00196\ \text{m}^2) = 120\ \text{kN}.$$

We can exploit Hooke's law and superpose the displacements of both portions of the bar:

$$\delta = \sum \frac{PL}{AE} = \left(\frac{PL}{AE}\right)_{St} + \left(\frac{PL}{AE}\right)_{Al}$$

$$= \frac{120{,}000 \text{ N}}{0.00196 \text{ m}^2}\left[\frac{1.5 \text{ m}}{200\times 10^9 \text{ N/m}^2}+\frac{0.5 \text{ m}}{70\times 10^9 \text{ N/m}^2}\right]$$

$$= 459 \times 10^{-6} \text{ m} + 437 \times 10^{-6} \text{ m} = 896 \times 10^{-6} \text{ m}$$

$$\delta = 896\ \mu\text{m, or } 0.896 \text{ mm.}$$

Note: The aluminum section is only a third as long as the steel, but it deforms nearly as much!

Example 2.6

A polystyrene bar consisting of two cylindrical portions AB and BC is restrained at both ends and supports two 26 kN loads as shown in Figure 2.35. Knowing that E is 3.1 GPa, determine (a) the reactions at A and C, and (b) the normal stress in each portion of the bar.

Given: Dimensions of and loading on composite polystyrene bar.

Find: Reactions and normal stresses.

Assume: Hooke's law applies. Neglect weight of polystyrene cylinders.

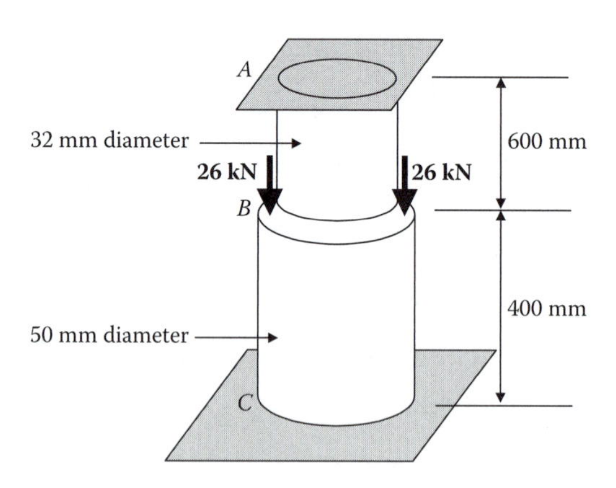

FIGURE 2.35

Solution

The first thing we need is an FBD (Figure 2.36):

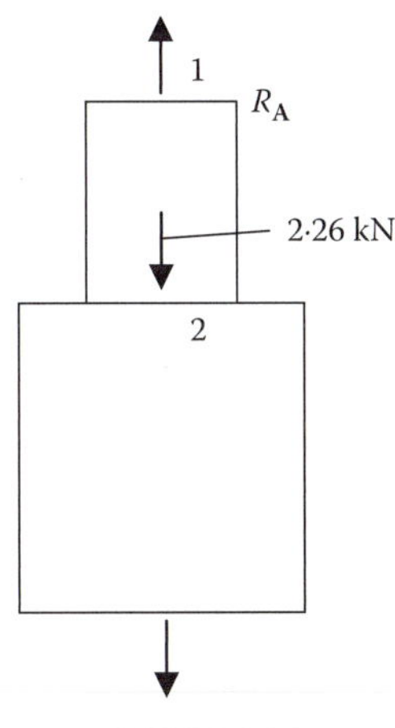

FIGURE 2.36

We ensure that this system is in equilibrium by stating,

$$\Sigma F_y = 0, \text{ or } R_A + R_B = 52 \text{ kN}.$$

This one equation contains two unknowns: The problem is statically indeterminate. So, what else do we know? Because both ends are fixed, the total elongation of the composite bar must be zero. So,

$$\delta_{AB} + \delta_{BC} = 0.$$

Using Hooke's law, we can write these displacements as

$$\delta_{AB} = \frac{P_{AB} L_{AB}}{A_{AB} E} \text{ and } \delta_{BC} = \frac{P_{BC} L_{BC}}{A_{BC} E}.$$

We then use the method of sections to find P_{AB} and P_{BC}, the internal forces in the two component sections (Figure 2.37).

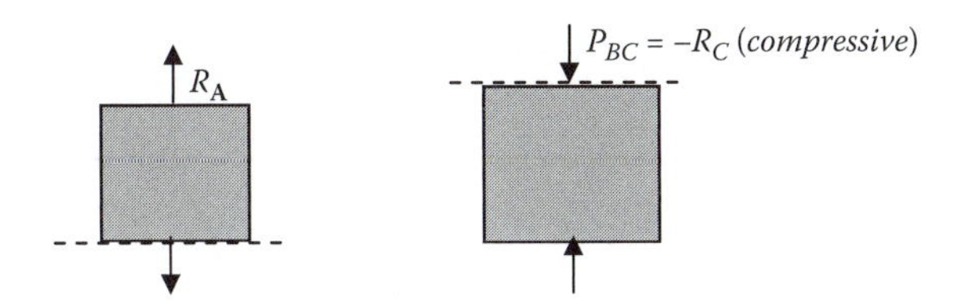

FIGURE 2.37
Measurement of tension in a string.

So, imposing the geometric constraint that the total elongation of the bar is zero, we have

$$\frac{R_A L_{AB}}{A_{AB}E} + \frac{-R_C L_{BC}}{A_{BC}E} = 0,$$

or

$$(2.407 \times 10^{-7})R_A - (6.572 \times 10^{-7})R_C = 0.$$

We also know $R_A + R_C = 52$ kN, so we can solve these two equations to get the reaction forces:

$$R_A = 11.2 \text{ kN}$$

$$R_C = 40.8 \text{ kN}.$$

Then we divide the internal forces by the cross-sectional areas they act on to obtain the normal stresses in both pieces:

$$\sigma_{AB} = \frac{P_{AB}}{A_{AB}} = \frac{R_A}{A_{AB}} = \frac{11.2 \times 10^3 \text{ N}}{\frac{\pi}{4}(0.032 \text{ m})^2} = 14.0 \text{ MPa}$$

$$\sigma_{BC} = \frac{P_{BC}}{A_{BC}} = \frac{-R_c}{A_{BC}} = \frac{-40.8 \times 10^3 \text{ N}}{\frac{\pi}{4}(0.05 \text{ m})^2} = -20.8 \text{ MPa}.$$

Example 2.7

Each bar in the truss shown in Figure 2.38 has a 2 in.[2] cross-sectional area, modulus of elasticity $E = 14 \times 10^6$ psi, and coefficient of thermal expansion $\alpha = 11 \times 10^{-6}$ (°F)$^{-1}$. If their temperature is increased by 40°F from their initial temperature T, what is the resulting displacement of point A? What upward force must be applied to prevent this displacement?

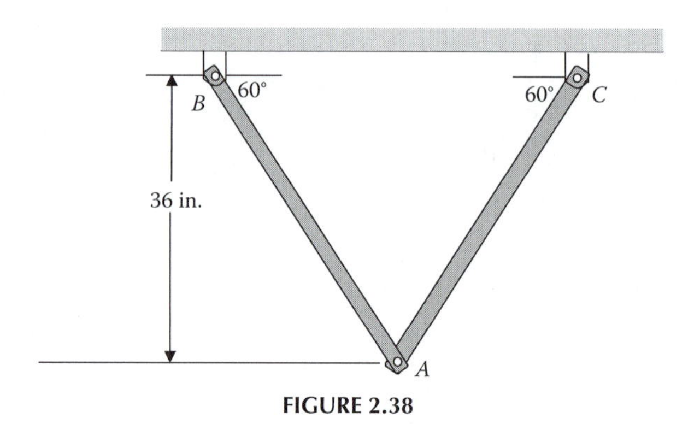

FIGURE 2.38

Given: Dimensions and properties of truss; imposed temperature change.
Find: Displacement of point *A*; force necessary to prevent this displacement.
Assume: Hooke's law applies. Neglect weight of bars.

Solution

The geometry of the problem allows us to find the original length of bars *AB* and *AC* (Figure 2.39):

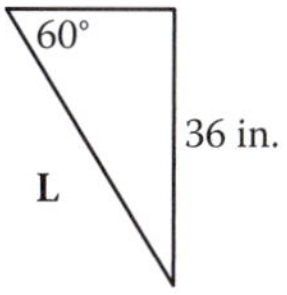

60°
36 in.
L

FIGURE 2.39

$$L = 36 \text{ in.}/\sin(60°) = 41.56 \text{ in.}$$

The change in the length of each bar due to the change in temperature ΔT is then

$$\delta_T = L\,\alpha\,\Delta T = (41.56 \text{ in.})(11 \times 10^{-6}\,(°\text{F})^{-1})(40°\text{F}) = 0.018 \text{ in.}$$

So, the new vertical distance from the fixed surface to point *A* is

$$(41.56 + 0.018) \cdot \sin(60°) = 36.008 \text{ in.}$$

The horizontal displacements of *AB* and *AC* will be equal and opposite, so the net displacement of point *A* is only vertical and is 0.008 in.

The upward force applied to prevent this must "undo" the thermal expansion of the two bars; it must induce a compressive axial load *P* in both bars such that, by Hooke's law,

$$\frac{PL}{AE} = -0.018 \text{ in., so } P = \frac{(-0.018 \text{ in.})(14 \times 10^6 \text{ psi})(2 \text{ in.}^2)}{41.56 \text{ in.}} = -12{,}127 \text{ lb (compressive!)}.$$

We construct an FBD and use equilibrium to find the force *F* necessary to induce this compressive load *P* in both bars (Figure 2.40):

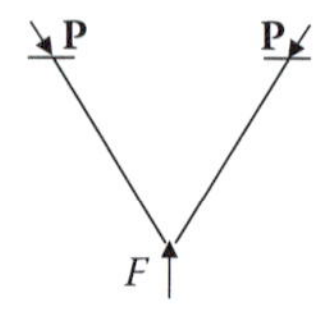

FIGURE 2.40

$$\Sigma F_y = 0 = F - 2P\sin(60^\circ),\text{ or } F = 2P\sin(60^\circ)$$
$$F = 21{,}327\text{ lb.}$$

Example 2.8

A steel railroad track (E = 200 GPa, $\alpha = 11.7 \times 10^{-6}/^\circ C$) was laid out at a temperature of 0°C. Determine the normal stress in a rail when the temperature reaches 50°C, assuming that the rails are (a) welded to form a continuous track, or (b) 12 m long with 6-mm gaps between them.

Given: Geometry of problem, material properties, imposed temperature change.

Find: Normal stress (a) when continuous or (b) when gaps are left.

Assume: Hooke's law applies.

Solution

Based on our understanding of thermal stresses, we expect the stress calculated in part (b) to be lower than that in part (a): We have learned that thermal stresses are induced only when a part is prevented from experiencing its natural thermal deformation, so the space left to accommodate thermal expansion in part (b) should help relieve the induced stress. We will see whether this expectation is met.

A schematic helps to illustrate the problem (Figure 2.41):

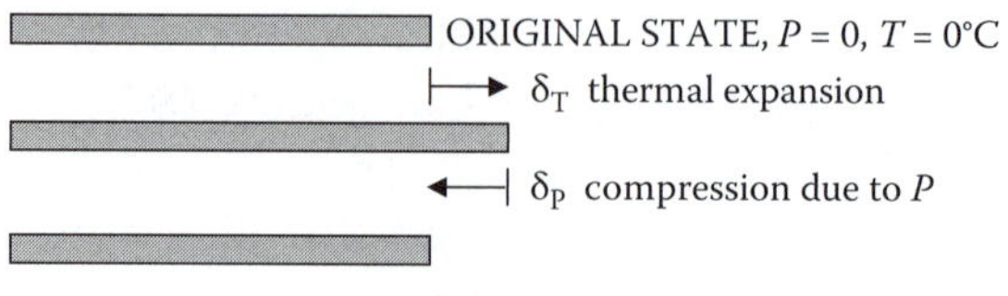

FIGURE 2.41

(a) The total deformation of a steel track segment is $\delta_T + \delta_P = 0$, as the welding allows no net change to the length of the segments. Hence, we add the deformations due to thermal effects and compressive forces:

$$0 = \underbrace{\alpha\,(\Delta T)L}_{\text{tends to stretch}} + \underbrace{\frac{-PL}{AE}}_{\text{tends to squash}},$$

so

$$\alpha\,\Delta T = \frac{P}{AE} = \frac{\sigma}{E}$$

$$\sigma = \alpha\,\Delta T\,E = (11.7 \times 10^{-6}\,(^\circ\text{C})^{-1})(50 - 0^\circ\text{C})(200 \times 10^9\text{ Pa})$$

$$\sigma = 117\text{ MPa (compressive)}$$

when welded.

(b) If a gap of 6 mm is left between rails, we allow each segment a net stretch of 6 mm:

$$+0.006\text{ m} = \alpha\,(\Delta T)L - \frac{PL}{AE},$$

so

$$\sigma = \frac{\alpha\,\Delta\,TL - 0.006\text{ m}}{L}E = \left(\frac{11.7\times10^{-6}(^\circ\text{C})^{-1}(50^\circ\text{C})(12\text{ m}) - 0.006\text{ m}}{12\text{ m}}\right)\left(200\times10^9\text{ Pa}\right)$$

$$\sigma = 17\text{ MPa (compressive)}$$

when a gap is left.

Example 2.9

A 6 mm × 75 mm plate, 600 mm long, has a circular hole of 25 mm diameter located at its center. Find the axial tensile force that can be applied to this plate in the longitudinal direction without exceeding an allowable stress of 220 MPa. How does the presence of the hole affect the strength of the plate?

Given: Dimensions of plate, limiting normal stress.

Find: Allowable axial load that can be applied to plate.

Assume: Hole is only feature that causes a stress concentration.

Solution

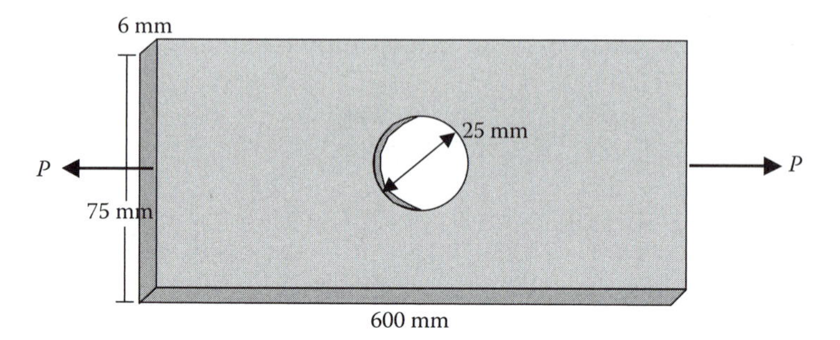

FIGURE 2.42

The cross-sectional area normal to an axial load P is $A_o = 6$ mm × 75 mm = 450 mm^2 (Figure 2.42). The average normal stress induced by such a load will be

$$\sigma_{ave} = P/A_o,$$

and due to the presence of the hole we must consider the effects of *stress concentration:*

$$\sigma_{\max} = K\sigma_{ave} = K\frac{P}{A_o}.$$

We can find K for this geometry using the graph in Figure 2.27:

$$\frac{r}{d} = \frac{(25 \text{ mm})/2}{75 \text{ mm } - 25 \text{ mm}} = \frac{1}{4},$$

$$K(\tfrac{r}{d} = 0.25) = 2.26$$

so

$$\sigma_{\max} = K\frac{P}{A_o} = 2.26\frac{P}{450 \text{ mm}^2} = 0.005P$$

and since we must have

$$220 \text{ MPa} \geq 0.005\,P$$

then

$$P \leq 43.8 \text{ kN}.$$

Note: If there were no hole in this plate, we would simply have

$$\sigma_{ave} = P/A_o,$$

and we could allow a force

$$P \le 99 \text{ kN}.$$

So with the hole, we can permit only 44% of the load we could have allowed without the hole.

2.14 Problems

2.1 In tissue engineering, biological materials are grown from seeded cells, so that artificial corneas, blood vessels, or other materials may be made from biological materials. Such materials are less likely than artificial parts made of plastic or metal to be rejected by the body. To engineer true replacement parts, it is necessary to understand the behavior of physiological systems and to match material properties such as elastic and shear moduli. It is impractical to construct a tension specimen like that in Figure 2.2 from soft tissues such as muscles, tendons, or blood vessels. What would you do instead?

2.2 Concrete, rocks, and bone are strong in compression and are usually designed for compressive loading. To test their strength in compression, what sort of test specimen would be useful?

2.3 The tension in your Achilles tendon is considerable when you stand on tiptoe or poise for a jump. Design a tension gauge that might be useful in measuring such tension, or the tension in a bow string or rubber slingshot. (*Hint*: after Fung 1994; Figure 2.43):

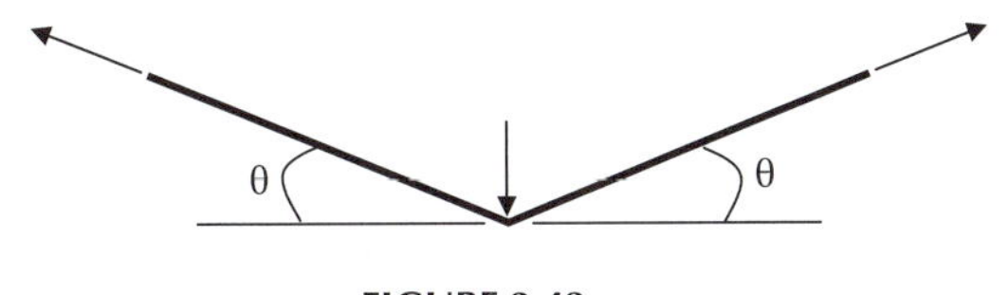

FIGURE 2.43

2.4 When a kangaroo switches from "pentapedal" (four limbs and tail) locomotion to hopping, its oxygen consumption drops, presumably because it then stores more energy in elastic tissues (Dawson and Taylor 1973). One of these elastic tissue "springs" in kangaroos (and other animals) is the Achilles tendon. A kangaroo's Achilles tendon was found to be 1.5 cm in diameter and 35 cm long. If each Achilles tendon has an elastic modulus of 1 GPa and is loaded to 2% strain (below its proportional limit), how much strain energy (i.e., stored potential energy) would both Achilles tendons contain? Based strictly on energy considerations, can you predict how high this amount of energy could lift a 40 kg kangaroo?

2.5 In Figure 2.44, the suspended mass $m = 20$ kg. Determine the axial force in the bar *AB*, and indicate whether it is in tension or compression. (*Hint*: Draw the free-body diagram of joint *B*.)

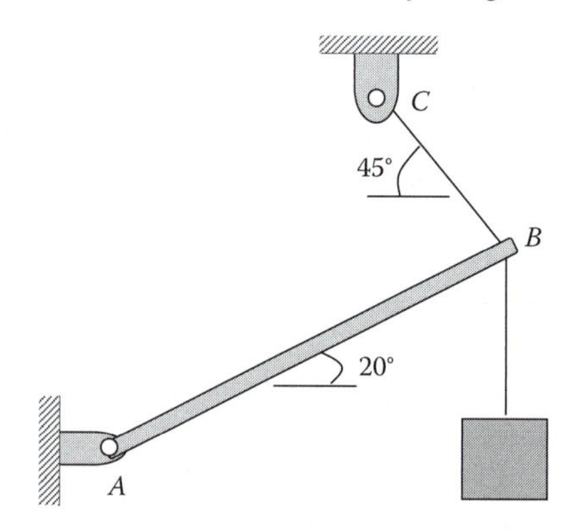

FIGURE 2.44

2.6 The bar shown in Figure 2.46 has a solid circular cross section, with a 2 in. radius. Determine the average normal stress (a) at plane P_1, and (b) at plane P_2.

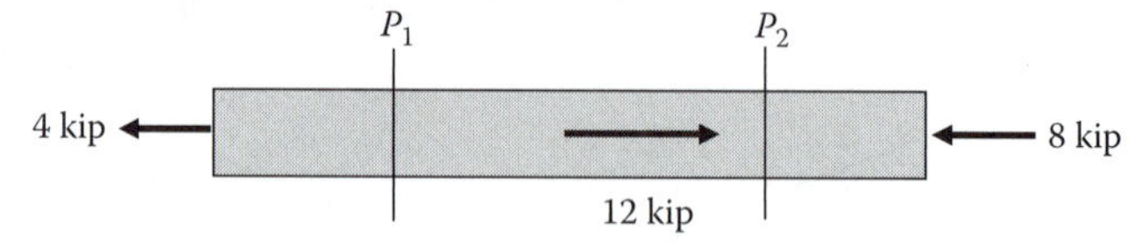

FIGURE 2.45

2.7 Suppose that a downward force is applied at point *A* of the truss, causing point *A* to move 0.360 in. downward and 0.220 in. to the left (Figure 2.46). If the resulting extensional strain ε_{AB} in the direction parallel to the axis of bar *AB* is uniform, what is ε_{AB}?

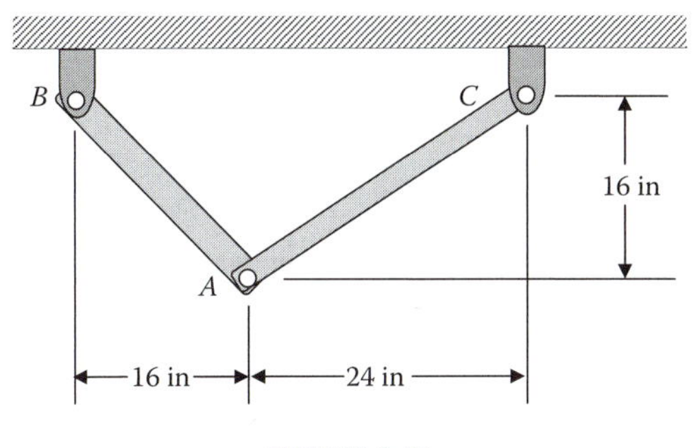

FIGURE 2.46

2.8 The jaws of the bolt cutter shown in Figure 2.48 are connected by two links *AB*. The cross-sectional area of each link is 750 mm². (a) What average normal stress is induced in each link by the 90 N forces exerted on the handles? (b) The pins connecting the links *AB* to the jaws of the bolt cutter are 20 mm in diameter. What average shear stress is induced in the pins by the 90 N forces exerted on the handles?

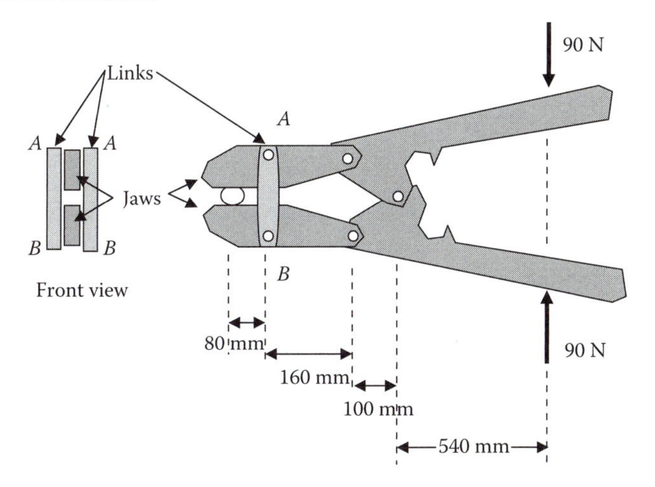

FIGURE 2.47

2.9 Determine the maximum allowable value of the force *F* if the tensile stress in segment *AB* must be less than 150 MN/m² (Figure 2.48). What are the changes in length of segment *BC* and of the entire bar for this value of *F*? The bar's cross-sectional area is 50 mm², and the bar is made of steel.

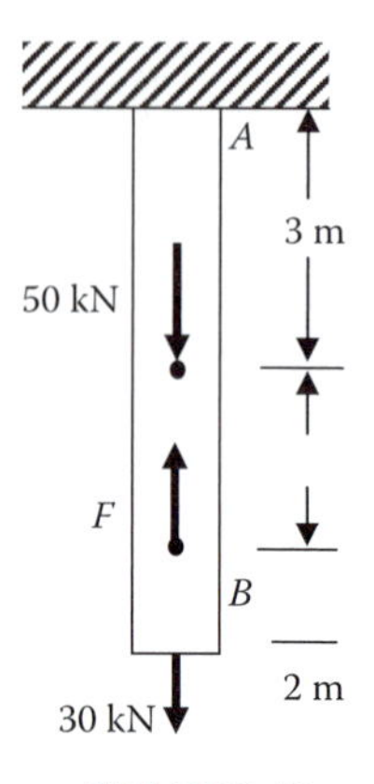

FIGURE 2.48

2.10 The bar shown in Figure 2.49 has a constant cross section and is fixed rigidly at both walls. Determine the reactions at both walls for the given applied load P.

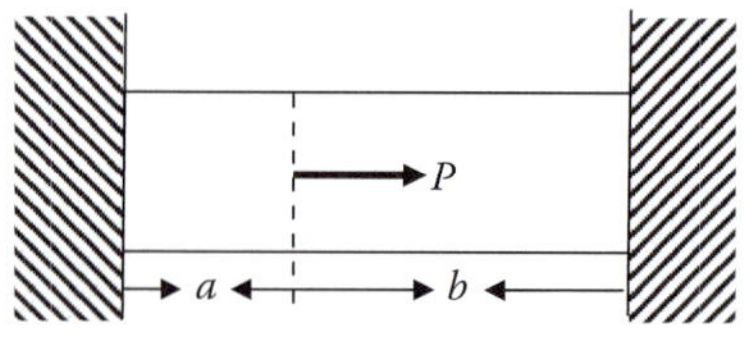

FIGURE 2.49

2.11 A rigid slab with mass m = 15,000 kg is supported by three columns, as shown in Figure 2.50. Determine the compressive force in each of the columns.

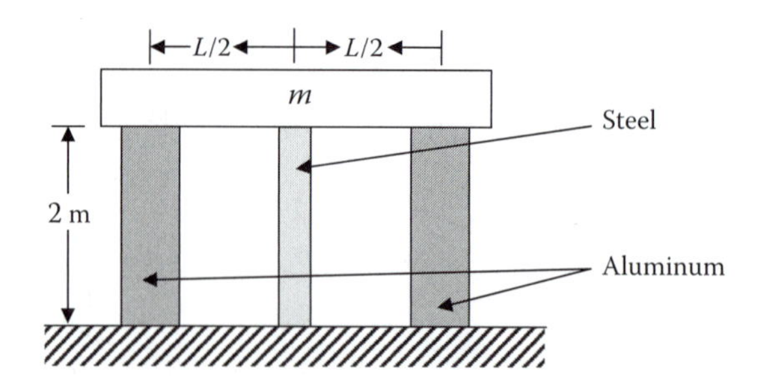

FIGURE 2.50

2.12 The bar shown in Figure 2.51 has a varying cross section and is fixed rigidly at both walls. The cross-sectional area of the narrower section is A; the cross-sectional area of the wider section is larger by a factor of m, or mA. Using the force (flexibility) method, determine (a) the reactions at both walls for the given applied load P; and (b) the displacement of the point D at which the load P acts.

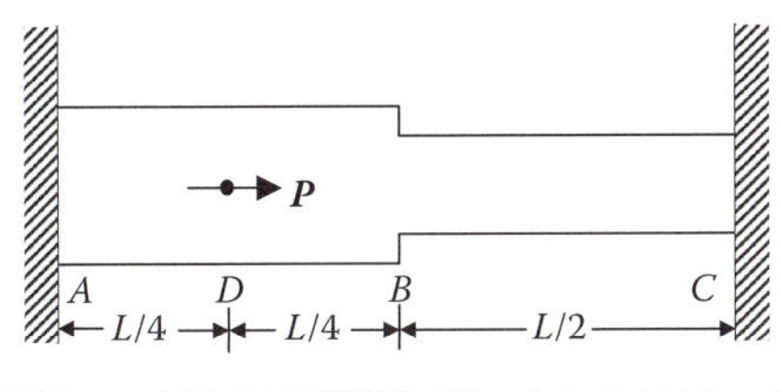

FIGURE 2.51

2.13 The bar shown in Figure 2.51 has a varying cross section and is fixed rigidly at both walls. The cross-sectional area of the narrower section is A; the cross-sectional area of the wider section is larger by a factor of m, or mA. Using the displacement (stiffness) method, determine (a) the reactions at both walls for the given applied load P; and (b) the displacement of the point D at which the load P acts.

2.14 Determine the movement δ of the tip of a dense, heavy uniform bar hanging vertically from one end. The bar has length L, modulus E, cross-sectional area A, and mass density ρ.

2.15 Your local lumber yard is providing a set of wooden 4 in. × 4 in. posts that you will mount on 6 in. × 6 in. concrete bases to support a section of roof. Handbooks provide the allowable compressive stresses: 1800 psi for wood and 1250 psi for concrete. The specific weight of concrete is also given as 150 lbf/ft^3, although the corresponding number for wood is not shown. For the postsupport configuration shown in Figure 2.52, determine (a) the specific weight of wood, given that a 2 ft section of the post weighs 21 lb and knowing that a 4 in. × 4 in. post is actually 3.5 in. × 3.5 in.; (b) the supported load P when the weights of the support and post are included; and (c) the supported load P when the weights of the support and post are not included.

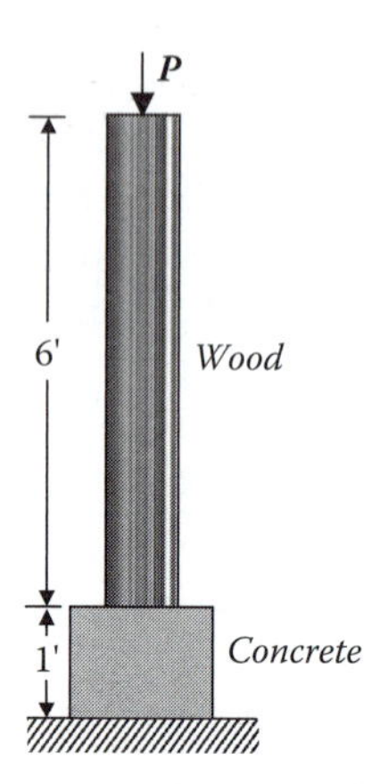

FIGURE 2.52

2.16 For the wood block shown in Figure 2.53, the allowable shear stress parallel to the grain is 1 MN/m^2, and the maximum allowable compressive stress in any one direction is 4 MN/m^2. Determine the maximum compressive force F that the block can support.

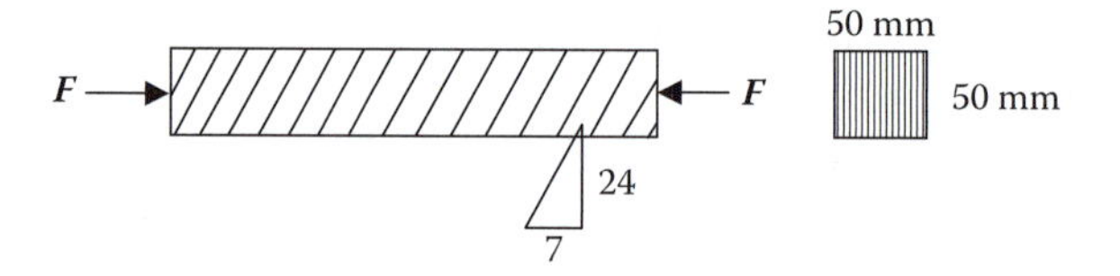

FIGURE 2.53

2.17 Two cylindrical bars with 30 mm diameters, one (ABC) made of yellow brass and the other (CDE) of stainless steel, are joined at C (Figure 2.54). End A of the composite bar is fixed, while there is a gap of 0.2 mm between end E and a vertical wall. A force of magnitude 40 kN and directed to the right is applied at B. Determine (a) the smallest force P needed at D to just close the gap without the steel bar actually coming into contact with the wall at E; (2) the reactions at A and E if a 40 kN force directed to the right is applied at D; and (3) the reactions at A and E if force P is twice the value you calculated in (a).

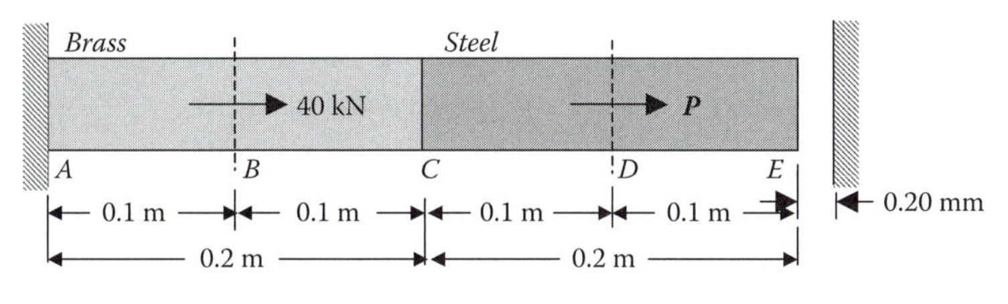

FIGURE 2.54

2.18 A bar consists of two portions BC and CD of the same material and same length L but of different cross sections (Figure 2.55). Determine the strain energy of the bar when it is subjected to an axial load P, expressing the result in terms of P, L, E, the cross-sectional area A of portion CD, and the ratio n of the two diameters.

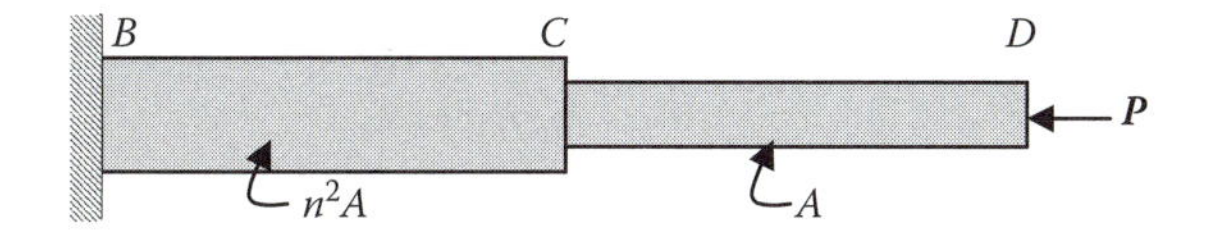

FIGURE 2.55

2.19 An electronic scoreboard is to be installed in a large stadium. Due to the design of the roof structure, the suspending cables will have different lengths, as is shown in Figure 2.56. (a) Determine a suitable cross-sectional area for each cable so that the scoreboard will hang level, accounting for the stretch in each cable. Use the data in the figure and the requirement that the cable's yield strength is 36 ksi. The modulus of elasticity for the cables is $E = 30{,}000$ ksi, and the weight of the scoreboard is $W = 10$ k. Remember, 1 kip (k) = 1000 lb. (b) The slope of the grain in the longer support cable has a maximum deviation from the cable's longitudinal axis of 15°, and there is some concern that the relatively low shear strength of the cable material along its grain could cause problems. Calculate both normal and shear stresses along this grain.

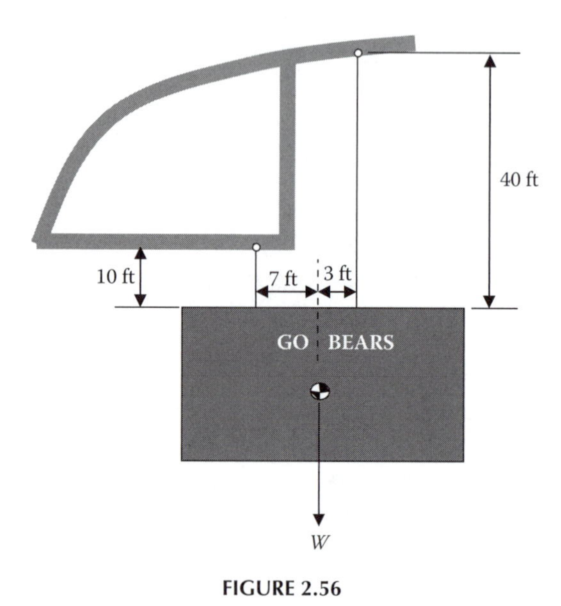

FIGURE 2.56

Case Study 1: Collapse of the Kansas City Hyatt Regency Walkways

On July 17, 1981, in the most damaging unforced structural failure in the history of the United States, two overhead walkways fell into the atrium lobby of the Hyatt Regency Hotel in Kansas City, Missouri. As a result of this collapse 114 people died, and millions of dollars of damage was sustained (Figure CS1.1).

The failure derived in large part from a key aspect of modern engineering design, which is that engineering designers do not, typically, build what they design. Rather, they produce a *fabrication specification,* a detailed description of the designed object that allows its assembly or manufacture by others. Separating the "designing" from the "making" means that such fabrication specifications must be complete and unambiguous.

Fabrication specifications are presented in drawings (e.g., blueprints, circuit diagrams, flow charts) and in text (e.g., parts lists, materials specifications, assembly instructions). Such traditional specifications can be complete and sufficiently specific, but they may not capture the designer's intent—and this can lead to catastrophe. The suspended walkways in the Hyatt Regency Hotel in Kansas City collapsed because a contractor fabricated the connections for the walkways in a manner different from the original design.

FIGURE CS1.1
The lobby of the Kansas City Hyatt Regency Hotel after the collapse of the second- and fourth-floor walkways on July 17, 1981. The devastation is evident. (Courtesy of Lee Lowry, Kansas City, MO.)

In the original design, walkways at the second and fourth floors were hung from the same set of 24-ft-long threaded rods that would carry their weights and loads to a roof truss (Figure CS1.2). The fabricator was unable to procure threaded rods sufficiently long to suspend the second-floor walkway from the roof truss, so instead, as shown in Figure CS1.3, he hung it from the fourth-floor walkway using shorter rods. (The original design would not have been easy to implement because of the difficulty involved in screwing on bolts over such long hanger rods and attaching walkway support beams.) The supports of the fourth-floor walkway were not designed to carry both the second-floor walkway and its own dead and live loads, resulting in the collapse. If the fabricator had understood the designer's intention to hang the second-floor walkway directly from the roof truss, this accident might have been avoided.

As Henry Petroski (1982) noted, the fabricator's redesign was akin to requiring that the lower of two climbers hanging independently from the same rope change his position so that he was grasping the feet of the climber above, causing the upper climber to carry the weights of both with respect to

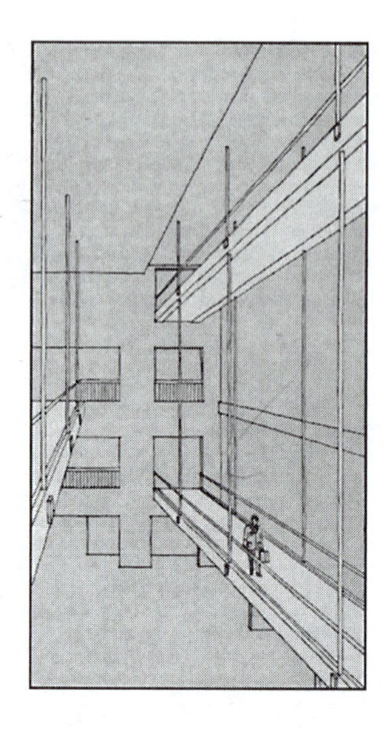

FIGURE CS1.2
An artist's sketch of the second- and fourth-floor walkways across the west side of the atrium of the Kansas City Hyatt Regency Hotel. The view looks southward and also shows a separate third-floor walkway that did not collapse but was taken down after a design review prompted by the collapse of the other two walkways on July 17, 1981. (From Pfrang, E. O. and Marshall, R., with E. J. Orwin and R. E. Spjut, *Civil Engineering*, pp. 65–68, July 1982. With permission.)

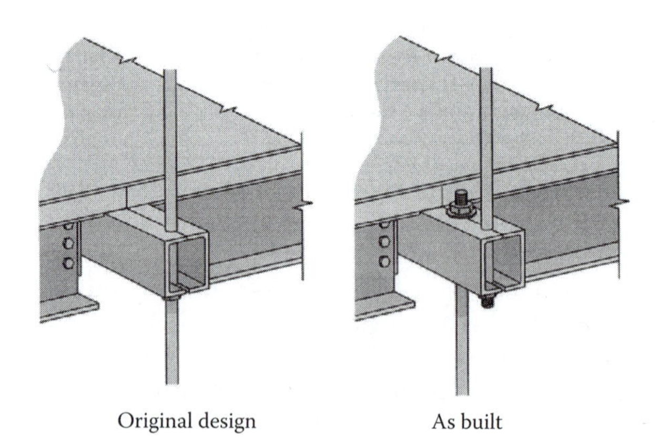

FIGURE CS1.3
The two hanger connections at the fourth-floor walkway: (a) The left sketch shows the configuration as designed, wherein the hanger rods went straight through the fourth-floor connection, down to the second floor, which these rods also supported. (b) The right sketch shows the configuration as built, with the hanger rods supporting the second floor now hung from the box beams that hold up the fourth-floor walkway. (From Dym, C. L. and Little, P., *Engineering Design: A Project-Based Introduction*, 3rd Ed., John Wiley & Sons, New York, 2008. With permission.)

the rope. The redesigned supports for the second-floor walkway were configured similarly.

Figure CS1.4 shows several sketches of the original design: (a) an elevation view of the second- and fourth-floor walkways, *each supported by the same pairs of hanger rods* (on east and west sides of the walkway) spaced at a distance L; and (b) an end view of the two walkways and FBDs of the supporting beam of each walkway. Consider now the lower, second-floor walkway. The load carried by each pair of its hanger rods can be estimated as the sum of the *dead load* of the walkway and its supporting beams and the *live load* of pedestrians likely to stand or walk across the walkways. Since the hanger rods are spaced a distance L apart we estimate the total force $2P$ needed to support a span of length $L/2$ on either side of a pair of hangers as

$$2P = (w + W)bL, \tag{CS1.1}$$

where w is the dead load per unit area, W the live load per unit area, and b the walkway width. In this instance, by both making calculations based on the design drawings and weighing pieces of the collapsed walkways, the engineers at the National Bureau of Standards (NBS)[9] who performed the forensic investigation of the walkway collapse determined that the combination of the dead and live loads, called the *design load*, was in this case $P = 90$ kN (20,300 lbf) per hanger rod. The analysis of the fourth-floor walkway based on the original design would be the same. Then the individual hanger rods needed to support both the second- and fourth-floor walkways as designed would each support a total load of $2P$ and would be sized accordingly.

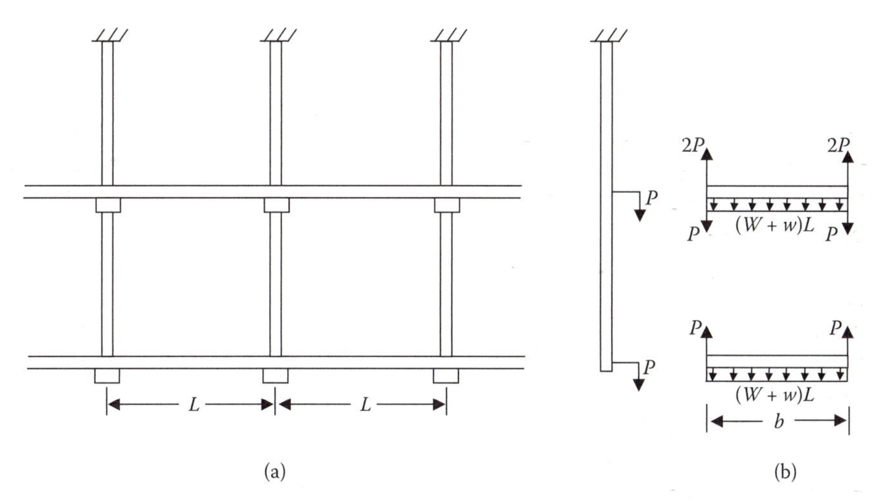

FIGURE CS1.4
Building a model of the walkways and their supports: (a) An elevation of the second- and fourth-floor walkways as originally designed. (b) An end view and free-body diagrams of the support beams. The forces carried by the hanger rods accumulate according to the number of walkways being supported below them.

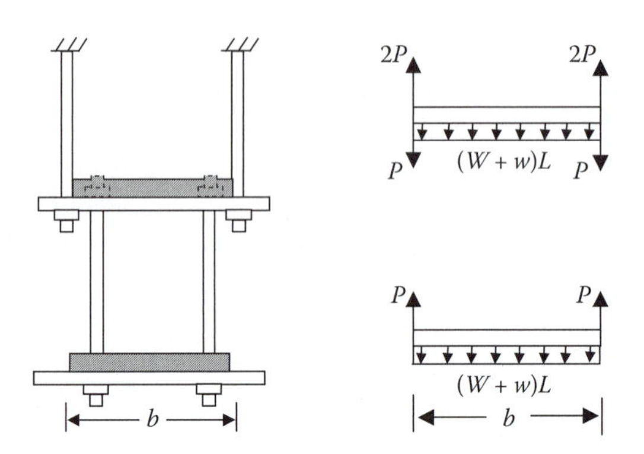

FIGURE CS1.5
Extending the model of the walkways and their supports to reflect the redesign. An end view of the second- and fourth-floor walkways designed so that the second-floor walkway hangs from the fourth-floor supporting beams, and free-body diagrams of a typical pair of supports. Note that the forces supported by the hanger rods are unchanged from the original design.

On the other hand, the end views of the walkways as built and their corresponding FBDs (Figure CS1.5) show that the rods would have to carry exactly the same loads at each level; that is, to support the lower walkway each rod carries a load equal to P, while above the fourth floor each rod would have to carry a load of $2P$ to support both the second- and fourth-floor walkways. So the rods in both designs would have equivalent designs with the same area, determined by equation (2.7),

$$A = \frac{2P}{\sigma_{\text{allow}}}, \quad \text{(CS1.2)}$$

where σ_{allow} is the allowable stress in the rod. In terms of the rope analogy, the part of the rope above the two climbers has to support the weight of both: It doesn't care whether each hangs directly from the rope or one climber hangs from the other.

So, why did the walkways collapse? They failed because an unanticipated connection was inserted into the design, the connection was not properly designed, and it failed (Figure CS1.6). As noted by respected engineers E. O. Pfrang and R. Marshall (1982), "With this modification the design load to be supported by each second floor ... connection was unchanged .... However, the load to be transferred from the fourth floor ... to the upper hanger rod under this arrangement was essentially doubled" (p. 68). Look again at the FBD in Figure CS1.5: It shows that the redesign required the nut under the fourth-floor supporting beam and its connection with the beam itself to support the transfer of twice the load that would have been transferred in the original design—which the fabricator's redesigned connection did not.

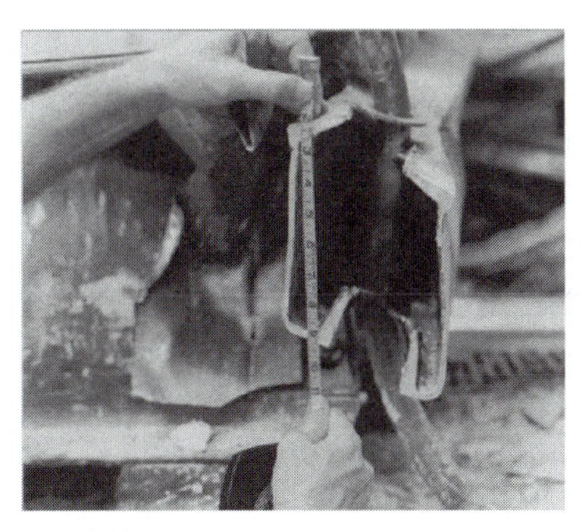

FIGURE CS1.6
Photographs of the failed connections that led to the collapse of the two walkways in the Kansas City Hyatt Regency Hotel. Compare it with Figure CS1.3 (b) and see that the outboard connection (on the right-hand edge) failed because the threaded nut and washer that went underneath the box beam pulled right through the box beam because that connection, designed originally to transmit a load of P, was actually carrying a load of $2P$. (Courtesy of Lee Lowry.)

Interestingly enough, it was also revealed in the subsequent forensic investigation that even the original design was only marginally safe. The NBS investigators found that the long-rod design would likely not have satisfied the Kansas City Building Code specifications. Further, it turned out that during construction, the building's construction workers had noticed that the walkways seemed flimsy and that they moved noticeably whenever workers moved wheelbarrows or the like across them. Their solution? Rather than report the problem and request a fix, they found other routes over which to transport their building materials!

The NBS official report issued in 1981 did not assign blame for this catastrophe. The essential problem was a lack of proper communication between the design engineers (Jack D. Gillum and Associates) and the manufacturers (Havens Steel). However, the NBS report's authors, Pfrang and Marshall, made it clear that responsibility lay primarily with the structural engineers. The Missouri licensing board and Court of Appeals agreed, finding that the design engineers should have noticed the difference between their design and what the contractor suggested and should have analyzed the redesigned connection. Basic calculations should have demonstrated the flaws in both the original design and in what was ultimately built. The principal structural engineers lost their Missouri engineer's licenses, and the firm, Jack D. Gillum and Associates, dissolved. The Hyatt Regency Crown Center lobby in Kansas City today features only one walkway, which is not suspended from the roof but instead rests on sturdy-looking columns that transmit its loads to the atrium floor.

Problems

CS1.1 If the Kansas City Building Code specified that a floor structure must support a live load of 4.79 kPa (100 psf), and if the walkway length $L = 9.1$ m = 30.0 ft and width $b = 2$ m = 6.56 ft, what contribution is made to the hanger rod load P?

CS1.2 If the design load is 90 kN (20,300 lbf), what is the dead load and what is the intensity of the dead load in the light of the live load calculation of Problem CS1.1?

CS1.3 Determine the specific weight of lightweight concrete and calculate its dead load intensity if it is used in an 80-mm (3.25 in.) cover of a formed steel deck walkway. Compare this result with that found in Problem CS1.2 and explain any differences.

CS1.4 Determine the stress induced in hanger rods carrying a design load of 90 kN (20,300 lbf), if their diameters are 32 mm (1.26 in.). Does that seem a reasonable stress level if the rods are made of mild steel? Explain your answer.

CS1.5 If the interfloor distance of the Hyatt Regency Hotel is 4.57 m (15 ft), how much does the second-floor walkway move with respect to the fourth-floor walkway?

Notes

1. Or, more generally, that Newton's second law is satisfied.
2. Carlson was a civil engineer investigating California dams; Simmons was an electrical engineer who first developed a way to manufacture a bonded-wire strain gauge and patented his design—though it took an 11-year court battle for him to win the patent rights for himself and not for Caltech, where he had been educated and continued to work.
3. Temperature considerations are important because the wires' electrical properties may be temperature dependent and also because temperature itself can result in deformation, as is quantified in Section 2.9.
4. Both P and F are used to represent forces in this textbook. The vector A is nA, where n is the outward normal vector of the area A.
5. From Gordon (1988) p. 45.
6. Such experiments were performed by Jacob Bernoulli (1654–1705) and J. V. Poncelet (1788–1867) in the quest to understand materials' response to loading.
7. Hooke (1635–1703) has never received due recognition for his scientific achievements. In addition to crafting what we know as Hooke's law, Hooke was an architect and geologist whose studies of microorganisms (using his friend Anton van Leeuwenhoek's newfangled microscopes) and fossils were seminal. Hooke's relative obscurity is largely a result of the efforts of his vindictive contemporary, Sir Isaac Newton, who used his own fame and influence to

diminish Hooke's accomplishments; it was his fear that Newton would steal or diminish *ut tensio, sic vis* that led Hooke to publish only his encrypted anagram for Hooke's law. He and Newton had had a feud over the inverse-square law of planetary motion, and Newton was so perturbed by it that he removed all traces of Hooke from his *Principia*. Hooke had even been prescient enough to anticipate the application of his observation of springs to material behavior, having stated that every kind of solid changes its shape when a mechanical force is applied and that it is this deformation that enables the solid to do what Gordon (1988) called "the pushing back." In so observing, Hooke anticipated the fields of continuum mechanics and elasticity. However, his intellectual heirs Thomas Young and Leonhard Euler were denied their inheritance by Newton, and Hooke remained obscure. Furthermore, we do not know what Hooke looked like, perhaps because—as he is often described as a "lean, bent, and ugly man"—he was unwilling to sit for a portrait.

8. If, instead, equilibrium is expressed in terms of displacements, the distinction between determinate and indeterminate problems vanishes, and we may apply the solution method developed in Section 2.7. However, it may prove useful (cf. Problem 2.8 and Problem 2.9) to work out a technique to resolve the indeterminacy of problems expressed in terms of stresses.
9. Since 1988 called the National Institute of Science and Technology (NIST).

3

Strain and Stress in Higher Dimensions

Now that we have constructed a foundation for our study of continuum mechanics, consisting of (1) kinematics or compatibility, (2) stress, (3) constitutive relationships, and (4) equilibrium, and have applied this to uniaxial loading and deformation, we are curious about the form this foundation will take in higher dimensions.

3.1 Poisson's Ratio

So far when we have discussed deformations of bodies in tension or compression, we have been referring to the deformation of a body in the direction of the applied uniaxial force. It is also true that in all solid materials, some deformation occurs at right angles to this force. That is, when a material is pulled along its axis, as shown in Figure 3.1a, it experiences some transverse (i.e., lateral) contraction. This is easily visualized using a rubber band. When pushed, the material feels transverse expansion (Figure 3.1b).

The deformations in Figure 3.1 are greatly exaggerated; in most engineering materials, this effect is small. One way to quantify material behavior, in fact, is to consider the relative *axial* and *lateral* strains due to axial loading. We do this by means of Poisson's ratio, first formulated by French scientist S. D. Poisson in 1828 and denoted by the Greek letter ν (nu):

$$\nu = \left| \frac{lateral\ strain}{axial\ strain} \right|, \tag{3.1a}$$

since

$$lateral\ strain = -\nu \cdot (axial\ strain). \tag{3.1b}$$

Poisson's ratio is a property of a material and can be found tabulated with other properties such as the modulus of elasticity E (e.g., in Appendix C of this book). Remember that the axial strains in question are caused by uniaxial stress only: by simple tension or compression. The value of ν varies for different materials; generally, it is on the order of 0.25 to 0.35 but can

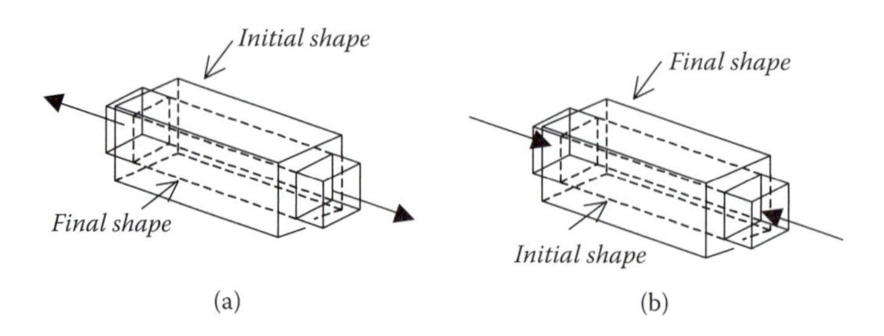

FIGURE 3.1
(a) Lateral contraction and (b) lateral expansion of solid bodies subjected to axial forces (Poisson's effect).

range from 0.1 (for some concretes) to 0.5 (for rubber).[1] Table 3.1 shows some of these values.

Note that the Poisson effect does not cause any additional stresses—*unless* the transverse deformation is inhibited or prevented. Incidentally, for Hookean solids it is possible to relate the three material properties we have so far discussed (elasticity modulus E, rigidity modulus G, and Poisson's ratio):

$$G = \frac{E}{2(1+\nu)}. \tag{3.2}$$

What Poisson's ratio reminds us is that our ideal situation of one-dimensional strain, considered in the previous chapter, is rarely physically realized. We must be conscious of a material's deformation in every dimension, even when loading is purely uniaxial. Although we will sometimes choose to neglect other dimensions, we should recognize that this is a choice to simplify our modeling and that we are leaving something out of our analysis.

TABLE 3.1
Poisson's Ratio for Common Materials

Material	ν
Rubber	0.50
Aluminum	0.35
Glass fiber-reinforced plastic	0.34
Mild steel	0.32
Wood	0.30
Cast iron	0.28
Stainless steel	0.15
Concrete	0.10

3.2 The Strain Tensor

The equations for strain as an average "percent deformation" presented in Chapter 2 (e.g., equations 2.1–2.4) are useful in a variety of straightforward loading conditions. However, in many cases we need to keep track of normal strains in multiple directions as well as shear strain. We can see how complicated the strain picture might become. There are three normal strains, in the x, y, and z directions, and in addition six shear strains, a pair in each plane. That's nine strain components in all. All of these "directions" or senses of strain are contained quite elegantly in the *strain tensor.*

Strain is a local property, and the values of each strain component may change dramatically within a material. And so we come to our mathematical definition of strain, which relates to relative deformations of an infinitesimal element.

If we consider the extensional strain in one direction of an original element AB with length Δx, as shown in Figure 3.2, we see that during straining, point A experiences a displacement u. This displacement is common to the whole element, a kind of "rigid-body displacement." A stretching Δu also takes place *within* the element, so that point B experiences a total displacement $u + \Delta u$.

Based on this situation, we define the extensional (normal) strain of this element as

$$\varepsilon = \lim_{\Delta x \to 0} \frac{\Delta u}{\Delta x} = \frac{du}{dx}, \tag{3.3}$$

(taking the limit as $\Delta x \to 0$ so that the expression will apply to any length Δx of the element). We see that this definition is independent of whatever rigid-body displacement occurred and is reminiscent of our third definition of normal strain, equation (2.4) (Chapter 2, Section 2.1.1).

If we extend our thinking to higher dimensions, as in Figure 3.3, we see that we now need to use subscripts to keep track of the direction of strain; we also need to use partial derivatives.

And so, if u, v, and w are the three displacements occurring in the x, y, and z directions, and if we again take the limits as dx, dy, and dz go to zero, we have three components[2] of normal strain:

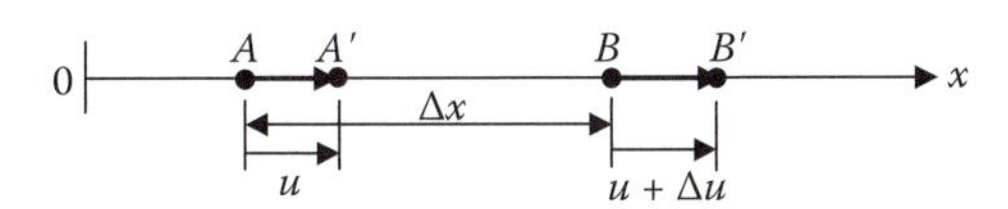

FIGURE 3.2
One-dimensional extensional strain.

$$\varepsilon_x = \varepsilon_{xx} = \frac{\partial u}{\partial x}, \tag{3.4a}$$

$$\varepsilon_y = \varepsilon_{yy} = \frac{\partial v}{\partial y}, \tag{3.4b}$$

$$\varepsilon_z = \varepsilon_{zz} = \frac{\partial w}{\partial z}. \tag{3.4c}$$

FIGURE 3.3
Two-dimensional extensional strain.

The analysis here shows that the three normal strains define the *change in shape* of a rectangular parallelapiped with initial volume $dxdydz$. A problem at the end of this chapter asks you to confirm that the normalized *change in volume* of the rectangular parallelapiped $dxdydz$ can be shown to be a function of the normal strains

$$\frac{\Delta V}{V} = \varepsilon_x + \varepsilon_y + \varepsilon_z \tag{3.5}$$

We can also encounter shear strain, as in Figure 3.4. After straining, the initially horizontal side with initial length dx has slope $\frac{dv}{dx}$, and the initially vertical side with initial length dy has slope $\frac{du}{dy}$. So, the initially right angle ACB is reduced by the amount $\frac{dv}{dx} + \frac{du}{dy}$. This is an angular deformation, or a shear strain, on the xy plane. We have

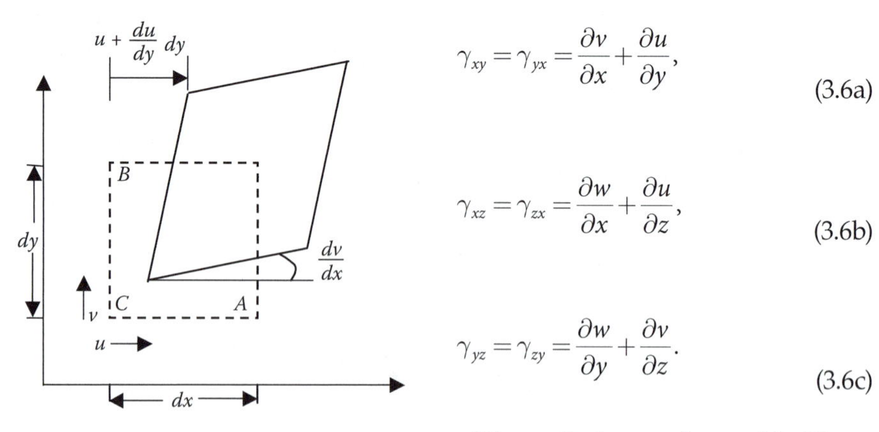

$$\gamma_{xy} = \gamma_{yx} = \frac{\partial v}{\partial x} + \frac{\partial u}{\partial y}, \tag{3.6a}$$

$$\gamma_{xz} = \gamma_{zx} = \frac{\partial w}{\partial x} + \frac{\partial u}{\partial z}, \tag{3.6b}$$

$$\gamma_{yz} = \gamma_{zy} = \frac{\partial w}{\partial y} + \frac{\partial v}{\partial z}. \tag{3.6c}$$

FIGURE 3.4
Shear strain in two dimensions.

We see that, as we learned in Chapter 2, Section 2.1.2, the engineering shear strain is equal to the change in the right angle between the two coordinates denoted in the subscripts.

By considering all the possible deformations of our parallelapiped, we have identified nine total, and six unique, strain components: ε_x, ε_y, ε_z, $\gamma_{xy} = \gamma_{yx}$, $\gamma_{yz} = \gamma_{zy}$, and $\gamma_{xz} = \gamma_{zx}$. Somehow all of these components together represent the total state of strain for our continuum. Strain is a *second-order tensor*, with one more level of sophistication than a vector.

Remember that we were able to write vectors, such as a force $\underline{P}$, as column vectors:

$$\begin{pmatrix} P_x \\ P_y \\ P_z \end{pmatrix}.$$

A vector is also known as a *first-order tensor.* It contains information about both magnitude and direction. For strain, we have magnitudes and directions as well as data relative to which our strain is quantified (e.g., the xy plane for the strain component γ_{xy}). The normal strain component ε_{xx}, for example, represents the magnitude of deformation in the x direction, relative to a reference length in the x direction. We are able to write the nine components of strain as a 3 × 3 matrix, which is one way to represent a second-order tensor.

Although this definition is physically motivated and mathematically sound, these engineering shear strains are not exactly the shear components of the strain tensor. For a reason having to do with the fact that elements do not truly behave as rigid bodies, the strain tensor components are actually defined using a factor of ½. This factor is necessary to make ε_{ij} behave mathematically as a proper tensor, as future study in continuum mechanics will show. We can write the strain tensor (using index notation) as

$$\varepsilon_{kl} = \frac{1}{2}(u_{l,k} + u_{k,l}) = \frac{1}{2}\left(\frac{\partial u_l}{\partial x_k} + \frac{\partial u_k}{\partial x_l}\right). \tag{3.7}$$

This equation defines six independent terms that form the components of a symmetric, second-order tensor—*symmetric* refers to the fact that each $\varepsilon_{ij} = \varepsilon_{ji}$ for $i \neq j$ (e.g., $\varepsilon_{xy} = \varepsilon_{yx}$) and of course also that $\gamma_{xy} = \gamma_{yx}$. Writing these out in longhand notation, we get terms like

$$\varepsilon_{xx} = \frac{\partial u}{\partial x} \tag{3.8}$$

and

$$\varepsilon_{xy} = \frac{1}{2}\left(\frac{\partial u}{\partial y} + \frac{\partial v}{\partial x}\right) \equiv \frac{1}{2}\gamma_{xy} \tag{3.9}$$

Or, we can write the strain tensor in its matrix form as

$$\varepsilon_{ij} = \begin{pmatrix} \varepsilon_x & \varepsilon_{xy} = \frac{\gamma_{xy}}{2} & \varepsilon_{xz} = \frac{\gamma_{xz}}{2} \\ \varepsilon_{yx} = \frac{\gamma_{xy}}{2} & \varepsilon_y & \varepsilon_{yz} = \frac{\gamma_{yz}}{2} \\ \varepsilon_{zx} = \frac{\gamma_{xz}}{2} & \varepsilon_{zy} = \frac{\gamma_{yz}}{2} & \varepsilon_z \end{pmatrix}. \tag{3.10}$$

3.3 Strain as Relative Displacement

Both normal and shear strain can also be characterized as *relative displacements* (Williams 2001) of segments of a body. For example, for small displacements, the axial extension or stretch of a body is simply the axial component of the relative displacement of the two ends of the bar. This way of thinking about strain lends itself nicely to the analysis of assemblies of bars known as trusses.

For the one-dimensional system in Figure 3.2, the quantity $\frac{du}{dx}(\Delta x)$ is identified as the relative displacement of the right end of the element *AB* from the perspective of the left end. This is consistent with our previous discussions, as $\frac{du}{dx}$ is the axial normal strain. By generalizing this thinking to higher dimensions, we can also discuss shear strains in terms of "relative displacements." Figure 3.5 illustrates a pair of line elements in the *xy* plane, before and after deformation of some body on which these elements have been etched. These line elements have initial lengths *dx* and *dy*.

When the body on which the elements *dx* and *dy* are etched is loaded, the line elements move. The arc lengths AA^*, BB^*, and CC^* result from the respective displacements $\underline{\delta}(A)$, $\underline{\delta}(B)$, and $\underline{\delta}(C)$ of the endpoints of the elements:

$$AA^* = \underline{\delta}(A) = u(x,y)\hat{\underline{i}} + v(x,y)\hat{\underline{j}} \tag{3.11}$$

$$BB^* = \underline{\delta}(B) = (u + \frac{\partial u}{\partial x}dx)\hat{\underline{i}} + (v + \frac{\partial v}{\partial x}dx)\hat{\underline{j}} \tag{3.12}$$

$$CC^* = \underline{\delta}(C) = (u + \frac{\partial u}{\partial y}dy)\hat{\underline{i}} + (v + \frac{\partial v}{\partial y}dy)\hat{\underline{j}} \tag{3.13}$$

where we have included the corresponding unit vectors to identify $u(x, y)$ and $v(x, y)$ as the displacements in the x and y directions, respectively. Then

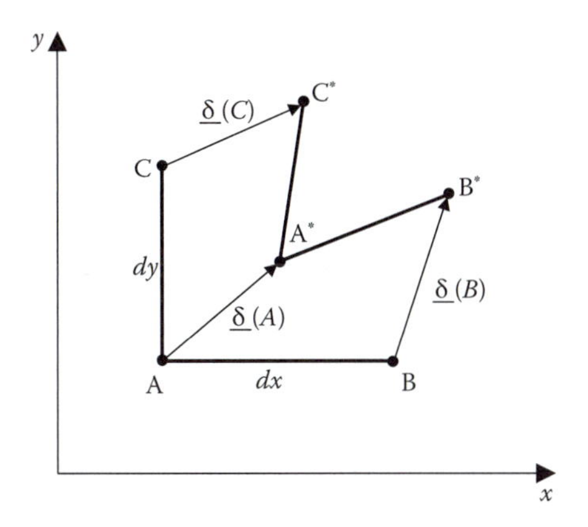

FIGURE 3.5
Pair of line elements before and after (*) deformation.

the relative displacements (with respect to the new location A^*) of the endpoints B^* and C^* of the two line segments dx and dy are

$$\underline{\delta}(B)-\underline{\delta}(A)=\left(\frac{\partial u}{\partial x}dx\right)\underline{\hat{i}}+\left(\frac{\partial v}{\partial x}dx\right)\underline{\hat{j}}$$

$$\underline{\delta}(C)-\underline{\delta}(A)=\left(\frac{\partial u}{\partial y}dy\right)\underline{\hat{i}}+\left(\frac{\partial v}{\partial y}dy\right)\underline{\hat{j}}. \tag{3.14}$$

Under this interpretation, the normal strain of a line element is the relative *longitudinal* or axial displacement of one end with respect to the other, divided by the original element length. Thus, here, the *normal strains* are just

$$\varepsilon_{xx}=\frac{(\vec{\delta}(B)-\vec{\delta}(A))\cdot\underline{\hat{i}}}{dx}=\frac{\partial u}{\partial x} \tag{3.15a}$$

$$\varepsilon_{yy}=\frac{(\vec{\delta}(C)-\vec{\delta}(A))\cdot\underline{\hat{j}}}{dy}=\frac{\partial v}{\partial y} \tag{3.15b}$$

The *engineering shear strain* is defined as the change in the (initially) right angle at the intersection of the two line elements caused by the relative *transverse* or normal displacement. Thus, here,

$$\gamma_{xy} = \frac{(\underline{\delta}(B) - \underline{\delta}(A)) \cdot \hat{\underline{j}}}{dx} + \frac{(\underline{\delta}(C) - \underline{\delta}(A)) \cdot \hat{\underline{i}}}{dy} = \frac{\partial v}{\partial x} + \frac{\partial u}{\partial y}. \tag{3.16}$$

These three equations provide the *strain–displacement* relations needed to analyze two-dimensional problems. It is not too difficult to extend these definitions to three dimensions, thus finding a complete set of six (symmetric) strains in terms of three displacement components (u, v, and w):

$$\varepsilon_{xx} = \frac{\partial u}{\partial x}, \quad \varepsilon_{yy} = \frac{\partial v}{\partial y}, \quad \varepsilon_{zz} = \frac{\partial w}{\partial z}$$

(3.17 a–f)

$$\gamma_{xy} = \frac{\partial v}{\partial x} + \frac{\partial u}{\partial y}, \quad \gamma_{xz} = \frac{\partial w}{\partial x} + \frac{\partial u}{\partial z}, \quad \gamma_{yz} = \frac{\partial w}{\partial y} + \frac{\partial v}{\partial z}$$

To reiterate the significance of this perspective on strain, the normal strain components represent the extensions of line elements originally in the direction indicated by a matched pair of subscripts. Similarly, the engineering shear strain is equal to the change in the right angle between the two coordinates denoted in the subscripts. Note that the definitions of strain in Chapters 2 and 3 are in agreement and are simply two ways to consider multidimensional deformations.

3.4 The Stress Tensor

We now understand stress as the intensity of an internal force (F/A), where both the force $\underline{F}$ (or, sometimes, $\underline{P}$) and the area $\underline{A}$ are vectors. This suggests to us that a full description of the stress distribution in a body will require a bit of careful bookkeeping.

As a starting point, we consider a section of a loaded body, as shown in Figure 3.6. On that section we identify a very small area, ΔAn, characterized by an outward normal n. This area contains the point O in which we are interested. We denote $\underline{\Delta F}$ as the net force acting on that small area, knowing that this is a contribution to the resultant force acting on the section to maintain equilibrium. (If we knew the particulars of the external loading on this body, we would have used the method of sections to calculate $\underline{\Delta F}$.)

Extending the idea of stress as the *intensity* of the distribution of force over area, we introduce the *stress vector* (also called a *traction vector*) $\hat{\sigma}_n$ as the resultant of all of the forces applied over an entire section per unit area of that section. We define the stress vector as the point function yielded by a limit process in which we divide the net force $\underline{\Delta F}$ acting at a point O by the area of the section ΔAn at the point O, and then let the area become vanishingly small:

$$\hat{\sigma}_n \equiv \lim_{\Delta A_n \to 0} \frac{\Delta F}{\Delta A_n}. \tag{3.18}$$

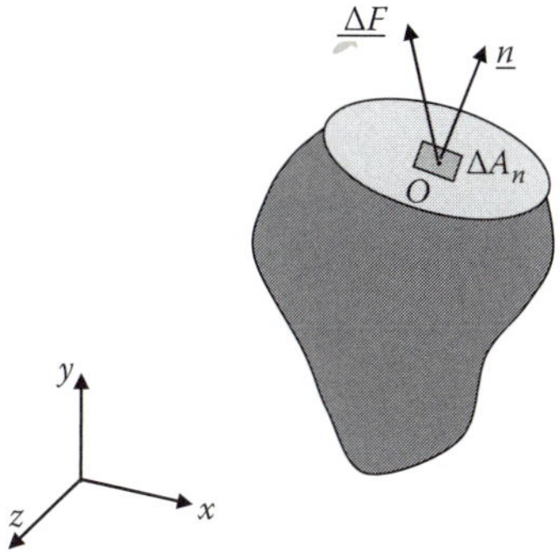

The stress vector $\hat{\sigma}_n$ is truly a vector because we are looking at the vector force $\underline{\Delta F}$ acting on a known area specified by both a normal, n, and a magnitude, ΔA_n.

FIGURE 3.6
The stress vector on a planar section through point O with outward normal n.

We can decompose the vector force $\underline{\Delta F}$ into components referred to a standard Cartesian system with unit vectors $(\hat{\underline{i}}, \hat{\underline{j}}, \hat{\underline{k}})$ in the (x, y, z) directions, respectively. Then we can write

$$\hat{\sigma}_n \equiv \lim_{\Delta A_n \to 0} \frac{\Delta F_x \hat{\underline{i}} + \Delta F_y \hat{\underline{j}} + \Delta F_z \hat{\underline{k}}}{\Delta A_n}. \tag{3.19}$$

If we take the section (area) to be perpendicular to the positive x, y, and z axes, respectively, the associated normals will align themselves with those axes. Then the limit formulas for the stress vectors normal to these three planes become

$$\hat{\sigma}_x \equiv \lim_{\Delta A_x \to 0} \frac{\Delta F_x \hat{\underline{i}} + \Delta F_y \hat{\underline{j}} + \Delta F_z \hat{\underline{k}}}{\Delta A_x}, \tag{3.20a}$$

$$\hat{\sigma}_y \equiv \lim_{A_y \to 0} \frac{F_x \hat{\underline{i}} + \ F_y \hat{\underline{j}} + \ F_z \hat{\underline{k}}}{A_y}, \tag{3.20b}$$

$$\hat{\sigma}_z \equiv \lim_{A_z \to 0} \frac{F_x \hat{\underline{i}} + \ F_y \hat{\underline{j}} + \ F_z \hat{\underline{k}}}{A_z}. \tag{3.20c}$$

We next want to simplify these equations for the stress vectors in each direction and to characterize the entire stress state in one elegant package. Perhaps the most straightforward approach is to use two subscripts, just as we did for multidimensional strain. For stress, the first subscript denotes the direction of the normal vector to the area in question, while the second denotes the direction of the component of force measured on that plane. Thus, we will define stress components such as

$$\sigma_{yx} \equiv \lim_{\Delta A_y \to 0} \frac{\Delta F_x}{\Delta A_y} \quad \text{and} \quad \sigma_{yy} \equiv \lim_{\Delta A_y \to 0} \frac{\Delta F_y}{\Delta A_y}$$

as the resulting limits of forces in the x and y directions divided by the very small area normal to the y axis. Then, the stress vector equations can be written as

$$\underline{\hat{\sigma}}_x \equiv \sigma_{xx}\underline{\hat{i}} + \sigma_{xy}\underline{\hat{j}} + \sigma_{xz}\underline{\hat{k}}, \tag{3.21a}$$

$$\underline{\hat{\sigma}}_y \equiv \sigma_{yx}\underline{\hat{i}} + \sigma_{yy}\underline{\hat{j}} + \sigma_{yz}\underline{\hat{k}}, \tag{3.21b}$$

$$\underline{\hat{\sigma}}_z \equiv \sigma_{zx}\underline{\hat{i}} + \sigma_{zy}\underline{\hat{j}} + \sigma_{zz}\underline{\hat{k}}, \tag{3.21c}$$

for a general stress state with any ΔF and ΔA. We see that there are nine different stress components, three of which have repeated subscripts, while the remaining six have mixed subscripts. The nine components can be displayed in the *stress tensor* array:

$$\underline{\underline{T}}: \begin{pmatrix} \sigma_{xx} & \sigma_{xy} & \sigma_{xz} \\ \sigma_{yx} & \sigma_{yy} & \sigma_{yz} \\ \sigma_{zx} & \sigma_{zy} & \sigma_{zz} \end{pmatrix}. \tag{3.22}$$

These components are often cast in mixed-format symbols: σ_{xx} for components with repeated subscripts (*normal stresses*) and τ_{zx} for components with mixed subscripts (*shear stresses*):

$$\underline{\underline{T}}: \begin{pmatrix} \sigma_{xx} & \sigma_{xy} & \sigma_{xz} \\ \sigma_{yx} & \sigma_{yy} & \sigma_{yz} \\ \sigma_{zx} & \sigma_{zy} & \sigma_{zz} \end{pmatrix} = \begin{pmatrix} \sigma_{x} & \tau_{xy} & \tau_{xz} \\ \tau_{yx} & \sigma_{y} & \tau_{yz} \\ \tau_{zx} & \tau_{zy} & \sigma_{z} \end{pmatrix}. \tag{3.23}$$

This does get a bit simpler; it turns out that, like the strain tensor, the stress tensor is symmetric (i.e., $\tau_{xy} = \tau_{yx}$). And in many cases we will only speak of a single component each of average normal stresses (σ) and shear stresses (τ). However, it's useful to know the big picture.

Visualizing the stress tensor as a matrix may cause us to wonder whether it is "diagonalizable," that is, whether a general stress state may be characterized by a plane on which only normal stress components are nonzero. We return to this issue in Chapter 4, Section 4.3.

Stress, strain, and strain rates are all second-order tensors, and as such they exhibit some common properties about their respective *principal values*. These principal values have to do with extreme values of stress and strain

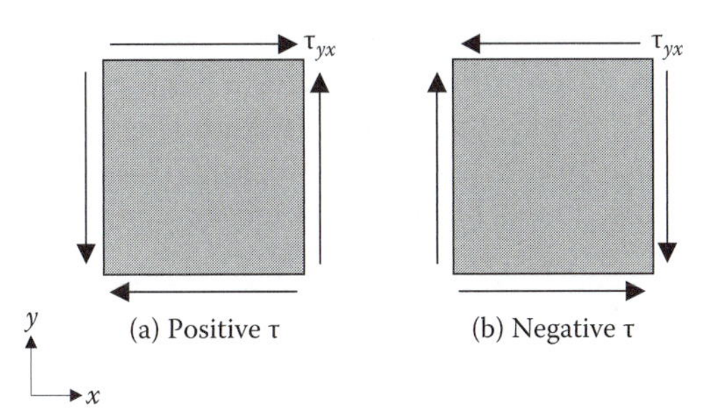

FIGURE 3.7
Sign convention for shear stress.

on planes at different orientations through a given point. By considering the stresses on particular inclined sections of an axially loaded bar, as we did in Chapter 2, Section 2.5, we were finding the *principal stresses* and the planes to which they corresponded.

A note on sign conventions: Normal stress is considered positive if it puts an element in tension and negative if it puts an element in compression. The convention for shear stress is somewhat more complex and is illustrated in Figure 3.7.

Decomposition of the formal definition of the stress vector (equation 3.18 into components in the usual Cartesian system has brought us to the set of the nine stress components displayed in equation (3.23). What is the physical meaning of these terms? How will this be useful to us?

First of all, in the light of our desire to know what's going on at any arbitrary point *O* within a loaded body—or, equivalently, at each and every point within the body—we now know that we can calculate the components of the forces on each of three perpendicular faces drawn through the point *O*. We now state the following (provable) mathematical assertion:

> If we pass three mutually orthogonal planes through a point O and find the stress vector on each of three mutually perpendicular faces drawn through O, then we have fully characterized the stress at point O.

If these faces or planes were themselves normal to the x, y, and z axes, we would have found the three stress vectors of equation (3.19) with stress components given in equation (3.20a), equation (3.20b), equation (3.20c), and (3.26). We are relieved to find that due to the symmetry of the stress tensor, we will only need six—not nine—components of the stress tensor to fully characterize stress at a point.

It bears repeating:

> Stress represents the intensity of internal forces on surfaces within a body subjected to loads.

At an imaginary cut or section, a vector sum of these forces (sometimes called stress resultants) keeps a body in equilibrium. The body we speak of could be a solid material, a liquid, or a gas. Although the internal forces in a gas more commonly arise from molecular collisions than from applied loads $\underline{F}$, we can represent these forces by a distributed load in the same way we built up the stress tensor for our potato from Figure 3.6. This will be useful to us throughout our study of continuum mechanics.

3.5 Generalized Hooke's Law

Although we have discussed the fact that stress and strain are second-order tensors, we looked in Chapter 2 at one-dimensional loading, for which we considered only one scalar component of stress and strain at a time, relating stress to strain by the one-dimensional form of Hooke's law, $\sigma = E\varepsilon$ or $\tau = G\gamma$. We can also write a more general form of Hooke's law, relating stress and strain in two and three dimensions.

If we think of the stress and strain tensors as, roughly, 3 × 3 matrices, it becomes clear that the constant of proportionality between them is a rather bulky, multicomponent construction. Remembering our index notation from Chapter 1, we write the general relation

$$\tau_{ij} = C_{ijkl}\, \varepsilon_{kl}. \tag{3.24}$$

C is an enormous, 3 × 3 × 3 × 3 tensor—a fourth-order tensor—whose components depend on E and G and ν. In practice, C is known as a material's "stiffness matrix" and its inverse as the "compliance matrix." It has 81 components. However, due to the symmetry of both stress and strain tensors, C is also symmetric, with only 36 independent components. The necessity of the existence of a strain energy function (U_0 from Chapter 2) adds some additional symmetry. Because of these symmetry conditions, there are only (!) 21 independent constants (assuming material homogeneity) needed to fully represent a linear elastic solid. However, this most general case, referred to as *anisotropic elasticity*, is also sufficiently unusual that we will simplify further.

In fact, for many, many practical materials,[3] Hooke's law can be written quite satisfactorily for the *isotropic* case, in which case the constants C_{ijkl} must be invariant with respect to coordinate rotations—that is, they will not change as we look in different (orthogonal) directions. In this case, there are, in fact, only two elastic constants (E and G), and Hooke's law for an isotropic elastic solid is written as

$$\varepsilon_x = \varepsilon_{xx} = \frac{\sigma_x}{E} - \nu\frac{\sigma_y}{E} - \nu\frac{\sigma_z}{E}, \tag{3.25a}$$

$$\varepsilon_y = \varepsilon_{yy} = -\nu\frac{\sigma_x}{E} + \frac{\sigma_y}{E} - \nu\frac{\sigma_z}{E}, \tag{3.25b}$$

$$\varepsilon_{zz} = \varepsilon_z = -\nu\frac{\sigma_x}{E} - \nu\frac{\sigma_y}{E} + \frac{\sigma_z}{E}, \tag{3.25c}$$

and

$$\gamma_{xy} = \frac{\tau_{xy}}{G}, \tag{3.25d}$$

$$\gamma_{yz} = \frac{\tau_{yz}}{G}, \tag{3.25e}$$

$$\gamma_{zx} = \frac{\tau_{zx}}{G}, \tag{3.25f}$$

Please remember that the equations here apply only to homogeneous isotropic materials: materials that have the same properties in all directions. Note that even for these ideal, simplest-case materials, the deformation in one direction depends on the normal stresses in all directions.

Finally, by summing equations (3.25a) through (3.25c) and noting the result in equation (3.5), we find that

$$\begin{aligned}\frac{\Delta V}{V} &= \varepsilon_{xx} + \varepsilon_{yy} + \varepsilon_{zz} \\ &= \frac{1-2\nu}{E}\left(\sigma_{xx} + \sigma_{yy} + \sigma_{zz}\right) \\ &= \frac{\sigma_{xx} + \sigma_{yy} + \sigma_{zz}}{K}\end{aligned} \tag{3.26}$$

where K is known as the *bulk modulus*. Note that both G and K are defined in terms of E and ν and that there are (still) only two independent constants for a homogeneous, linearly elastic, isotropic solid.

3.6 Limiting Behavior

Let's look more closely at the stress–strain diagram we referred to in our discussion of Hooke's law (Chapter 2, Figure 2.16). Hooke's law, as we already know, governs the "early" (low strain) regime of the diagram, where stress

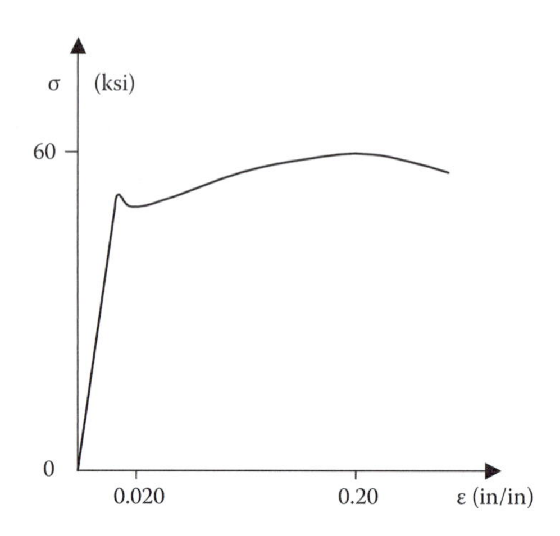

FIGURE 3.8
Idealized stress–strain diagram.

and strain are linearly related. But what's going on "later" in the picture, at higher stress and strain, as the curve bends and ultimately terminates?

Here again is an idealized stress–strain diagram (Figure 3.8). This diagram represents the behavior of mild steel. We are familiar with the Hookean regime of behavior, the linear region with a slope equal to the material's Young's modulus E. The point at which the curve is no longer linear, often a plateau on the stress–strain diagram, is called the material's *yield point,* defining a *yield stress.* Generally, beyond the yield point, it takes much less stress to cause much higher strains than in the Hookean region, and of course the relationship between stress and strain is no longer linear. In some materials, a maximum stress may be reached just before fracture. This is called the *ultimate strength* of the material. Finally, we see that the curve ends abruptly at a certain stress point. This point represents the stress that would cause the material to rupture or fracture. From Chapter 2, Figure 2.11, we observe that ductile materials can withstand much more strain but much less stress than brittle materials, while brittle materials can sustain great stress but very little strain. We classify materials as *ductile* or *brittle* based on their behavior at room temperature; at elevated temperatures, an otherwise brittle material may behave more like a ductile one, while at lowered temperatures, a ductile material may behave like a brittle one.[4]

Stress–strain curves for various materials are obtained through rigorous testing of the materials' behavior in tensile, compressive, and bending tests. The tensile tests are performed in a setup like that shown in Chapter 2, Figure 2.2. Near the breaking point, a specimen of mild steel may resemble the sketch in Figure 3.9. The narrowing of the specimen at its midpoint is called *necking* and is the physical manifestation of the specimen's Poisson's ratio.

Necking occurs in ductile materials, or in materials in ductile states. Figure 3.10 contains photographs of a ductile material undergoing necking (Figure 3.10a) and after failure (Figure 3.10b). Brittle materials, such as cast iron, glass, and stone, experience rupture without any observable change in deformation rate and with no necking. A photograph of a brittle material after failure is shown in Figure 3.10c.

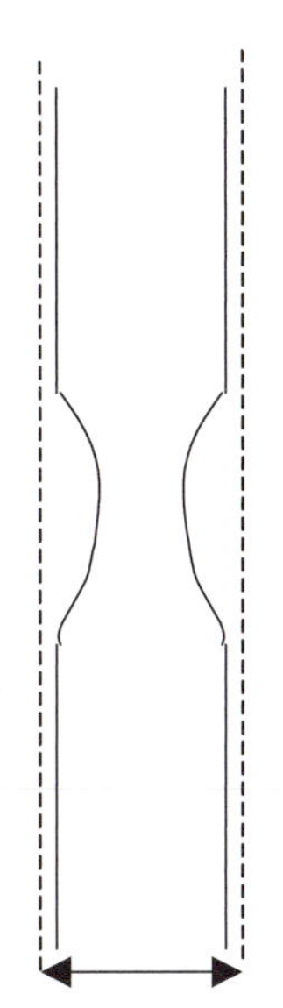

FIGURE 3.9
Necking of a material during tensile testing.

The testing, necking phenomenon, and modulus values discussed here are all uniaxial, or one-dimensional, in nature. Values for Young's modulus and ultimate tensile strength are obtained by stretching a specimen (as shown in Figure 2.8) until it fails. Values for shear modulus and ultimate shear strength are obtained by applying purely shearing deformations to a specimen. We are beginning to realize that real-world loading is not as simple as these testing conditions. Because of the complexities of real loading and real materials, various criteria are used to predict failure of structures. The simplest is the "maximum normal stress criterion," in

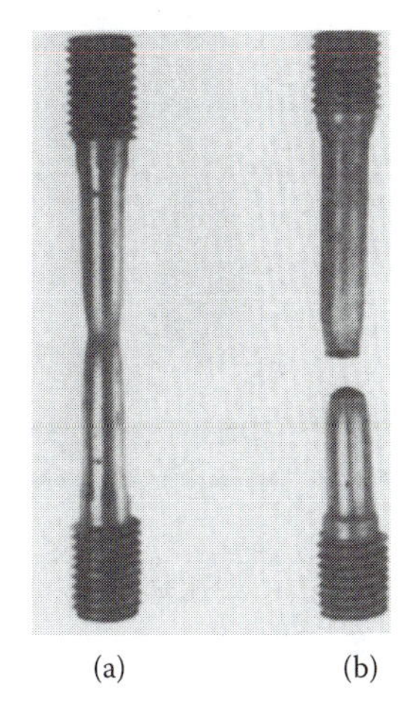
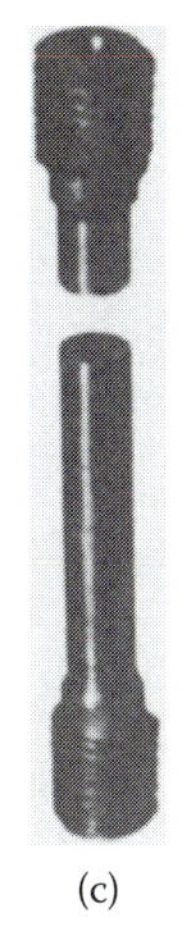

(a) (b) (c)

FIGURE 3.10
Ductile material (a) experiencing necking and (b) after failure, and (c) brittle material after failure in uniaxial tension test.

which failure is predicted to occur when the maximum of the three normal stresses reaches the material's ultimate tensile or compressive stress. This criterion would clearly be well suited to a brittle material like that shown in Figure 3.10b. However, for other materials, subject to different loading, other criteria are useful. We will keep this need for a reliable predictor of structural failure in mind as we continue our study of continuum mechanics.

The yield and ultimate strengths of a range of engineering materials are shown in Table 3.2. The fracture strength of a solid depends on the strength of intermolecular bonds in the material. Based on this reasoning, the theoretical cohesive strength of a brittle elastic solid can be estimated to be approximately one tenth the value of E. However, experimentally determined fracture strengths of most engineering materials lie between 10 and 1000 times below this theoretical value. In the 1920s, A. A. Griffith proposed a rationale for this discrepancy that has now become widely accepted: the presence of microscopic flaws or cracks that exist under normal conditions on surfaces of and within a body of material. Griffith found that these flaws lower a material's overall strength due to

TABLE 3.2

Ultimate and Yield Properties of Common Engineering Materials

Material		Ultimate Strength, MPa			Yield Strength, MPa	
		Tensile	Compressive	Shear	Tensile	Shear
Aluminum alloy	2014-T6	414	—	220	300	170
	6061-T6	262	—	165	241	138
Cast iron	Gray	210	825	—	—	—
	Malleable	370	—	330	250	165
Concrete	—	20–35	—	—	—	
Magnesium alloy	AM100A	275	—	145	150	—
Steel	0.2% Carbon, hot-rolled	450	—	330	250	165
	0.6% Carbon, hot-rolled	690	—	550	415	250
	0.6% Carbon, quenched	825	—	690	515	310
	3.5% Ni, 0.4% C	1380	—	1035	1035	620
Wood	Douglas Fir	—	51	7	—	—
	Southern Pine	—	58	10	—	—

stress concentration (as in Chapter 2, Section 2.10) at the cracks. The local amplification of stress accelerates the growth of the crack, accelerating the material's failure. Using strain energy analysis (as in Chapter 2, Section 2.11), Griffith showed that the critical stress required for crack propagation in a brittle material depended on the material's modulus of elasticity and the specific surface energy and was inversely proportional to the initial size of the crack. A study of materials science would address the microscopic issues involved in stress concentration and crack propagation; in this text we are concerned with the macroscopic implications for our structures.

You may sometimes be asked to include a *safety factor* in your designs; this is a margin of insurance against unforeseen conditions. The allowable stress in your design must be less than the failure or (more conservatively) the yield stress. The safety factor is simply the ratio of failure/yield stress to the allowable stress in the current loading conditions (a limit determined from several factors, including material properties, confidence in load prediction, type of loading, possible deterioration, and design life of the structure). Safety factors should have values over 2.0 in robust designs. That is, our analysis should assure us that the stress will never exceed half of the reference (failure or yield) value.

3.7 Properties of Engineering Materials

By performing a tension test, we obtain values for the proportional limit (the end of Hookean behavior), yield stress, ultimate stress, and rupture stress; we also determine the modulus of elasticity, percent elongation, and percent reduction in cross-sectional area. These values provide quantifiable definitions for the vocabulary generally used to discuss the way a material responds to loading and deformation.

Stiffness: is the property that enables a material to withstand high stress without great strain. In other words, stiffness is a resistance to any sort of deformation. As we've seen, the stiffness of a material is a function of its modulus of elasticity E.

Strength: refers to the greatest stress a material can withstand before failure. This may be quantified by the proportional limit, the yield stress (or yield strength), or ultimate stress (ultimate strength), depending on the type of material and loading being considered. Materials that are very stiff are generally very strong as well, as both properties are related to the strength of atomic bonds.

Elasticity: is what enables a material to regain its original dimensions after a deforming load is removed. No known material is completely elastic in all ranges of stress. However, most engineering materials are elastic over large ranges of stress. Steel, for example, is elastic up to its proportional limit. Deformations beyond the elastic region are referred to as *plastic deformations* and cannot be completely recovered.

Ductility: is what allows a material to undergo considerable plastic deformation under *tensile* load before rupture—to "bend before it breaks," like soft metals and rubber. We can see this on the stress–strain curve for ductile materials (Figure 3.8): The curve features a sizeable, near-flat region beyond the Hookean limit, in which stress increases very little while deformation increases. Ductility is characterized by the percent elongation of the specimen during the tensile test and by the percent reduction in area of the cross section (due to necking—Poisson again!) at the plane of fracture. A high-percent elongation indicates a highly ductile material; a material is considered ductile if elongation is greater than 5%.

Brittleness: implies the absence of any plastic deformation before abrupt failure. It exhibits no necking and has a rupture strength roughly equal to its ultimate strength. This is reflected in a stress–strain curve that ends rather suddenly after the proportional limit. Brittle materials, such as cast iron, concrete, and stone, are relatively weak in tension and are usually tested in compression.

Malleability: is what enables a material to undergo considerable plastic deformation under *compressive* load before rupture. Most ductile materials are also quite malleable. When processing includes hammering or rolling of a metal, malleable materials are the best choice, because they are able to withstand the large compressive deformation that accompanies these processes.

Toughness: enables a material to endure high-impact loads or shock loads. In a high-impact load, some of the energy of the blow is transmitted to and absorbed by the body. Toughness is a measure of the energy required to crack the material, and, in general, increasing a material's strength will decrease its toughness.

Resilience: enables a material to endure high-impact loads without inducing a stress above the elastic limit. In a resilient material, the energy absorbed during the blow is stored and recovered when the body is unloaded. Resilient materials are well suited to applications like baseball bats. We can measure resilience by calculating the area under the elastic portion of the stress–strain curve from the origin through the elastic limit. This is the strain energy U_o, as we remember, and this is used to calculate a "modulus of resilience" that is the strain energy per unit volume, or $\sigma_y^2/2E$.

As has been suggested in the descriptions of these properties, treatments or manufacturing techniques that change one of these properties will also affect the others. For example, quenching carbon steel makes it harder and tougher but less strong and more brittle than it was before quenching. As designers we must make trade-offs and optimize the combination of material behaviors in our systems.

Metals are typically categorized as ferrous (i.e., iron containing) or nonferrous. Ferrous metals are, at the present time, the primary materials used in engineering structures.

Ferrous Metals

The iron in ferrous metals must be extracted from iron ores, which often contain impurities such as phosphorous and silica that must be removed during production. Cast iron, wrought iron, and steel are the three most common forms of ferrous metals. All three are fundamentally iron-carbon alloys containing small amounts of sulfur, phosphorous, silicon, and manganese. Other elements such as nickel and chromium may be added to alter physical and mechanical properties.

Cast iron is a generic name for a group of alloys of carbon and silicon with iron. Most have at least 3% total carbon. The graphite flakes in cast iron act like tiny cracks, making the cast iron as a whole pretty brittle. Wrought iron is a soft, easily worked material with a high resistance to corrosion; its carbon content is typically less than 0.1%. Steel is an alloy consisting almost entirely of iron, and its properties may be changed dramatically by varying the composition. Up to a point, increasing the carbon content increases the hardness, strength, and abrasion resistance of steel. However, ductility, toughness, impact properties, and machinability will be decreased.

Nonferrous Metals

The mechanical properties of the primary nonferrous metals depend on their principal element and the quantity and type of alloying elements and on the method of manufacturing and the heat-treating process.

Aluminum's basic raw material is bauxite ore. High-purity aluminum is soft, weak, and ductile with an ultimate tensile strength of approximately 10,000 psi. Aluminum is lightweight and highly resistant to corrosion under most conditions. It also has good thermal conductivity and high electrical conductivity. Despite its advantages, though, aluminum also has a high coefficient of thermal expansion (see Chapter 2, Section 2.9) and a modulus of elasticity of only 10,000,000 psi, approximately one third that of steel.

Titanium and its alloys have attractive engineering properties. They are about 45% lighter than steel and also possess very high strength, up to twice that of aluminum. This combination of moderate weight and high strength gives titanium alloys the highest strength-to-weight ratio of any structural metal. Titanium alloys also have excellent corrosion resistance, low coefficient of thermal expansion, high melting point (higher than iron), and high electrical resistivity. The modulus of elasticity, a measure of stiffness, is 16,000,000 psi. However, titanium's high cost has limited its utility and range of applications.

Copper's most significant properties are its high electrical conductivity, high thermal conductivity, good resistance to corrosion, and good malleability and strength. These properties are exploited in heat-exchange equipment and many other applications but most of all in the electrical field.

Nonmetals

Concrete consists mainly of a mixture of cement, fine and coarse aggregates (e.g., sand, gravel, crushed rock), and water to harden the mixture. The compressive strength of concrete is relatively high, but it is a fairly brittle material with low tensile strength. Steel reinforcing rods are often used in combination with concrete; the steel resists tension, and the concrete resists compression. Under favorable conditions the strength of concrete increases with its age; the tabulated values for strength are usually those occurring 28 days after the placing of the concrete.

Wood, one of the oldest natural construction materials, is a cellular organic material. We divide wood into two classes: hardwood and softwood. These are somewhat misleading terms in that there is no direct relationship between these designations and the actual hardness or softness of the wood. Softwood comes from conifers (i.e., trees with needlelike or scalelike leaves), and hardwood comes from deciduous trees. Most of the wood used in the United States for structural purposes is softwood, most often Douglas fir and southern pine. Allowable stresses for lumber must take into account species and grade (quality), as well as conditions under which the lumber is to be used, such as load duration and moisture conditions.

Plastics are a group of synthetic organic materials derived by a process called polymerization. Generally, plastics may be either thermoplastics or thermosetting plastics. Thermoplastic material can be repeatedly softened and made to flow by heating. Some thermoplastics are made to be capable of large plastic deformation; others, such as polyvinyl chloride (PVC) and polystyrene, are rigid. Thermosetting plastics have no melting or softening point, though they may be damaged by heat. All thermosetting plastics are brittle, hard, and strong, while thermoplastics are typically ductile, low in strength, and resistant to impact. The thermal expansion of most thermoplastics is about 10 times that of steel.

The modern usage of the phrase *engineering materials* also includes both natural and synthetic biomaterials, whose properties can be quite complex. Please see Case Study 5 for a more involved discussion of the mechanics of biomaterials.

3.8 Equilibrium

We have now addressed the first three items in our continuum mechanics checklist. We have developed ways to talk about (1) deformation or strain,

(2) stress, and (3) constitutive laws or stress–strain relationships in multiple dimensions. The fourth item that concerns us is equilibrium, our governing principle. Often, we will be able to tackle equilibrium using statics and the method of sections. Please see the worked example problems at the end of this chapter for illustrations of this. It is also useful to recognize that we can formulate equilibrium as an *elasticity problem*.

3.8.1 Equilibrium Equations

Let's derive the "microscopic" equations of equilibrium in three dimensions, starting with the six components that fully characterize stress at any arbitrary point (recall our assertion that the stress tensor is symmetric (cf. Chapter 4, equation 4.8). We use the notation displayed in the stress tensor array of equation (4.7) (Chapter 4) to denote stress components by the symbols σ_{xx} and τ_{yx}.

Consider an element of volume *dxdydz* in which we look at the changes in the stress as we sum forces in three independent directions (Figure 3.11). We assume that the components of stress are known at the left, bottom, and rear faces and use the first term of the Taylor expansion to approximate the values of these components at the right, top, and front faces, distances *dx*, *dy*, and *dz* away, as illustrated in Figure 3.11.

For example, forces in the *x* direction result from stresses on all six faces:

$$
\begin{aligned}
&(\sigma_{xx} + \frac{\partial \sigma_{xx}}{\partial x} dx)dydz - \sigma_{xx} dydz \\
&+(\tau_{yx} + \frac{\partial \tau_{yx}}{\partial y} dy)dxdz - \tau_{yx} dxdz \\
&+(\tau_{zx} + \frac{\partial \tau_{zx}}{\partial z} dz)dxdy - \tau_{zx} dxdy \\
&+B_x dxdydz = 0.
\end{aligned}
\tag{3.27}
$$

In equation (3.27) we have once again introduced a *body force*, per unit volume, whose *x* component is *Bx*. After canceling terms appropriately and dividing through by the element volume, we find the following equation of equilibrium:

$$\frac{\partial \sigma_{xx}}{\partial x} + \frac{\partial \tau_{yx}}{\partial y} + \frac{\partial \tau_{zx}}{\partial z} + B_x = 0 . \tag{3.28a}$$

Similarly, in the *y* and *z* directions,

$$\frac{\partial \tau_{xy}}{\partial x} + \frac{\partial \sigma_{yy}}{\partial y} + \frac{\partial \tau_{zy}}{\partial z} + B_y = 0 \tag{3.28b}$$

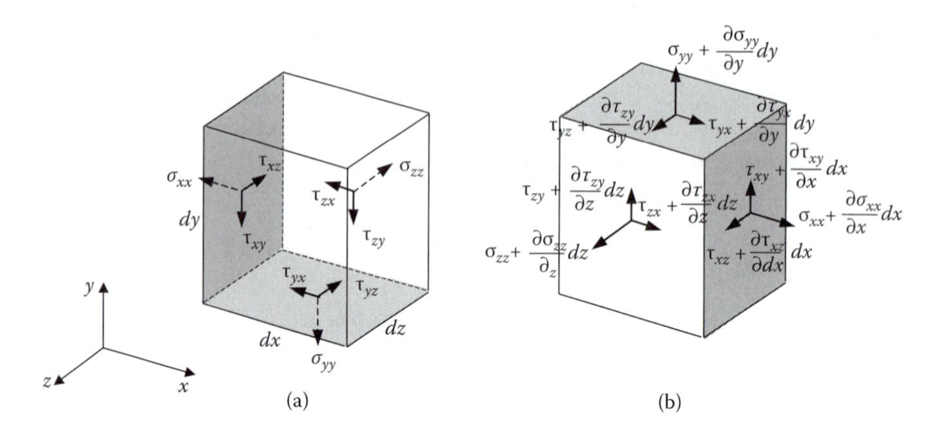

FIGURE 3.11
An infinitesimal element with a three-dimensional stress state. For clarity, stress components on three faces at a time are shown: (a) stress components on left, bottom, and rear faces; (b) stress components on right, top, and front faces.

$$\frac{\partial \tau_{xz}}{\partial x} + \frac{\partial \tau_{yz}}{\partial y} + \frac{\partial \sigma_{zz}}{\partial z} + B_z = 0 . \tag{3.28c}$$

Thus, equations (3.28a) through (3.28c) represent three equations of equilibrium from which we must determine six components of stress. Interestingly, we reduced the number of stress unknowns to six from nine by effectively using three *moment equilibrium* equations, conserving angular momentum.

We can also use *indicial* or *index notation* from Chapter 1, Section 1.5 to write these equations in a very elegant form. We use coordinates *xi*, $i = 1, 2, 3$, to stand for x, y, and z, respectively. Further, we remember the notation that

$$\frac{\partial(\ )}{\partial x_i} \equiv (\ )_{,i} .$$

And we recall the *summation convention*, by which any repeated subscript means we are summing over all values of that subscript. For example,

$$(\ )_{,ii} = \frac{\partial^2(\ )}{\partial x_i \partial x_i} \equiv \sum_{i=1}^{3} \frac{\partial^2(\ )}{\partial x_i \partial x_i}$$

$$(\ )_{i,i} = \frac{\partial(\ )_i}{\partial x_i} \equiv \sum_{i=1}^{3} \frac{\partial(\ )_i}{\partial x_i} .$$

Then the equations of equilibrium can be written simply as

$$\sigma_{ij,j} + B_i = 0 \quad \text{or} \quad \tau_{ij,j} + B_i = 0, \qquad i, j = 1, 2, 3, \tag{3.29}$$

where σ or τ is used interchangeably for all stress components, and the range of the indices is normally not written out as it is understood from the context.

3.8.2 The Two-Dimensional State of Plane Stress

In many circumstances we can simplify the analysis of stress by recognizing that a structure is *thin in one dimension* (often, the dimension along the z axis) in comparison with its dimensions in the other two (x and y) directions. This class of problems is called *plane stress*, and it includes the analysis of aircraft and spacecraft structures, pressure vessels, and similar *thin-walled structures.* In this case, because we assume that

$$\sigma_{zz}, \tau_{xz}, \tau_{yz} \cong 0, \tag{3.30}$$

there are only three stress components to worry about: $\sigma_{xx}, \tau_{xy}, \text{and}\, \sigma_{yy}$.

In this context, we need solve only two equilibrium equations (see also Figure 3.12):

$$\frac{\partial \sigma_{xx}}{\partial x} + \frac{\partial \tau_{yx}}{\partial y} + B_x = 0$$

$$\frac{\partial \tau_{xy}}{\partial x} + \frac{\partial \sigma_{yy}}{\partial y} + B_y = 0, \tag{3.31}$$

It is important to note the following:

- Plane stress in the z direction does not mean or imply that there are no loads applied in that direction. Indeed, the loads in the thickness or z direction in a thin-walled structure cause *membrane stresses* of significant magnitude in the *in-plane* directions. These stresses are significantly larger than the stress in the thickness direction.
- Plane stress in the z direction does not mean or imply that the deflection (or displacement, w) is zero in that direction. Again, for thin-walled structures, it is the deflection in the direction of the thickness that is usually the most prominent and visible deformation.

For the state of plane stress in the z direction, as defined in (3.31a, b), the constitutive law, or appropriate form of generalized Hooke's law, for the remaining two normal stresses becomes

$$\sigma_{xx} = \frac{E}{(1-\nu^2)}\left(\varepsilon_{xx} + \nu\varepsilon_{yy}\right) \tag{3.32a}$$

$$\sigma_{yy} = \frac{E}{(1-\nu^2)}\left(\nu\varepsilon_{xx} + \nu\varepsilon_{yy}\right). \tag{3.32b}$$

3.8.3 The Two-Dimensional State of Plane Strain

Problems in *plane strain* occur when there is reason to believe that there is no appreciable variation or deformation in a direction. In such instances, the movement of any point is likely to be very small in one direction. We could look at every plane perpendicular to this direction, assuming also that the loading doesn't vary appreciably along this direction, and expect to realize the same behavior in each plane. We realize plane strain in objects that are both very long in one direction and loaded (relatively) uniformly in that direction so that

$$\varepsilon_{zz} = \varepsilon_{xz} = \varepsilon_{yz} = 0. \tag{3.33}$$

Equation (3.33) indicates that when the conditions of plane strain are judged to apply, we simply set the corresponding strain components to zero. However, this does not mean that the corresponding stresses are zero. For the case of plane strain in the z direction, as defined in (3.32a, b), the normal stress in the z direction is not zero. In fact, it is the stress required to maintain the rigid planes that can be said to characterize plane strain. Thus,

$$\sigma_{zz} = \nu\left(\sigma_{xx} + \sigma_{yy}\right) \neq 0, \tag{3.34}$$

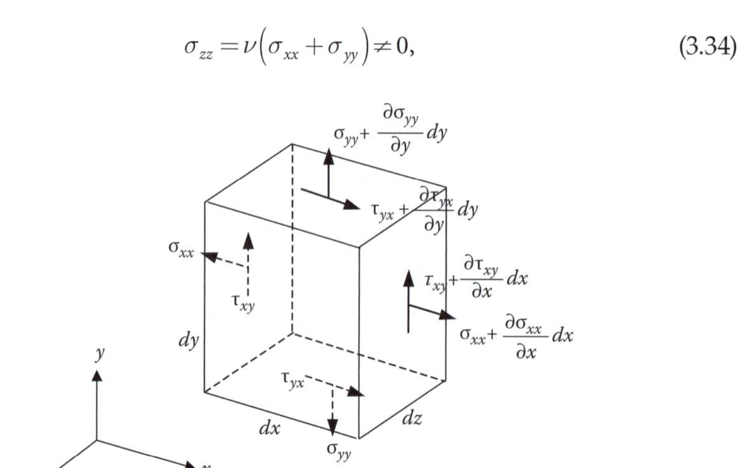

FIGURE 3.12
A three-dimensional element in plane stress.

so that the constitutive laws for plane strain become

$$\sigma_{xx} = \frac{E}{(1+\nu)(1-2\nu)}\left[(1-\nu)\varepsilon_{xx} + \nu\varepsilon_{yy}\right] \tag{3.35a}$$

$$\sigma_{yy} = \frac{E}{(1+\nu)(1-2\nu)}\left[\nu\varepsilon_{xx} + (1-\nu)\varepsilon_{yy}\right]. \tag{3.35b}$$

The equilibrium equations and the relations between strain and displacement are the same in both plane stress and plane strain. Incidentally, for plane strain conditions, just as for plane stress, we will often be interested in knowing just how the two-dimensional strain field varies as we rotate a plane through a point (see Chapter 4, Section 4.3).

3.9 Formulating Two-Dimensional Elasticity Problems

In this section we briefly describe how a two-dimensional problem is formulated in the theory of elasticity. Bearing in mind the notation often used in structural mechanics, we now place our problems in the (x, z) plane so that we are considering plane stress or plane strain in the y direction.

The starting point is equilibrium, and the two-dimensional version (3.31a, b) is repeated here with a body force due to gravity in the vertical or z direction:

$$\frac{\partial \sigma_{xx}}{\partial x} + \frac{\partial \tau_{zx}}{\partial z} = 0, \tag{3.36a}$$

$$\frac{\partial \tau_{xz}}{\partial x} + \frac{\partial \sigma_{zz}}{\partial z} + \rho g = 0. \tag{3.36b}$$

If we were to integrate these partial differential equations, we could then algebraically calculate the corresponding strains, depending on which planar model we were applying. For plane stress, for example, and from either equations (3.25a–f) or the inversion of equations (3.35a, b), we could find the engineering strains

$$\varepsilon_{xx} = \frac{1}{E}(\sigma_{xx} - \nu\sigma_{zz}) \tag{3.37a}$$

$$\varepsilon_{zz} = \frac{1}{E}(\sigma_{zz} - \nu\sigma_{xx}) \tag{3.37b}$$

$$\gamma_{xz} = \frac{\tau_{xz}}{G}. \tag{3.37c}$$

Next we can determine the two meaningful displacements in this plane stress model—that is, $u(x, z)$ and $w(x, z)$—by integrating the relevant strain–displacement equations (a subset of equations 3.17a–f):

$$\varepsilon_{xx} = \frac{\partial u}{\partial x}, \quad \varepsilon_{zz} = \frac{\partial w}{\partial z}$$
$$\gamma_{xz} = \left(\frac{\partial u}{\partial z} + \frac{\partial w}{\partial x}\right). \tag{3.38}$$

Summarizing what we've said so far, in tabular form:

Equations	Number	Unknowns	Number
Equilibrium (3.45)	2	$\sigma_{xx}, \sigma_{zz}, \tau_{xz}$	3
Hooke's law (3.46)	3	$\varepsilon_{xx}, \varepsilon_{zz}, \gamma_{xz}$	3
Strain–displacement (3.47)	3	$u(x,z), w(x,z)$	2
Totals	8		8

On the surface, this seems copasetic: In equations (3.36a, b) through (3.38) we clearly have a *system* of eight equations involving eight unknowns. But some questions remain. First and foremost, are there ways to restructure the problem to make it seem less onerous? We espy a glimmer of hope, recalling that a footnote in Chapter 2, Section 2.8 hinted that by formulating a problem in terms of displacements rather than stresses we would be able to erode the distinction between statically indeterminate and determinate problems.

3.9.1 Equilibrium Expressed in Terms of Displacements

There are two ways to structure solution processes for elasticity problems, and choosing between the two depends on whether one wants to get directly to displacements or directly to stresses. The approach to calculating displacements directly requires a short chain of straightforward substitutions: First, the equations in (3.46) are substituted into the equations in (3.45) to cast equilibrium in terms of strain components. Second, the strain–displacement relations (3.47) are used to eliminate the strains from the intermediate results just found. As Problem 3.11 asks you to confirm, the equations of equilibrium cast in terms of displacements are, for plane stress,

$$G\nabla^2 u + \frac{1+\nu}{1-\nu} G \frac{\partial}{\partial x}\left(\frac{\partial u}{\partial x} + \frac{\partial w}{\partial z}\right) = 0$$

$$G\nabla^2 w + \frac{1+\nu}{1-\nu} G \frac{\partial}{\partial z}\left(\frac{\partial u}{\partial x} + \frac{\partial w}{\partial z}\right) = -\rho g. \tag{3.39}$$

This is a system of (just!) two equations for the two unknown displacements, and an elegant system at that. Notice the symmetry of the differential operators[5] and of the groupings of elastic constants. In fact, the equations in (3.39) are a *subset* (remember, we are talking about plane stress) of the well-known Navier equations of elasticity theory. In indicial notation, the three-dimensional Navier equations are

$$G\nabla^2 u_i + \frac{1}{1-2\nu} G \frac{\partial \varepsilon_{\alpha\alpha}}{\partial x_j} + B_i = 0, \tag{3.40}$$

where B is the net body force acting on the body with modulus G and Poisson's ratio v: In almost all cases, the only body force that cuts any ice is gravity. The parallels between the equations in (3.39) and (3.40) are unmistakable. The difference in the elastic constants results from the plane stress assumption in the equations in (3.39). Note, also, that both equations in (3.39) include an "abbreviated," two-dimensional version of the volume change, or *dilatation* (cf. equation 3.5):

$$\frac{\partial u}{\partial x} + \frac{\partial w}{\partial z} = \varepsilon_{xx} + \varepsilon_{zz}. \tag{3.41}$$

This term also clearly reflects the two-dimensional nature of this discussion.

3.9.2 Compatibility Expressed in Terms of Stress Functions

How would the solution process be different if we wanted to find stresses directly? It is a neat piece of arithmetic to show that if we can identify some function φ from which we can calculate the stress components by performing the following derivatives,

$$\sigma_{xx} = \frac{\partial^2 \varphi}{\partial z^2} \tag{3.42a}$$

$$\sigma_{zz} = \frac{\partial^2 \varphi}{\partial x^2} - \rho g z \tag{3.42b}$$

$$\tau_{xz} = -\frac{\partial^2 \varphi}{\partial x \partial z}, \tag{3.42c}$$

then these stress components will identically satisfy the two-dimensional equations of equilibrium. All well and good, but what is this function φ, and how do we find and calculate it?

The mysterious function, φ, is called a *potential function*, and in this instance it derives from the stress–strain and strain–displacement relations, equations (3.37) and (3.38). Starting with the equations in (338), we seem to have three strain–displacement relations for determining (only) two displacements. In fact, these three equations are themselves related; that is, they are not entirely independent. We can see this by eliminating the displacements u and v among the three equations in (3.38), that is, by noting that

$$\frac{\partial^2 \varepsilon_{xx}}{\partial z^2} + \frac{\partial^2 \varepsilon_{zz}}{\partial x^2} = \frac{\partial^3 u}{\partial x \partial z^2} + \frac{\partial^3 w}{\partial z \partial x^2} \equiv \frac{\partial^2 \gamma_{xz}}{\partial x \partial z}, \tag{3.43a}$$

or

$$\frac{\partial^2 \varepsilon_{xx}}{\partial z^2} + \frac{\partial^2 \varepsilon_{zz}}{\partial x^2} = \frac{\partial^2 \gamma_{xz}}{\partial x \partial z}. \tag{3.43b}$$

Equation (3.43) is a *compatibility condition* that the strains must satisfy so that displacements obtained by integrating the equations in (3.38) are continuous and single valued. That compatibility condition can then be straightforwardly expressed in terms of the potential function—also called the *Airy stress function* after its originator—by substituting for the strains using the stress–strain law (3.37) and for the stresses from the definition of the potential function (3.42). Then, as Problem 3.15 asks you to confirm, the resulting form of the compatibility equation expressed in terms of the Airy stress function[6] is

$$\nabla^4 \varphi = \nabla^2 \nabla^2 \varphi = 0. \tag{3.44}$$

3.9.3 Some Remaining Pieces of the Puzzle of General Formulations

We can summarize our two formulations so far as follows. In the first instance we used constitutive and kinematics relations to write equilibrium entirely in terms of displacements (viz., 3.39). The solutions to these equations can be inspected to ensure that the resulting displacements are continuous, single valued, and consistent with any constraints. From these displacements, we can calculate strain and then stress.

In the second instance, we used equilibrium and a constitutive law to write compatibility entirely in terms of a (single) stress function (3.44). A solution to equation (3.44) automatically satisfies equilibrium and will produce "compatible" displacements (although it is always a good idea to inspect displacement results).

How and where do the actual loads come into the picture? And what are the correct boundary conditions corresponding to equations (3.39) and (3.44)? These two questions—and their answers—are related, in part because of issues raised in Chapter 2. Remember that there are two kinds of external or applied loads that are applied to solids or structures. One kind of applied load comes through *body forces* that typically reflect response on a *specific*, "per unit" basis to a field, such as gravity or electromagnetic radiation. As we have seen, body forces appear in our formulations of equilibrium. The second kind of external load results from *surface loading*, that is, the distribution of forces (and moments) on the surface of the solid or structure, including points on the solid's bounding surface at which the structure is supported (or "grounded"). Surface loads may appear in equations of equilibrium (as they did for axially loaded bars, and as they will in our discussion of beams in later chapters), as well as in appropriate boundary conditions, as we now discuss.

We begin by rewriting our definition of the stress vectors (e.g., equations 3.21a–c) in terms of the familiar stress components:

$$\underline{\hat{\sigma}}_x \equiv \sigma_{xx}\underline{\hat{i}} + \sigma_{xy}\underline{\hat{j}} + \sigma_{xz}\underline{\hat{k}}, \tag{3.45a}$$

$$\underline{\hat{\sigma}}_y \equiv \sigma_{yx}\underline{\hat{i}} + \sigma_{yy}\underline{\hat{j}} + \sigma_{yz}\underline{\hat{k}}, \tag{3.45b}$$

$$\underline{\hat{\sigma}}_z \equiv \sigma_{zx}\underline{\hat{i}} + \sigma_{zy}\underline{\hat{j}} + \sigma_{zz}\underline{\hat{k}}. \tag{3.45c}$$

Then, if we simply apply equations (3.45a) through (3.45c) at points on a surface bounding the solid or structure of interest, with the various T's taken as *known* or *prescribed* forces, then equations (3.45a) through (3.45c) serve as the boundary conditions on the corresponding stresses at those points on the surface. In fact, written in indicial form, equations (3.45a) through (3.45c) are the famous Cauchy equations of the theory of elasticity, relating stress vectors to the stress tensor:

$$\hat{\sigma}_i = \sigma_{ij} n_j. \tag{3.46}$$

Of course, there are problems where we also know or prescribe the displacements at points. These seem easier to express because we are simply equating displacement components to specified values or functions. However, it is also the case that we cannot prescribe both a force and a displacement in the same direction at the same point. We will see how that plays out in detail when we talk about engineering beam theory in Chapter 5. In the meantime, we leave it as an assertion that should have at least intuitive appeal. Would it make sense to prescribe the force we might apply to one end of a spring and, at the same time, prescribe independently how far that end should move?

Finally, we note that the previous formulations of the plane stress problem can be duplicated for plane strain, although the final details may differ. Plane stress and plane strain are important concepts that find frequent use in elasticity theory and in structural mechanics. And while their mathematics may be quite similar, their applications are rather different. Thus, it is important to remember the domain of each. The plane stress model is valid for solids or structures that are both thin and unloaded through that thickness, while the plane strain model is valid for a thin slice of a solid that is very long in one direction, along which there is no (or very little) variation in load or geometry.

This introduction to elasticity has been just that—an introduction. The important concepts here should not surprise you: kinematic description of displacements (strain), internal loading (stress), equilibrium, and compatibility. We've packed a full bag of mathematical tools for problems in continuum mechanics. In the following chapter, we investigate applications of these tools to problems in torsion and pressure vessels and return to the question of transforming our descriptions of stress and strain.

3.10 Examples

Example 3.1

A rectangular copper alloy block as shown in Figure 3.13 has the following dimensions: $a = 200$ mm, $b = 120$ mm, and $c = 100$ mm. This block is subjected to a triaxial loading in equilibrium having the following magnitude: $\sigma_x = +2.40$ MPa, $\sigma_y = -1.20$ MPa, and $\sigma_z = -2.0$ MPa. Assuming that the applied forces are uniformly distributed on the respective faces, determine the size changes that take place along a, b, and c. Let $E = 140$ GPa and $\nu = 0.35$.

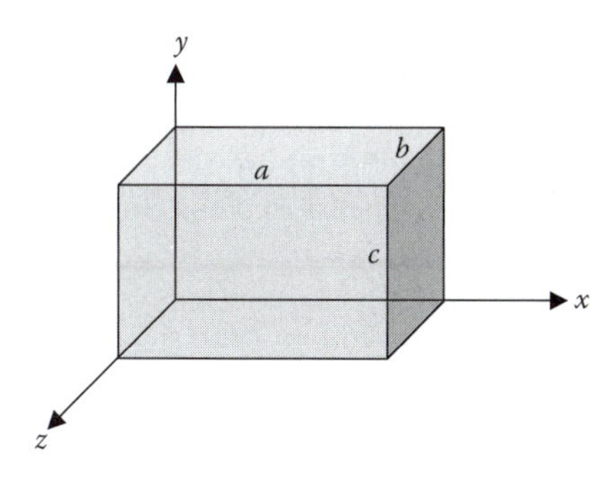

FIGURE 3.13

Given: Stress state, dimensions, and properties of copper block.

Find: Size changes in each direction.

Assume: Homogeneous and isotropic; Hooke's law applies.

Solution

We need the generalized form of Hooke's law, since we have stresses and deformations in multiple directions.

$$\varepsilon_x = \frac{\sigma_x}{E} - \frac{\nu\,\sigma_y}{E} - \frac{\nu\,\sigma_z}{E}.$$

$$\varepsilon_y = \frac{\sigma_y}{E} - \frac{\nu\,\sigma_x}{E} - \frac{\nu\,\sigma_z}{E}.$$

$$\varepsilon_z = \frac{\sigma_z}{E} - \frac{\nu\,\sigma_y}{E} - \frac{\nu\,\sigma_x}{E}.$$

Plugging in the given values of each normal stress component, Poisson's ratio, and E,

$$\varepsilon_x = 25\times10^{-6} = \frac{\delta_a}{a}, \text{ so } \delta_a = (25\times10^{-6})(200\text{ mm}) = 5.029\times10^{-3}\text{mm}.$$

Similarly,

$$\delta_b = -9.575\times10^{-4}\text{mm}.$$

$$\delta_c = -2.075\times10^{-3}\text{ mm}.$$

so that the new dimensions of the copper block are $a' = 200.005$ mm, $b' = 119.999$ mm, and $c' = 99.9979$ mm.

Example 3.2

A rectangular block is compressed by a uniform stress σ_0 as it sits between two rigid surfaces with the gap a shown in Figure 3.14. Determine (a) the stress σ_{yy}; (b) the change in the length along the x axis as the gap a is closed; and (c) the minimum value of σ_0 need to close the gap.

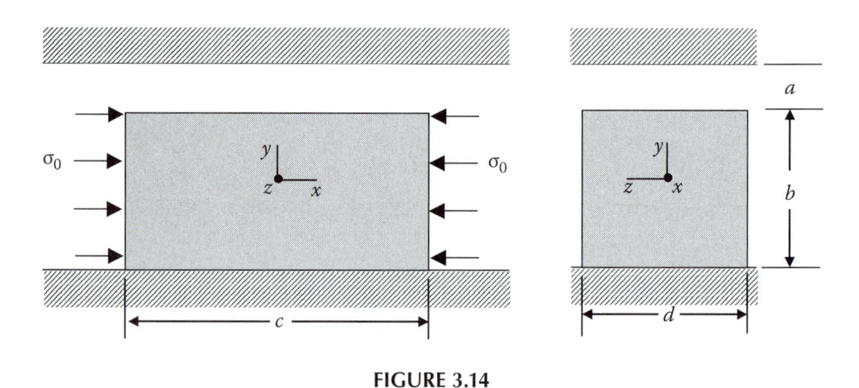

FIGURE 3.14

Given: Rectangular block under uniform stress.

Find: Normal stress in y direction; deformation in x direction; applied stress needed to close gap.

Assume: Material is homogeneous and isotropic; Hooke's law applies.

Solution

Here, $\sigma_{xx} = -\sigma_0, \sigma_{zz} = 0\,(d << c)$, so $E\varepsilon_{yy} = \sigma_{yy} + \nu\sigma_0$.

As the block expands in the y direction, $\varepsilon_{yy} = \delta a / b$ where $0 \leq \delta a \leq a$,

(a) $\delta a \leq a:\ \sigma_{yy} = 0$ and $\sigma_0 = E(\delta a)/(\nu b)$; $\delta a = a:\ \sigma_{yy} = (Ea/b) - \nu\sigma_0$.

(b) Need $\sigma_{yy} = 0.$ to just close the gap as $\delta a = a$.

(c) $\sigma_0 = (Ea/\nu b)$.

3.11 Problems

3.1 Verify that the results $\tau_{xy} = \tau_{yx}$ $\tau_{yz} = \tau_{zy}$ $\tau_{xz} = \tau_{zx}$ are correct by satisfying moment equilibrium for an infinitesimal volume element about each of three orthogonal axes.

3.2 Show that the change in volume of a solid body element whose initial, unstrained volume is $V = l_x l_y l_z$ is given (to first order in the normal strains) by

$$\frac{\Delta V}{V} = \varepsilon_{xx} + \varepsilon_{yy} + \varepsilon_{zz} \, .$$

3.3 Perform the indicated matrix multiplication in the following equation and determine the explicit formulas for the three stress or traction vectors:

$$\begin{Bmatrix} \hat{\sigma}_x \\ \hat{\sigma}_y \\ \hat{\sigma}_z \end{Bmatrix} = \begin{bmatrix} \sigma_{xx} & \sigma_{xy} & \sigma_{xz} \\ \sigma_{yx} & \sigma_{yy} & \sigma_{yz} \\ \sigma_{zx} & \sigma_{zy} & \sigma_{zz} \end{bmatrix} \begin{Bmatrix} \hat{i} \\ \hat{j} \\ \hat{k} \end{Bmatrix}.$$

3.4 Determine, for the following three-dimensional state of stress,

$$\begin{bmatrix} 1{,}000 & 200 & 200 \\ 200 & 1{,}000 & 200 \\ 200 & 200 & 1{,}000 \end{bmatrix} \text{psi}$$

(a) The components of the surface traction vector acting on an element of surface that has a normal vector $\hat{\boldsymbol{n}} = 0.50\hat{i} + 0.50\hat{j} + 0.707\hat{k}$.

(b) The component of this surface traction vector in the direction of the unit vector $\hat{\boldsymbol{\lambda}} = 0.25\hat{i} + 0.935\hat{j} + 0.25\hat{k}$.

3.5 Calculate all of the normal and shear strains for the following displacement field:

$$u(x,y,z) = -z\frac{\partial w(x)}{\partial x}, \quad v(x,y,z) = 0, \quad w(x,y,z) = w(x)\, .$$

Which of these results are most relevant for a planar system of coordinates with x positive to the right and z positive downward? (*Note*: These equations will reappear when we study beams.)

3.6 Calculate all of the normal and shear strains for the following displacement field:

$$u(x,y,z) = -\alpha yz, \quad v(x,y,z) = \alpha xz, \quad w(x,y,z) = 0\, .$$

(*Note*: These equations will reappear when we study torsion.)

3.7 Calculate all of the normal and shear strains for the following displacement field:

$$u(x,y,z) = -\alpha yz, \quad v(x,y,z) = \alpha xz, \quad w(x,y,z) = \kappa(x,y).$$

Compare and contrast these results with those of Problem 3.6.

3.8 Two small cubes of equal size but made of different materials are stacked (as shown in Figure 3.15) so they just fit between two rigid surfaces. The bottom cube is subjected to a uniform pressure p on each of its exposed surfaces. Find the *contact* or *interfacial stress* on the connecting plane, expressed in terms of p and the two sets of material properties.

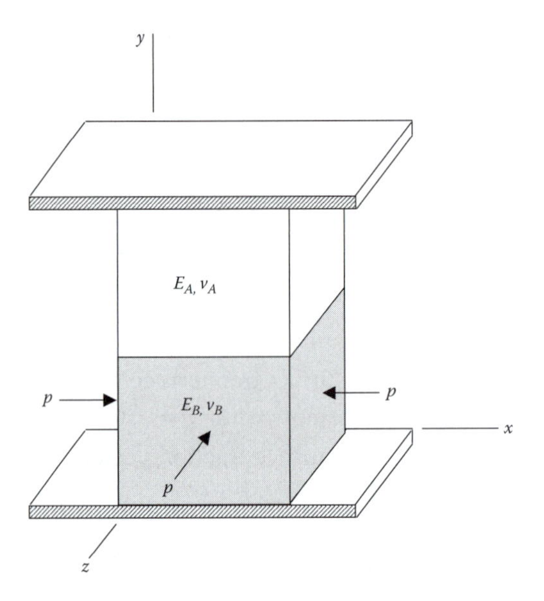

FIGURE 3.15

3.9 The two small cubes of Problem 3.8 are each subjected to a separate uniform pressure, pb, on the exposed faces of the bottom cube and pu on the exposed surfaces of the top cube. What is the ratio of these two pressures, expressed in terms of the two sets of material properties, such that the volume changes of each cube are the same?

3.10 A circle of diameter d is inscribed on the surface of an unstressed metal (Young's modulus E and Poisson ratio ν) square plate of thickness t and side length l (as pictured in Figure 3.16). If the plate is subjected to planar stresses $\sigma_{xx} = 82.7$ MPa and $\sigma_{yy} = 137.8$ MPa and has properties $E = 200$ GPa, $\nu = 0.30$, $t = 2.00$ cm, and $l = 40.0$ cm, find the changes in (a) the length of the diameter AB; (b) the length of the diameter CD; (c) the thickness of the plate; and (d) the volume of the plate.

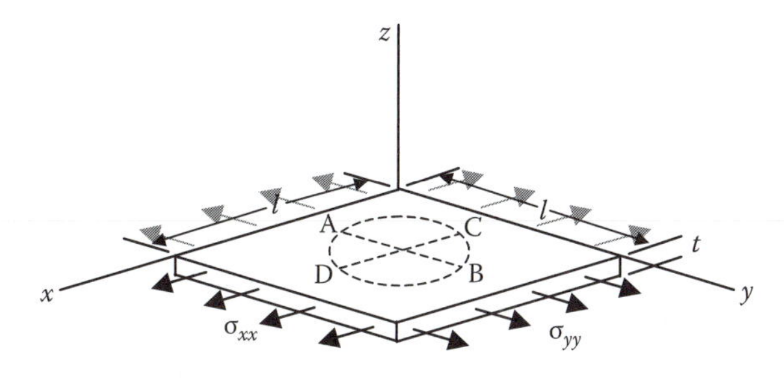

FIGURE 3.16

3.11 Verify the following equilibrium equations for plane stress in the xz plane:

$$G\nabla^2 u + \frac{1+\nu}{1-\nu} G \frac{\partial}{\partial x}\left(\frac{\partial u}{\partial x} + \frac{\partial w}{\partial z}\right) = 0$$

$$G\nabla^2 w + \frac{1+\nu}{1-\nu} G \frac{\partial}{\partial z}\left(\frac{\partial u}{\partial x} + \frac{\partial w}{\partial z}\right) = -\rho g.$$

3.12 Verify the following compatibility equation for plane stress in the xz plane:

$$\nabla^4 \varphi = \nabla^2 \nabla^2 \varphi = 0 .$$

3.13 Explain the notation and meaning of Cauchy's formula:

$$\hat{\sigma}_i = \sigma_{ij} n_j .$$

3.14 Derive counterparts of the following equilibrium equations for plane strain in the xz plane:

$$G\nabla^2 u + \frac{1+\nu}{1-\nu} G \frac{\partial}{\partial x}\left(\frac{\partial u}{\partial x} + \frac{\partial w}{\partial z}\right) = 0$$

$$G\nabla^2 w + \frac{1+\nu}{1-\nu} G \frac{\partial}{\partial z}\left(\frac{\partial u}{\partial x} + \frac{\partial w}{\partial z}\right) = -\rho g.$$

How do these results differ the equilibrium equations for plane stress?

3.15 Derive counterparts of the following compatibility equation for plane strain in the xz plane:

$$\nabla^4 \varphi = \nabla^2 \nabla^2 \varphi = 0\,.$$

How do these results differ from the compatibility equations for plane stress?

3.16 Consider a 4-in.-square steel bar subjected to transverse biaxial tensile stresses of 20 ksi in the x direction and 10 ksi in the y direction. (a) Assuming the bar to be in a state of plane stress, determine the strain in the z direction and the elongations of the plate in the x and y directions. (b) Assuming the bar to be in a state of plane strain, determine the stress in the z direction and the elongations of the bar in the x and y directions. Let $E = 30 \times 10^3$ ksi and $v = 0.25$.

3.17 A piece of 50 × 200 × 10 mm steel plate is subjected to loading along its edges, as shown in Figure 3.17. (a) If $P_x = 100$ kN and $P_y = 200$ kN, what change in thickness occurs due to the application of these forces? (b) For P_x alone to cause the same change in thickness as in part (a), what must be the magnitude of P_x? Let $E = 200$ GPa and $v = 0.25$.

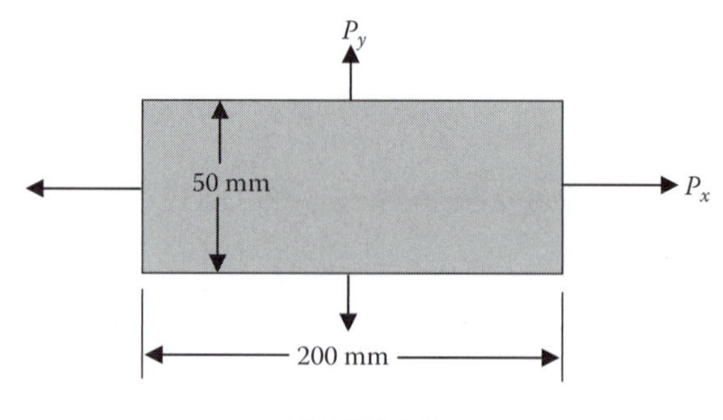

FIGURE 3.17

Notes

1. An elegant demonstration of the Poisson's ratio effect can be seen by stretching a swatch of chicken wire: The wire mesh visibly expands in the direction you're pulling and contracts in the transverse direction. Rod Lakes (1987) created polymer foams that exhibit negative Poisson's ratios: When pulled, they expand in the transverse direction as well as the axial. Some materials composed of fibrous networks (e.g., textiles, biomaterials) have also exhibited this "antirubber" or "auxetic" negative Poisson's behavior (Evans 1989).
2. The subscripts on epsilon provide directional orientation. One subscript, j, tells us we're considering deformation in the jth direction. The other subscript, i, tells us what to compare that deformation to. When $i = j$, we typically use only one subscript i. This is normal strain: The deformation is in the reference direction, making strain a change in length with respect to a length in the same direction. For shear strain, $i \neq j$, and the displacement of interest is perpendicular to the reference direction.
3. One prominent exception to this optimistic assumption of homogeneity and isotropy arises in the consideration of biological materials. Arteries and other biological structures have varying properties in different directions, and this variation serves them well. (It also complicates the modeling and mimicking efforts of engineers and biologists.) Please see Case Study 5 for further discussion of the mechanics of biomaterials.
4. One example of each of these transitions has achieved notoriety. Combustion heating of the steel support members of the World Trade Center in 2001 caused the steel to become more ductile and to lose strength, contributing to the progressive failure of the towers. Also, the infamous O-ring seal on the Space Shuttle Challenger in 1986 became brittle due to cold weather and allowed hot gas to escape from the solid rocket booster, leading to the destruction of the vehicle.
5. The del-squared ∇^2 operator, or Laplacian, is reviewed in Appendix B.
6. See note 5.

4

Applying Strain and Stress in Multiple Dimensions

In both one and multiple dimensions, we have now considered how continuum mechanics will help us create and analyze effective designs. We made sure to include (1) kinematics, or strain; (2) stress; (3) constitutive laws, or how strain and stress are related; and (4) equilibrium as we developed general results to help us analyze the internal response of continuous materials to external loading. In Chapter 3 we recognized that strain, stress, and the modulus that relates them are each *tensors* (of second, second, and fourth order, respectively) and that working with tensors involves some bookkeeping. The somewhat involved mathematics of Chapter 3 should not have distracted us from our goal to obtain useful results that we will apply to the design and analysis of structures. In Chapter 4, we apply the formulations and results of Chapter 3 to several canonical types of external loading, and we return to the question of how strain and stress depend on the reference plane in question.

4.1 Torsion

In the previous sections, we discussed primarily axial loading conditions and how to determine stresses and deformations under these conditions. We now turn our attention to bodies subjected to a twisting action caused by a torque or a twisting moment. As before, we look at the isolated effects of this one type of loading; we will later be able to combine multiple loading configurations to address more realistic, real-world problems. One example of purely twisting external load is in the tightening of a vise grip; the user applies a torque to the threaded screw of the vise, turning it, which in turn causes the jaws to tighten. In practice, members for transmitting torque, such as motor shafts, are generally circular or tubular in cross section. Most of our examples and applications, therefore, involve circular sections.

4.1.1 Method of Sections

What happens when a member in static equilibrium is subjected to a twisting motion? If the member is free, this means a pair of externally applied, equal and oppositely directed couples acting in parallel planes. If the member is fixed, this means a single external couple is applied and that the fixed end supplies an internal resisting torque. The portion of the members between these two external, or between the external and internal, torques is said to be in *torsion*, or under torsional load. For example, the screw of the bench vise previously mentioned is in torsion when the jaws are fully tightened and forces still applied to the handle.

Generally, only one equation of statics will be relevant:

$$\Sigma M_x = 0,$$

where the x axis is directed along the member in question. So, when we apply the method of sections, the internal torque must balance the externally applied torque: It must be equal but have opposite sense. For an example of this, see Figure 4.1.

The torque, being the product of force and lever arm, has units of in.-lb. in the U.S. customary system (USCS) and N·m in the International System of Units (SI). Note that the terms *torque, moment,* and *couple* are used interchangeably in modern engineering practice.

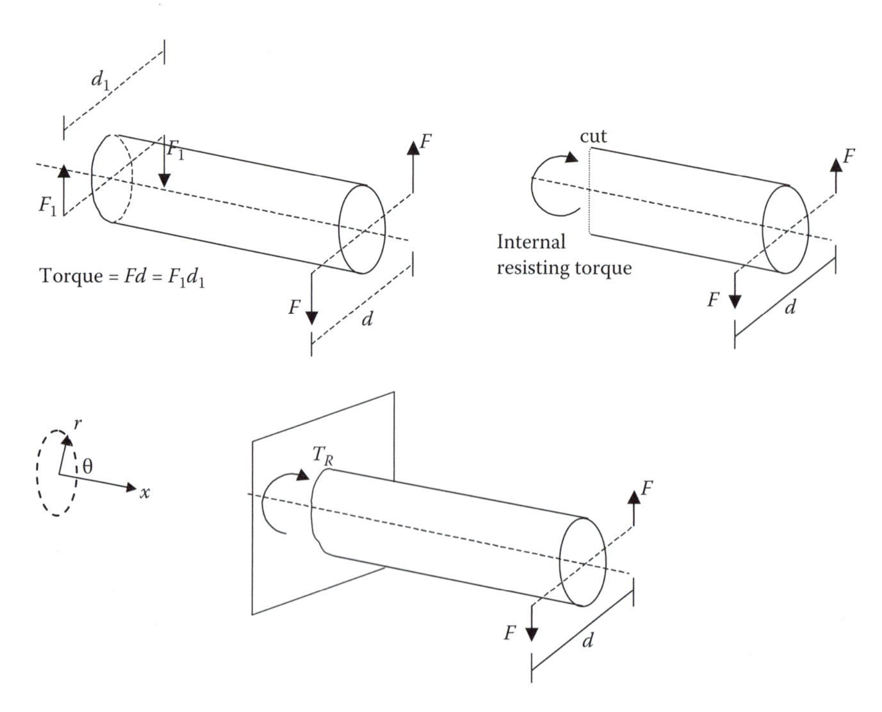

FIGURE 4.1
Sketches showing (clockwise from top left): equal and opposite torques; free-body diagram of section; rigidly fixed bar.

To relate the internal torque and the stresses it sets up in members with circular solid and tubular cross sections, we make the following assumptions, all of which are rooted in and validated by copious experimental data:

1. A plane section of material initially perpendicular to the member's axis remains plane after torques are applied; that is, no warpage or distortion of parallel planes takes place. Imagine a cylinder composed of very thin disks, like a roll of pennies, if the reader will indulge this anachronism. When you twist the roll, the pennies are each displaced but are not warped out of plane.
2. Shear strains γ vary linearly from the central axis, reaching γ_{max} at the periphery. That is, shear strain varies linearly with radial coordinate r. The radius itself remains straight. (On any cross section or penny, the outer edges are deformed the most.)
3. And, *if* the material composing the member is linearly *elastic*, we may apply Hooke's law, from which it follows that shear stress is proportional to shear strain, as we have seen: $\tau = G\gamma$.

Thus, we expect τ, like γ, to increase with r. Sliding motion in the θ-r plane will cause stresses $\tau_x\theta$, and by symmetry $\tau_x\theta = \tau\theta_x$. And, since there is no direct tension or compression on the θ-r plane, we will drop the subscripts on $\tau_x\theta$ and $\gamma_x\theta$, and simply use τ and γ.

4.1.2 Torsional Shear Stress: Angle of Twist and the Torsion Formula

Consider a torsionally loaded member like the bottom sketch in Figure 4.1. A circular member is fixed against rotation at one end and subjected to a torque at the other end. Since torques cause neither direct tension nor compression, this loading develops pure shear stresses on each cross-sectional plane between the torque and the fixed end.

If we imagine that the member is made up of a series of ultrathin plates bonded together (a roll of micropennies), we can visualize each thin plate tending to slide by, or shear, across the contact surface with the adjacent plate. Since the member is in equilibrium (and does not fracture), some internal resistance must develop that prevents any such slippage. This internal resistance (per unit area) is called the *torsional shear stress.* The resultant of these resisting stresses on any cross-sectional plane is an internal resisting torque.

Since we are by now well aware that all materials have limited (tensile, compressive, and shear) strength, we desire a mathematical relationship among torsional shear stress, applied torque, and the physical properties of the member. As always, we seek to understand the internal response of a material to (in this case, torsional) loading.

The free end of the member in Figure 4.1 (bottom) will rotate slightly when a torque is applied. This is shown in Figure 4.2. The shaft radius c will be

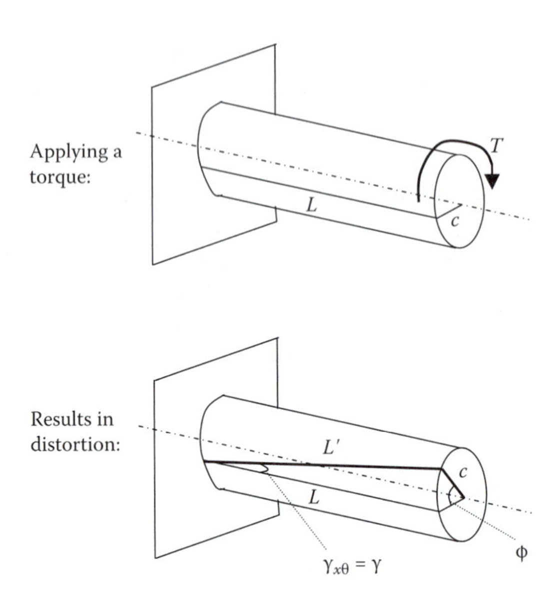

FIGURE 4.2
Circular shaft in torsion.

rotated an angle ϕ, called the *angle of twist*, and line L will become L', actually part of a helical curve. So, the shear distortion of line L is equal to the arc length subtended by this twisting ϕ. We are interested in finding an expression for this arc length, with which we can write the shear strain. (The shear strain, the change in the initially right angle between line L and the vertical, may be approximated by its tangent: the arc length divided by the original length L.) If we look closely at the circular front face of the shaft, we can write the arc length as ϕc; if we look at the length of the shaft, we see that (assuming small deformations) the arc length is γL. These two expressions for arc length must be equivalent, so we must have

$$\gamma L = \phi c. \tag{4.1}$$

Remember, this shear strain is at the outer radius, and so is the maximum possible shear strain (since shear strain increases linearly with r). In other words, the maximum shear strain γ depends on the angle of twist of the shaft, ϕ:

$$\gamma_{\max} = \frac{\phi c}{L}, \tag{4.2}$$

where both γ and ϕ are expressed in radians. In general, the shear strain is given by

$$\gamma(r) = \frac{\phi r}{L}, \tag{4.3}$$

where both γ and ϕ are in radians.

In the previous section we listed the assumptions made for a circular member in torsion. The first was that a plane cross section will remain a plane after the shaft has twisted; also, a straight line radius such as c will remain a straight line as the shaft is twisted. Our second assumption, that shear strains vary linearly with r, tells us that halfway between the center of the shaft and its outer edge the shear strain will have half its value at the outer surface. We write this statement, which readily follows from equations (4.2) and (4.3), as

$$\gamma = \tfrac{r}{c}\gamma_{\max}. \tag{4.4}$$

If Hooke's law $\tau = G\gamma$ applies, our third assumption lets us use a similar distribution for shear stress, as shown graphically in Figure 4.3.

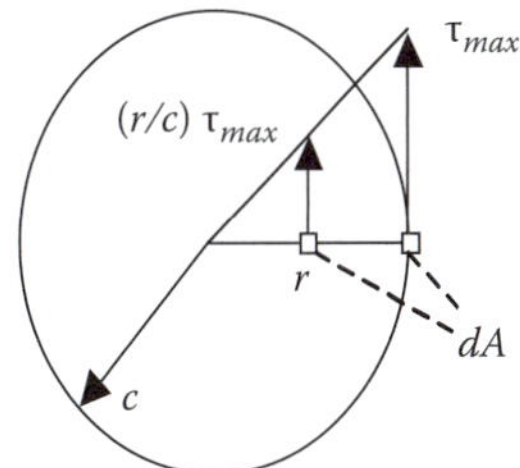

FIGURE 4.3
Shear stress variation on a plane.

Once the stress distribution at a section is established, the resisting torque in the member can be expressed in terms of stress. Remember that stress is the internal resistance to applied loads. To satisfy equilibrium, this internal resisting torque must balance the externally applied torque T. Hence, we have

$$\underbrace{\int_A \underbrace{\underbrace{\frac{r}{c}\tau_{\max}}_{\text{stress}}\ \underbrace{dA}_{\text{area}}}_{\text{force}}\ \ \underbrace{r}_{\text{moment arm}}}_{\text{torque}} = T, \tag{4.5}$$

where the integral sums all torques developed on the section in question by the infinitesimal internal forces acting at some distance r from a member's axis, over the whole area A of the cross section. At any given section, $\tau_{\max}$ and c are constant; therefore, we take them out of the integral and write the expression as

$$\frac{\tau_{\max}}{c}\int_A r^2 dA = T. \tag{4.6}$$

We notice that the integral $\int r^2 dA$ is the polar second moment of area A, denoted by J.[1] For a circular cross section, $dA = 2\pi\, r dr$, and we can calculate J as

$$J = \int_A r^2 dA = \int_0^c 2\pi r^3 dr = 2\pi \left[\frac{r^4}{4}\right]_0^c = \frac{\pi c^4}{2} = \frac{\pi d^4}{32}, \tag{4.7}$$

where d is the diameter of the solid circular shaft in question. If c or d is expressed in millimeters, J has units of mm^4. We can now rewrite our expression for the internal torque as

$$\tau_{\max} = \frac{Tc}{J}, \tag{4.8}$$

which is the well-known *torsion formula* for circular shafts, giving us τ_{max} in terms of the resisting torque and the member's dimensions. More generally, we can find the shear stress τ at any point a distance r away from the center of a section from

$$\tau = \frac{r}{c}\tau_{\max} = \frac{Tr}{J}. \tag{4.9}$$

If our shaft is not solid but a tube of some thickness, similar expressions can be derived. The limits of integration in this case are not 0 and c but are b and c where b marks the inner radius of the tube, so the polar second moment of area becomes

$$J = \int_A r^2 dA = \int_b^c 2\pi r^3 dr = \frac{\pi c^4}{2} - \frac{\pi b^4}{2}. \tag{4.10}$$

And for very thin tubes, where $b \sim c$, and $c - b = t$, the thickness of the tube, J, reduces to

$$J \approx 2\pi R_{av}{}^3 t, \tag{4.11}$$

where

$$R_{av} = (b + c)/2.$$

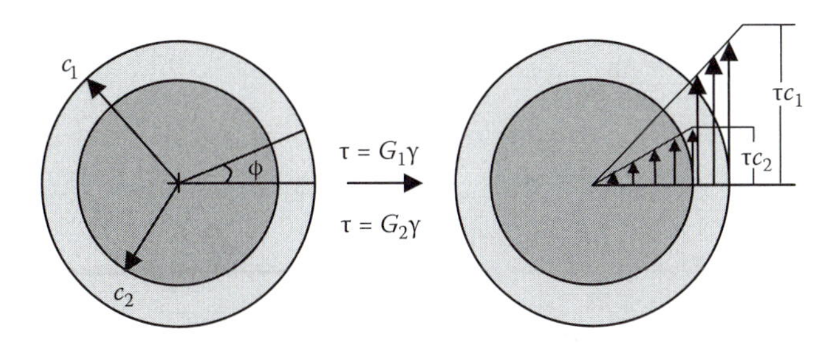

FIGURE 4.4
Elastic behavior of circular member in torsion having an inner core of soft material.

If a circular shaft is made from two different materials bonded together as in Figure 4.4, our original *strain* assumption applies. Through Hooke's law, the shear stress distribution will be found to be more like that in Figure 4.4.

Incidentally, the angle of twist can now be related to the shear stress and, hence, to the torque. (The internal resisting torque must balance the external torque: $T = T$.) Recall that

$$\gamma_{\max} = \frac{c\phi}{L}.$$

If Hooke's law applies, we can use our equation for τ_{max} to write

$$\gamma_{\max} = \frac{\tau_{\max}}{G} = \frac{Tc}{JG}, \tag{4.12}$$

so equating these two expressions for maximum shear strain, we have

$$\phi = \frac{TL}{JG}. \tag{4.13}$$

This expression suggests the technique used for measuring a material's rigidity modulus G in a torsion testing machine. In torsion testing, a known torque T is applied, the resulting deformation ø measured, and the slope of the plotted data is JG/L. Since the geometric parameters of the sample are known, J and L are known constants, yielding an experimental value of G.

Equation (4.13) gives the angle of twist for a shaft of uniform J, G, and L. In the case of adjoining sections of differing geometries, we are able to superpose the angles of twist by integrating or merely summing over the different components:

$$\phi = \int_0^L \frac{Tdx}{JG} \tag{4.14}$$

for continuous changes in diameter or properties, or

$$\phi = \sum_i \frac{T_i L_i}{J_i G_i} \tag{4.15}$$

for abrupt changes or stepped shafts.

4.1.3 Stress Concentrations

The equations we have so far developed for stresses and strains in circular shafts apply to solid and tubular circular shafts while the material behaves elastically and while the cross-sectional areas along the shaft remain reasonably constant. These equations also give acceptable results when changes in the cross-sectional area are gradual. But for stepped shafts where the diameter changes abruptly, large stress concentrations are possible. High local shear stresses occur at sites far from the center of the shaft. In this textbook, we do not calculate these local stress concentrations, but we use a torsional stress-concentration factor to estimate these effects. This method is completely analogous to that discussed in Section 2.10 for axially loaded members, and again the factors depend only on the member geometry. Figure 4.5 shows the stress-concentration factors for various proportions of stepped round shafts. The factor obtained from the chart is then used to adjust the value of maximum shear stress:

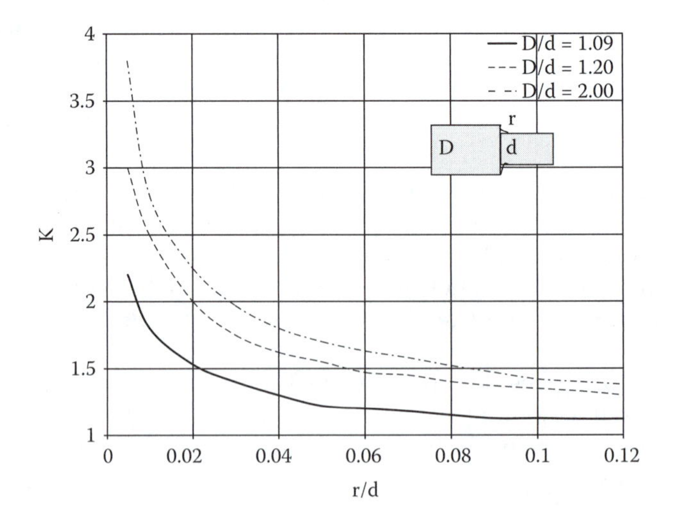

FIGURE 4.5
Torsional stress-concentration factors in circular shafts of two diameters. (After Jacobsen, L. S., *Transactions of the American Society of Mechanical Engineering* 47, pp. 619–638, 1925.)

$$\tau_{max} = K\frac{Tc}{J}, \tag{4.16}$$

where the shear stress Tc/J is obtained for the smaller shaft. It should be clear from the extreme curvature at low r in Figure 4.5 that it is desirable to have a large fillet radius r at all sections where a transition in shaft diameter is made.

4.1.4 Transmission of Power by a Shaft

Rotating shafts are commonly used to transmit power. If an applied torque turns a shaft, work is done by the torque. Work, you may recall, is defined as the energy developed by a force acting through a distance against a resistance. When the distance is linear, we express the work as force × distance. For a rotating shaft, the applied torque turns the shaft through a circular distance, so work is expressed as torque × angular distance = T_θ.

We express the rotation angle θ in radians; if a shaft is rotated at constant speed against some resistance, the work done in one revolution is $2\pi\ T$. The units of work are (N·m), (ft·lb) or (in·lb).

Power is defined as the work done per unit time, work/time. We therefore want to talk about the shaft rotation per unit time, or the shaft's angular speed. We use ω to represent the shaft's angular velocity ($\dot{\theta}$) in radians per second. (Often, we are given a shaft's angular velocity in revolutions per minute, or rpm; to convert this to radians per second we must multiply by 2 π and divide by 60.) Power can then be written

$$P = \omega\ T. \tag{4.17}$$

The unit conventionally used in the United States is the horsepower (hp). In the SI, the unit used to express (N·m/s) is the watt (W). It was the Scottish inventor James Watt who, having refined the Newcomen pump to create the useful steam engine, needed a standard to which to compare his new technology. The "industry standard" at the time was what a millhorse could produce, so Watt tested a brewery horse turning a mill wheel and found that the horse output 33,000 ft·lb/min, a number that became known as "1 horsepower." Some useful facts for dealing with these units are as follows:

$$1\text{ hp} = 33{,}000\ \frac{\text{ft}\cdot\text{lb}}{\text{min}} = 550\frac{\text{ft}\cdot\text{lb}}{\text{sec}} = 6{,}600\frac{\text{in}\cdot\text{lb}}{\text{sec}} = 745.7\text{W}.$$

4.1.5 Statically Indeterminate Problems

Just as in the case of axially loaded bars, there are times when we cannot determine the internal torques from statics alone. It's necessary to complement the equilibrium equations with relations involving the shaft deformations and considering the geometry of the problem. And, as before, several techniques are available to help us.

If we are within the elastic/Hookean regime of behavior, with one degree of *external* indeterminacy (i.e., one more reaction than we can solve for), the force or flexibility method is a good choice to resolve the indeterminacy. In the force method for bars in torsion, we first remove one of the redundant reactions and calculate the rotation ϕ_o at the released support. Then we restore the required boundary conditions by twisting the member at the released end through an angle ϕ_1 such that the sum $\phi_o + \phi_1 = 0$. This analysis does not depend on the number or kind of applied torques or on variations in shaft size of material.

It is also possible to encounter *internal* statical indeterminacy in composite shafts built up from two or more tubes or materials. In these cases, the angle of twist ϕ is the same for each member, and the displacement or stiffness method is the best choice to resolve the indeterminacy. The strategy is to consider the torque T_i for the ith shaft component as $T_i = (k_t)_i\,\varphi$ and then to obtain the total applied torque from the sum of its parts:

$$T = \sum_i (k_t)_i \phi.$$

In these expressions, k_t is the torsional stiffness of a member, again analogous to a spring constant, and we define it as

$$k_t = \frac{T}{\phi} = \frac{JG}{L},$$

with units of [(in·lb)/rad] or [(N·m)/rad]. And, as in the case of axial loads, we can define the reciprocal of stiffness to be the torsional flexibility, which we use in the force method for indeterminate problems.

We also use the displacement method for *externally* statically indeterminate problems with several kinematic degrees of freedom. As before, we write the equations of global equilibrium, and of geometric compatibility, and assume elastic (Hookean) behavior. That is to say, for a two-member bar in torsion, we write equilibrium as

$$T_1 + T_2 + T = 0$$

and impose geometric compatibility as

$$\phi_1 = \phi_2,$$

where these are the twists of the two bar segments, assuming that both far ends are fixed. For linearly elastic behavior, this equation for geometric compatibility becomes

$$\frac{T_1 L_1}{J_1 G_1} = \frac{T_2 L_2}{J_2 G_2}$$

So, just as we did for axially loaded bars that were statically indeterminate, we must ensure (1) equilibrium, (2) geometric compatibility, and (3) consistency of material properties, using constitutive laws such as Hooke's law, in any order we find convenient, until we can solve for all the unknowns in the problem.

4.1.6 Torsion of Inelastic Circular Members

We based our derivation of the *torsion formula* on Hooke's law, so the expressions developed for shear stress and angle of twist in a bar in torsion are only relevant when loads are under the proportional limit. If the yield strength is exceeded somewhere in the shaft, or if the material involved is a brittle material with a nonlinear shear stress–strain diagram, these relations are invalidated. It would be useful if we had a more general method, which we could use when Hooke's law did not apply, to find the stress distribution in a solid circular shaft and to calculate the torque required to produce a given angle of twist.

We made no assumptions of elastic behavior to say that the shear strain γ varies linearly with distance r from the central axis of the shaft. So even in the inelastic case, we may write

$$\gamma = \frac{r}{c}\gamma_{\max}, \tag{4.18}$$

where c is the radius of the shaft.

What we can't yet say is how shear stress varies with r, because we no longer have Hooke's law at the ready. But we do have a shear stress–strain diagram for the given material, and if we don't we can perform some torsion tests on a sample and obtain one. From this diagram, we find the value of γ_{max}, which corresponds to the local maximum τ_{max}. For each value of r, we can find the corresponding value of γ from equation (4.18) and from the stress–strain diagram can find the value of shear stress τ corresponding to this γ. We can then plot τ as a function of r to see the local distribution of stresses. Figure 4.6 shows a range of stress–strain diagrams and the resulting distributions.

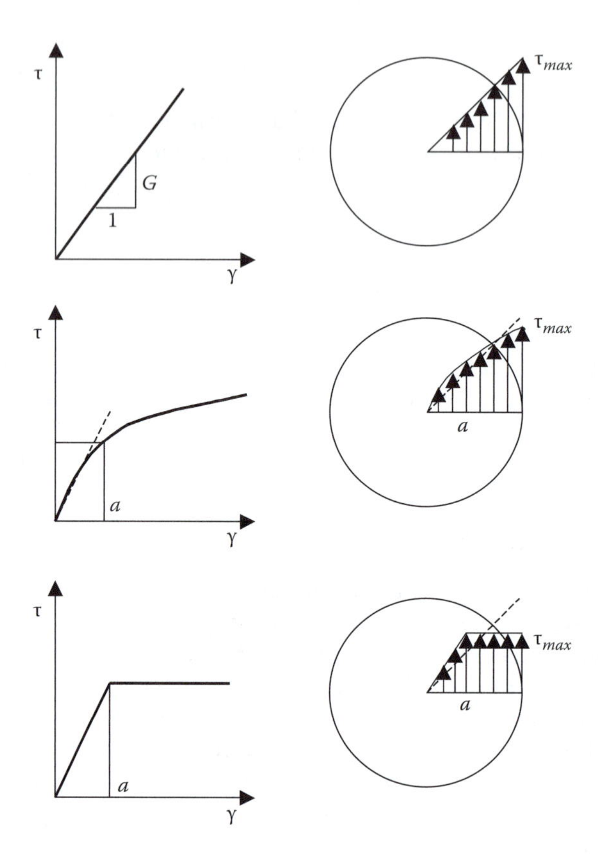

FIGURE 4.6
Stress–strain diagrams (left) and corresponding stress distributions (right).

Once we know the stress distribution, we can integrate over the area of a given cross section to find the internal resisting torque carried by these stresses:

$$T = \int_A (\tau dA) r, \tag{4.19}$$

which, for a circular member, may be simplified as

$$T = 2\pi \int_0^c r^2 \tau dr, \tag{4.20}$$

where in these integrals we are using $\tau(r)$ as obtained from the stress–strain diagram. If $\tau(r)$ is an analytical function, we may proceed with the integration. Otherwise, we must obtain the torque T through a numerical integration.

A significant value is the ultimate torque, T_U, which causes shaft failure. This value may be determined from the ultimate shearing stress of the material by setting $\tau_{max} = \tau_U$ and carrying out the computations for T. It is more convenient to determine T_U experimentally by twisting a specimen of a given material until it breaks. Assuming a (fictitious) linear stress distribution, it is then possible to find the maximum shearing stress R_T:

$$R_T = \frac{T_U c}{J}. \tag{4.21}$$

R_T is purely a reference value, because it is based on a linear stress distribution that is known to be false. It is called the *modulus of rupture in torsion,* or simply *modulus of rupture,* of the material and is always larger than the actual ultimate shearing stress τ_U.

It is also useful to be able to determine the stress distribution and torque T corresponding to a given angle of twist φ. To do this, we recall that

$$\gamma = \frac{r\phi}{L}, \tag{4.22}$$

so that for a given φ and L, we are able to find the shear strain γ at any value r. We can then use the shear stress–strain diagram to find the corresponding τ and can then plot $\tau(r)$. Once we have this distribution, we can proceed with the integral for T as described earlier.

4.1.7 Torsion of Solid Noncircular Members

Everything we have said about torsion has applied to members with circular cross sections. We assumed early and often that plane sections (such as the cross section itself) remained plane. This assumption depends on the *axisymmetry* of the member—that is, upon the fact that it appears the same when viewed from a fixed position and rotated about its axis through an arbitrary angle.

In a square bar, however, because of the lack of axisymmetry, most lines drawn through a cross section will deform when the bar is twisted, and the cross section itself will be warped out of its original plane. See Figure 4.7 for an illustration of this behavior, or draw an even grid on a rubber eraser and apply a twisting moment to see the irregularity of the grid under torsion.

Disappointingly, then, our equations for strain and stress distribution in elastic circular shafts are nontransferable to noncircular shafts. It would be wrong to assume that shear stress in a square bar varied linearly with

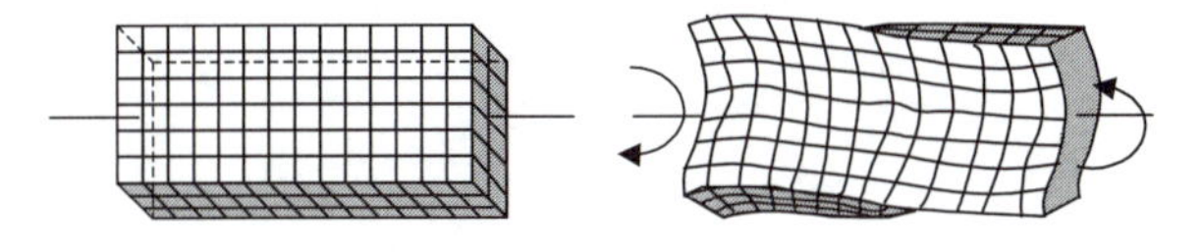

FIGURE 4.7
Rectangular bar (a) before and (b) after a torque is applied.

distance from the axis of the bar; under this assumption, shear stress would be highest at the corners, and it is actually zero at these points.

The mathematical computation of the stresses and strains in noncircular bars in torsion is quite complex. In fact, it was the French elastician Adhémar Barré de Saint-Venant (of the eponymous principle in Chapter 2, Section 2.10) who developed the solution in 1853. This solution is somewhat beyond our scope. However, we can gain some intuition about these problems from the final results of his analysis.

For straight bars with a uniform rectangular cross section, of length L and with a and b denoting the wider and narrower sides of the cross section as in Figure 4.8, the maximum shear stress occurs along the center line of the wider face of the bar and is equal to

$$\tau_{\max} = \frac{T}{C_1 ab^2}, \tag{4.23}$$

and the angle of twist may be expressed as

$$\phi = \frac{TL}{C_2 ab^3 G}. \tag{4.24}$$

In these expressions, the coefficients C_1 and C_2 depend only on the ratio a/b and are given in Table 4.1 for a range of values of this ratio. Both these expressions are valid only within the elastic regime. Similar results for different types of cross sections are available in books such as W. C. Young's *Roark's Formulas for Stress and Strain* (1989).

It is also possible to recast the equation for angle of twist to express the torsional stiffness k_t for a rectangular section:

$$k_t = \frac{T}{\phi} = C_2 ab^3 \frac{G}{L}. \tag{4.25}$$

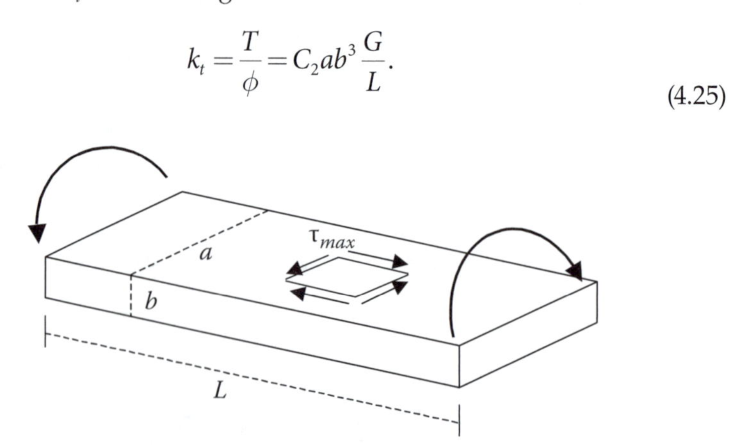

FIGURE 4.8
Generic rectangular bar in torsion.

TABLE 4.1

Coefficients for Rectangular Bars in Torsion

a/b	*C1*	*C2*
1.0	0.208	0.1406
1.2	0.219	0.1661
1.5	0.231	0.1958
2.0	0.246	0.229
2.5	0.258	0.249
3.0	0.267	0.263
4.0	0.282	0.281
5.0	0.291	0.291
10.0	0.312	0.312
∞	0.333	0.333

An elegant *membrane analogy* provides a way to visualize the shear stress distribution in noncircular members. This analogy was introduced by the prolific German scientist Ludwig Prandtl in 1903. The idea comes from the fact that the partial differential equation governing the shear stress in a bar in torsion is the same equation that governs the deformation of an elastic membrane (e.g., a soap film) attached to a fixed frame and subjected to a uniform pressure on one of its sides. For the equations to be mathematically identical, the frame must be the same shape as the bar cross section. The solution of this equation shows the following:

1. The shear stress at any point is proportional to the slope of the stretched membrane at the same point, as illustrated in Figure 4.9.
2. The direction of a particular shear stress at a point is normal to the slope of the membrane at the same point, as also illustrated in Figure 4.9.

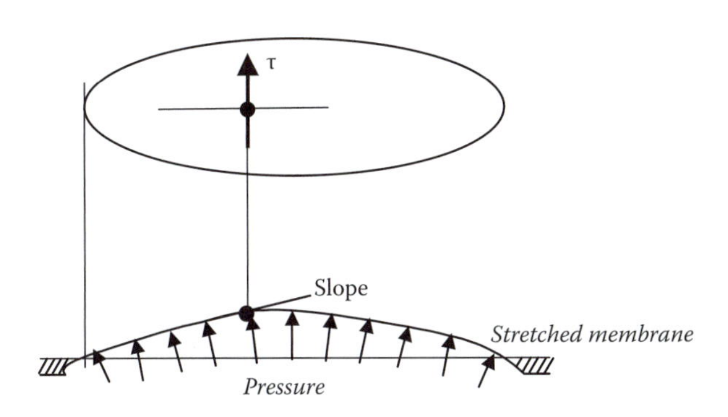

FIGURE 4.9
Membrane analogy for bars in torsion.

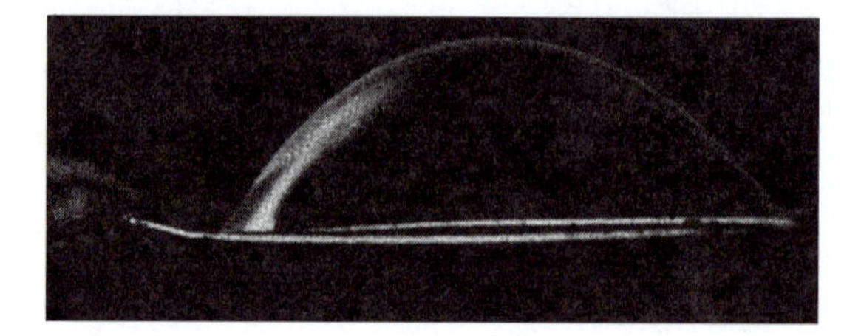

FIGURE 4.10
Soap film on circular frame. (From Isenberg, C., *The Science of Soap Films and Soap Bubbles,* Dover, Mineola, NY 1992. With permission.)

3. Twice the volume enclosed by the membrane is proportional to the torque carried by the section.

If you simply imagine blowing on a soap film, too gently to detach the film and blow a bubble, you should be able to picture the places at which the film will distort and where its slope will be greatest. The membrane analogy tells you that these points correspond to the locations of highest shear stress in a cross section in torsion. For example, in Figure 4.10, a circular soap film is shown being deformed by uniform pressure on its upper surface. Observe that it is nearly flat at the center, where we know the shear stress to be at its minimum value, and that the curvature is very steep at the outer edge, where we already know that due to its radial dependence the shear stress will be maximized.

4.1.8 Torsion of Thin-Walled Tubes

The procedure for obtaining the shear stress distribution and angle of twist for thin-walled tubes is much less complex than that for solid noncircular members. If we consider a hollow cylindrical member of noncircular section subjected to torsion, as in Figure 4.11, we will find the relevant relations.

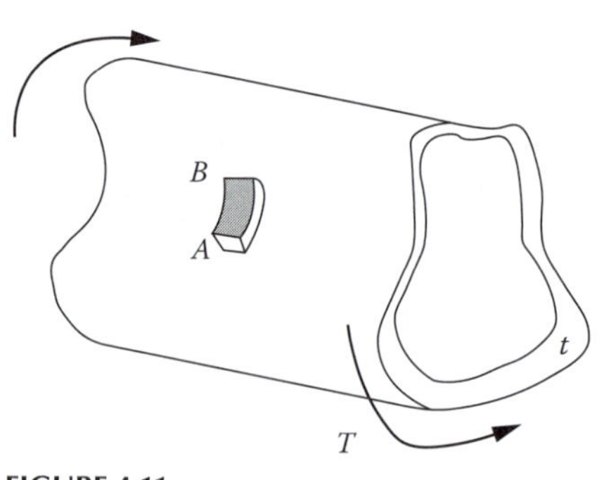

FIGURE 4.11
Thin-walled tube in torsion.

The thickness t of the wall may vary but is assumed to be small relative to the member's other dimensions. We look first at a small section of the wall, as in Figure 4.12. This section is in equilibrium, so there must be no net force acting on it. The only

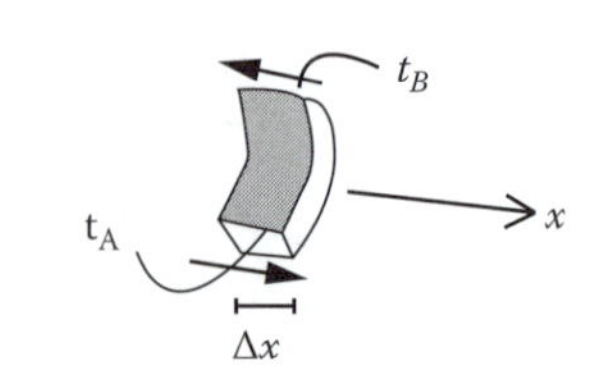

FIGURE 4.12
Segment of tube.

forces acting on it are the shear forces $\boldsymbol{F}_A$ and $\boldsymbol{F}_B$, exerted on the ends of portion AB. Therefore, $\Sigma F_x = 0$ gives us

$$F_A - F_B = 0. \tag{4.26}$$

We can express F_A as the product of the shear stress on the small face A and of the area, $t_A \Delta x$, of that face:

$$F_A = \tau_A t_A \Delta x. \tag{4.27}$$

Shear stress is independent of the x coordinate of the point in question but may vary across the wall, so in this expression we are using for τ_A the average shear stress along the wall. We can express F_B in a similar way and then can write the equilibrium requirement as

$$\tau_A t_A \Delta x - \tau_B t_B \Delta x = 0, \text{ or}$$

$$\tau_A t_A \Delta x = \tau_B t_B \Delta x. \tag{4.28}$$

Since A and B were chosen arbitrarily, this must be true for any two faces of the member, for any Δx, and so we must have the product τt constant throughout the member. If we denote this product by q, we have

$$q = \tau t = \text{constant}. \tag{4.29}$$

This quantity q is commonly called the *shear flow* in the wall of the hollow shaft, because of a parallel that can be drawn to the problem of water flowing in a channel of varying width. In this case, as we see later, conservation of mass requires that the product of water velocity and channel width must be constant. The requirement that the product of shear stress and tube thickness must be constant is thought to be sufficiently similar to justify calling q the shear flow.

We can relate the shear flow q to the torque T applied to our hollow member (Figure 4.13). To do this, we consider a cross section of the tube as shown in Figure 4.14. The force per unit distance (along the perimeter), by the previous argument, is constant and equal to q. This shear flow multiplied by the infinitesimal length ds of a segment of the perimeter gives a force qds per differential length. The product of this force qds and the perpendicular distance r from the centroid of the cross

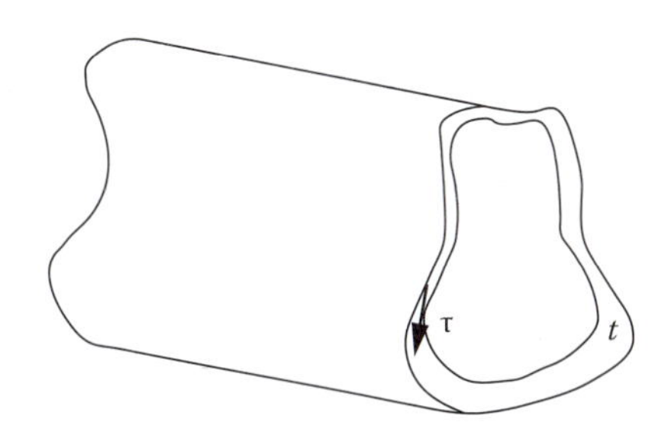

FIGURE 4.13
Shear stress at any point of a transverse section is parallel to the wall surface.

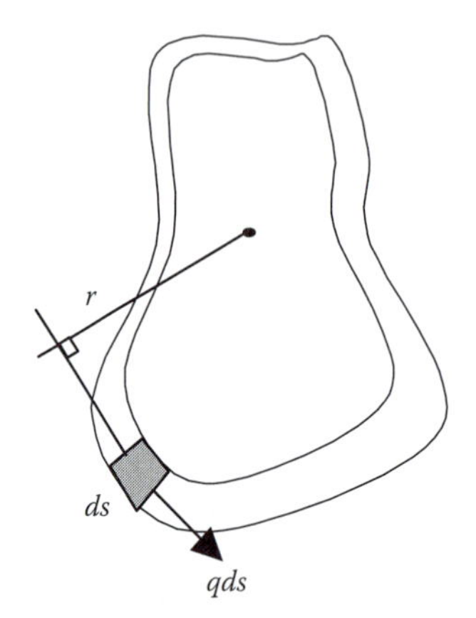

FIGURE 4.14
Cross section of thin-walled tube.

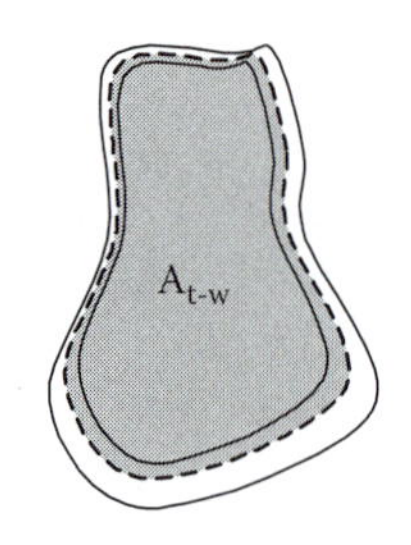

FIGURE 4.15
Definition of $A_{t\text{-}w}$.

section (or another convenient point near the center) to the segment ds gives us the contribution of this element to the resistance of applied torque T. Adding or integrating this around the entire cross section, we get

$$T = \oint rqds, \tag{4.30}$$

where the integration is carried out around the tube along the center line of the perimeter. Since q is constant, we can take it out of the integral and write

$$T = q\oint rds, \tag{4.31}$$

To simplify life, we do not carry out this integration but instead rely on an approximation. The value rds is twice the value of an infinitesimal triangle with base ds and altitude r, and the complete integral is therefore twice the area bounded by the centerline of the tube perimeter. This is sketched in Figure 4.15. We call this area $A_{t\text{-}w}$ and obtain

$$T = 2A_{t-w}q \qquad \text{or} \qquad q = \frac{T}{2A_{t-w}}. \tag{4.32}$$

The t-w subscript on A reminds us that this approximation holds only for thin-walled cylinders. $A_{t\text{-}w}$ is an estimate of the average of the two areas enclosed by the inner and outer surfaces of the tube. Since q is constant, we can determine the shear stress at any point where the wall thickness is t from $\tau = q/t$.

For linearly elastic materials, we can find an expression for angle of twist by applying the principle of energy conservation. Assuming the thin-walled shaft in question has length L and modulus of rigidity G, its angle of twist under torque T is

$$\phi = \frac{TL}{4A_{t-w}^2 G} \oint \frac{ds}{t}, \tag{4.33}$$

where the integral, once again, is computed along the centerline of the wall section. And we can talk about the torsional stiffness k_t of our tube:

$$k_t = \frac{T}{\phi} = \frac{4A_{t-w}^2 G}{\oint \frac{ds}{t} L}. \tag{4.34}$$

4.2 Pressure Vessels

Pressure vessels are generally spheres, cylinders, ellipsoids, or some combination of these, with the goal of containing liquids and gases under pressure. Examples of pressure vessels include boilers, fire extinguishers, shaving cream cans, and pipes, as well as the oxygen tanks carried by scuba divers such as those installing the artificial reef components in our motivating example from Chapter 1.

Actual vessels are usually composed of a complete pressure-containing shell with flange rings and fastening devices for connecting and securing mating parts. At this point, we are interested in the stresses developed in the walls of simple spheres and cylinders, two shapes that are widely used in industry. To perform our stress analysis, we employ a generalized form of Hooke's law.

Thin-walled pressure vessels are those that have a wall thickness t not more than a tenth of the internal radius r_i of the vessel ($t \leq 0.1\ r_i$). The walls of an ideal thin-walled pressure vessel act as a membrane, experiencing no bending. The internal pressures within such vessels are relatively low. Thick-walled vessels such as gun barrels or high-pressure hydraulic presses, on the other hand, have $t > 0.1\ r_i$ and experience dramatic variations in stress from the inner to the outer surface. In this section we consider the simpler thin-walled situation.

Cylindrical and spherical thin-walled pressure vessels are generally subjected to some level of internal fluid (gas or liquid or both) pressure. As a result of the internal pressure, tensile stresses are developed in the vessel walls. These stresses may not exceed specified allowable tensile stresses. Internal pressure tends to rupture the vessel along a joint.

Consider first the cylindrical pressure vessel shown in Figure 4.16a. If we take a section by passing a cutting plane through the pressure vessel, we obtain a plane as in Figure 4.16b, a typical cross section of a cylindrical

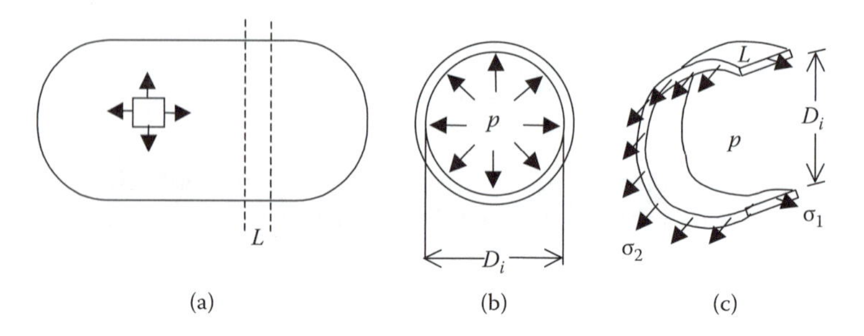

FIGURE 4.16
(a) Cylindrical pressure vessel; (b) cross section; and (c) section of thickness L. As an exercise, label the stresses in (a) as σ_1 or σ_2, so that they are in agreement with (c).

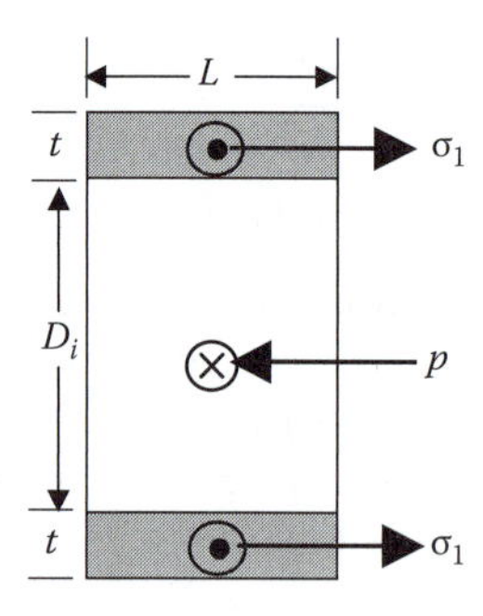

FIGURE 4.17
Projected area of pressure vessel.

thin-walled pressure vessel subjected to an internal pressure p. The internal pressure at any point acts equally in all directions and is always perpendicular to any surface on which it acts. This is reflected in Figure 4.16b.

As mentioned already, the radially acting internal pressure tends to rupture the vessel's joints. To resist this tendency, tensile stresses are developed in the walls of the pressure vessel. These are called *circumferential* or *hoop stresses*. In conventional cylindrical coordinates, these are normal stresses in the θ direction, or σθθ, although in the context of pressure vessels σ_1 is used to represent hoop stress. If we perform a force balance on the element in Figure 4.16c, we can obtain an estimate of these stresses.

As should be clear from Figure 4.16c, the two hoop stresses σ_1 resist the force developed by the internal pressure p, which acts normal to a projected area $D_i L$. Figure 4.17, which shows only the projected area $D_o L$, should serve to make this even clearer. (Remember that pressure, like stress, is a force per unit area.)

The hoop stresses σ_1 act on a combined area $2L(r_o - r_i) = 2\,L\,t$. Balancing the forces, we have

$$p\,D_i L = 2\,\sigma_1\,L\,t, \tag{4.35}$$

which neatly simplifies to an expression for hoop stress σ_1:

$$\sigma_1 = \frac{pD_i}{2t} = \frac{pr_i}{t}. \tag{4.36}$$

This is an expression for the *average* circumferential stress and is valid only for thin-walled cylindrical pressure vessels. In these vessels, in fact, it is often estimated that $r_o \approx r_i$, and so the subscript on r is omitted. And incidentally, this expression can also be arrived at by examining an infinitesimal slice of the cylindrical vessel and integrating over it.

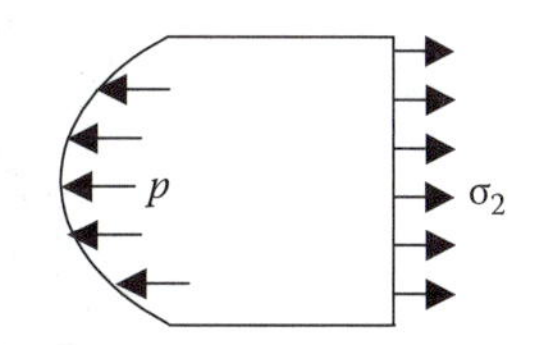

FIGURE 4.18
Longitudinal stress.

The other normal stress σ_2 acting in a cylindrical pressure vessel acts *longitudinally* and may be determined by the solution of an axial-force problem. Conveniently it is called the *longitudinal stress*. In cylindrical coordinates we would call it σ_{xx}, although it is more commonly known simply as σ_2. To find its value, we "slice" the body perpendicular to its axis and obtain the section shown in Figure 4.18.

As indicated in the figure, the internal pressure develops a force $p\,\pi r_i^2$, and this must be balanced by the force developed by the longitudinal stress σ_2 in the walls, $\sigma_2\,(\pi r_o^2 - \pi r_i^2)$. If we equate these two forces and solve for σ_2,

$$p\pi r_i^2 = \sigma_2(\pi r_o^{\ 2} - \pi r_1^{\ 2})$$

$$\sigma_2 = \frac{pr_i^2}{r_o^2 - r_i^2} = \frac{pr_i^2}{(r_o + r_i)(r_o - r_i)}. \tag{4.37}$$

But we have $r_o - r_i = t$, the thickness of the cylindrical wall, and since we are considering thin-walled vessels, we take $r_o \approx r_i \approx r$, so we may use

$$\sigma_2 = \frac{pr}{2t}. \tag{4.38}$$

We also notice that for thin-walled cylindrical pressure vessels, $\sigma_2 \approx \sigma_1/2$. That the hoop stresses are twice the longitudinal may be appreciated by cooking a hot dog until it "plumps" (deforms by expanding in response to rising internal pressure) and bursts—the tears in its casing will be along the longitudinal direction, because it will fail in the circumferential or hoopwise direction.

For thin-walled spherical pressure vessels, a similar method may be employed. In Figure 4.19 we see a sample vessel and a free-body diagram (FBD) that combines elements of Figure 4.16c and Figure 4.17. For a sphere, any section we take passing through the center of the sphere will yield the same result, whatever the inclination. So the maximum membrane stresses for thin-walled spherical pressure vessels are

$$\sigma_1 = \sigma_2 = \frac{pr}{2t}. \tag{4.39}$$

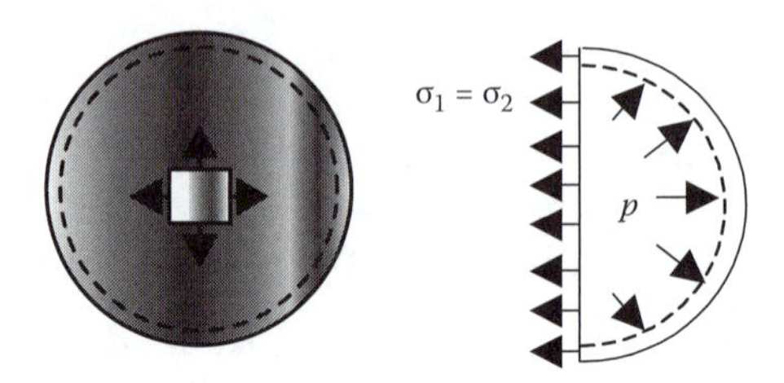

FIGURE 4.19
Spherical pressure vessel.

For either cylindrical or spherical pressure vessels, the effects of internal pressure may be combined with other loading conditions. Just as in the case of thermal effects, the hoop and circumferential stress are simply added to the other stresses in the corresponding directions. For further discussion of pressure vessels, please see Case Study 2 at the end of this chapter.

> Example: Aneurysm, a ballooning or dilation of a blood vessel, often afflicts the abdominal aorta, a large vessel supplying blood to the abdomen, pelvis, and legs. While aneurysms can develop and grow gradually, the rupture (rapid expansion and tearing) of an aneurysm is usually catastrophic. Although the healthy abdominal aorta has a diameter of 1.2 to 2 cm, an aneurismal abdominal aorta may have a diameter up to 6 to 10 cm. Figure 4.20 shows a rough sketch of this anatomy.

We'd like to model the artery as a pressure vessel, despite the many differences between a physiologically realistic blood vessel and the idealization we have just studied. Anatomy textbooks give a range of values for the thickness of artery walls, from which we choose a median value of 0.1 cm. If we choose a radius of 1 cm for our model healthy abdominal aorta, we can call our vessel thin-walled.

The pressure inside the artery varies from a low (diastolic) to high (systolic) value over each heartbeat. Using a typical healthy systolic pressure of

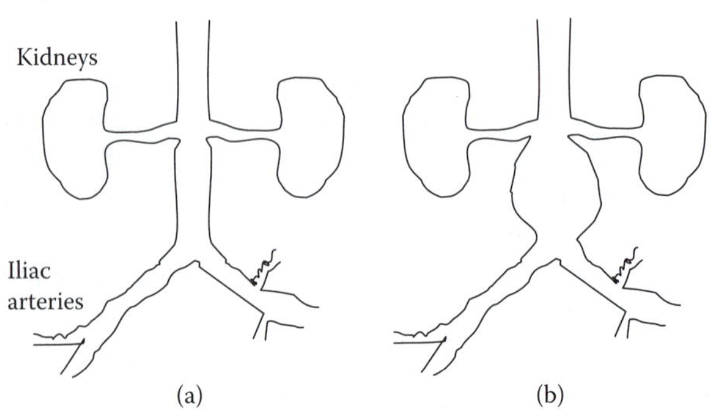

FIGURE 4.20
The abdominal aorta: (a) healthy and (b) affected by aneurysm.

120 mm Hg (1.6 N/cm²), we can calculate the circumferential or hoop stress in a healthy abdominal aorta:

$$\sigma_1 = \frac{pr}{t} = \frac{1.6\,{}^{\mathrm{M}}\!/_{\mathrm{m}^2} \cdot 1\ \mathrm{cm}}{0.1\ \mathrm{cm}} = 16\,{}^{\mathrm{N}}\!/_{\mathrm{cm}^2}.$$

If the vessel grows to a diameter of 5 cm, the hoop stress in a cylindrical vessel becomes

$$\sigma_1 = \frac{pr}{t} = \frac{1.6\,{}^{\mathrm{M}}\!/_{\mathrm{m}^2} \cdot 2.5\ \mathrm{cm}}{0.1\ \mathrm{cm}} = 40\,{}^{\mathrm{N}}\!/_{\mathrm{cm}^2}.$$

If, however, the abdominal aorta remodels itself into a more spherical shape, the hoop stress will be reduced:

$$\sigma_1 = \frac{pr}{2t} = \frac{1.6\,{}^{\mathrm{M}}\!/_{\mathrm{m}^2} \cdot 2.5\ \mathrm{cm}}{2(0.1\ \mathrm{cm})} = 20\,{}^{\mathrm{N}}\!/_{\mathrm{cm}^2}.$$

This crude calculation suggests that the aorta may change its shape in part to reduce the stress induced by internal (blood) pressure. It's worth noting again that this pressure pulses, too, resulting in a cyclic loading and unloading of the vessel. Other factors contributing to aneurysm development include elastin degradation, atherosclerosis, and genetics—but continuum mechanics is certainly part of the package.

4.3 Transformation of Stress and Strain

So far we have considered the isolated effects of normal stresses and shear stresses due to various loading by axial and shear forces, bending moments, and torques. We have seen that when stresses acting on an element are collinear (such as a pressure vessel's longitudinal normal stress, and the normal stress due to an axial load on the vessel), we can simply add them. When stresses are not collinear, we must consider all of the stress tensor components. In some cases, the combinations of stresses produce critical conditions worthy of more detailed examination.

In the previous sections, we have been able to calculate the stress state on a lateral cross section of a component. However, as we remember from our study of axially loaded bars, the stresses on an *inclined* cross section may be quite different. In designing a system, we might prefer to know the stress state at some other orientation—for example, if we were using a mate-

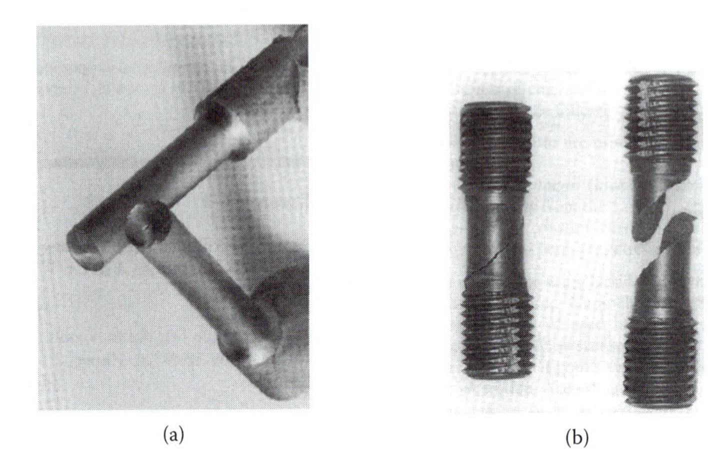

FIGURE 4.21
Failure of circular shafts due to torsion: (a) ductile failure, (b) brittle failure along 45° helix.

rial (e.g., wood, fiber-reinforced concrete) with a grain or with anisotropic properties or if a weld or bolt were inclined at some angle from our usual axes. Consider the failure of a material under torsion—some materials do fail along the interfaces between the imaginary pennies being twisted, in a "clean break" along the cross section (Figure 4.21a). But more brittle materials tend to fail in a different way so that the cleavage surface is inclined at an angle of about 45° (Figure 4.21b). Twist a piece of chalk in your hands to see this type of failure. To understand these failures, and to create robust designs, we want to develop a way of calculating the stress state on axes that are oriented at an arbitrary angle to our reference axes.

We are aware that the most general state of stress at a given point L may be represented by six unique components of a *stress tensor.* Three of these components, σ_x, σ_y, and σ_z, are the normal stresses exerted on the faces of a small cube-shaped element centered at point L and the shear components τ_{xy}, τ_{xz}, and τ_{yz} on the same element. (We remember that the stress tensor is symmetric; therefore, $\tau_{xy} = \tau_{yx}$, $\tau_{xz} = \tau_{zx}$, and $\tau_{yz} = \tau_{zy}$.) If the element is rotated from the standard coordinate axes, we will have to transform the stress components (as shown in Figure 4.22). The same goes for the six independent components of the *strain tensor.*

In this section we focus primarily on *plane stress,* a state in which two faces of the cubic element are stress free. This two-dimensional case, itself of significant use in practice, is easily extended to three dimensions once it is understood.

4.3.1 Transformation of Plane Stress

In Chapter 2, Section 2.5, we considered the stresses on an inclined plane in an axially loaded bar. A similar technique may be used to find the stress

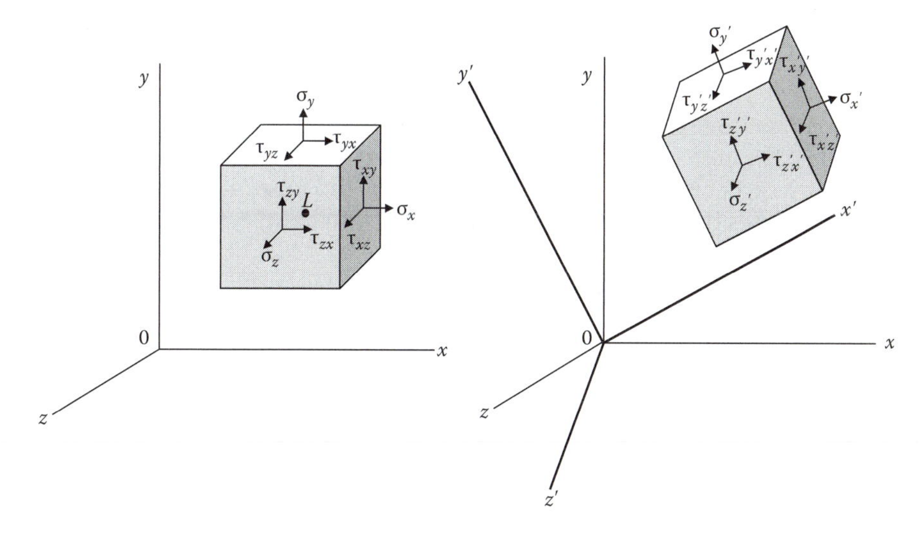

FIGURE 4.22
Stress components on a cubic element.

state on planes and cubic elements that have been rotated. To remind ourselves that we do in fact know how to transform the stresses on an inclined plane, especially now that multiple components of stress are in the picture, let's work out an example:

> Example: If the state of stress for an element is shown in Figure 4.23a, we may also express the state of stress on a wedge of angle $\alpha = 22.5°$. Because this wedge (ABC) is part of the original element, the stresses on faces AC and BC are known. They appear again on the wedge in Figure 4.19b. The

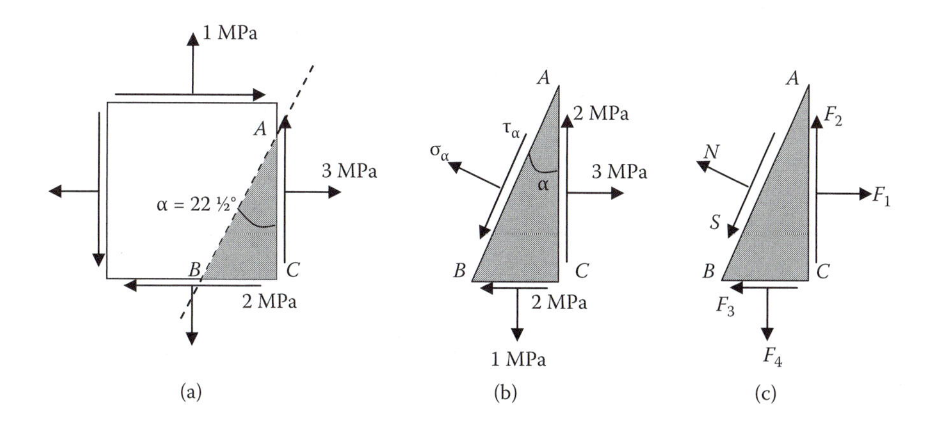

FIGURE 4.23
Finding stresses on an inclined plane. Element has unit depth into the page. (After Popov, E. P., *Engineering Mechanics of Solids*, Prentice Hall, 1998.)

unknown normal and shear stresses acting on face *AB*, σα and τα are what we want to find.

Face *AB* has area *dA*, in m². The area corresponding to line *AC* is $dA \times \cos\alpha = 0.924dA$; the area corresponding to line BC is $dA \times \sin\alpha = 0.383dA$.

Next we can obtain the forces on the faces, F_i in Figure 4.23c, by multiplying the stresses by their respective areas. All F_i are in MN.

$$F_1 = 3 \text{ MPa} \times 0.924dA = 2.78dA.$$

$$F_2 = 2 \text{ MPa} \times 0.924dA = 1.85dA.$$

$$F_3 = 2 \text{ MPa} \times 0.383dA = 0.766dA.$$

$$F_4 = 1 \text{ MPa} \times 0.383dA = 0.383dA.$$

To keep the wedge in equilibrium, the unknown forces due to unknown normal and shear stresses must balance these forces:

$$\Sigma F_N = 0 \qquad N = F_1\cos\alpha - F_2\sin\alpha - F_3\cos\alpha + F_4\sin\alpha = 1.29dA.$$

$$\Sigma F_S = 0 \qquad S = F_1\sin\alpha + F_2\cos\alpha - F_3\sin\alpha - F_4\cos\alpha = 2.12dA.$$

Since forces *N* and *S* act on the plane defined by *AB*, whose area is *dA*, we divide these values by *dA* to find the stresses. Thus, $\sigma_\alpha = 1.29$ MPa and $\tau_\alpha = 2.12$ MPa, in the directions shown in Figure 4.23b.

Conceptually, this is all we are doing in this section. This approach is the starting point for all of the seemingly more sophisticated analyses to follow.

We want to generalize the approach of the example to any initial element and to any inclined wedge. This is illustrated in Figure 4.24. Again, we want to determine the transformed stresses (in the "prime" directions, as in Figure 4.22); again, we apply the equations of equilibrium to our wedge.

Equilibrium in the x' and y' directions requires (check these results as an exercise)

$$\sigma_{x'} = \frac{\sigma_x + \sigma_y}{2} + \frac{\sigma_x - \sigma_y}{2}\cos 2\theta + \tau_{xy}\sin 2\theta, \qquad (4.40)$$

$$\tau_{x'y'} = -\frac{\sigma_x - \sigma_y}{2}\sin 2\theta + \tau_{xy}\cos 2\theta. \qquad (4.41)$$

These are the general expressions for the normal and shear stress on any plane located by the angle θ. Clearly, we must know the state of stress in the

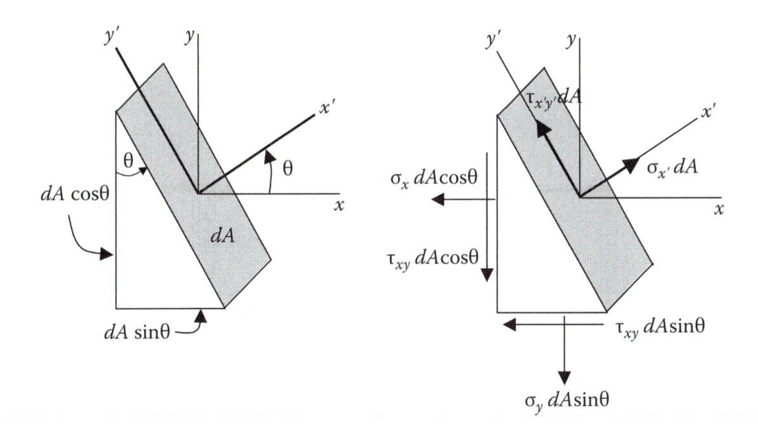

FIGURE 4.24
Derivation of stress transformation on an inclined plane.

initial (x, y, z) orientation to find these transformed stresses. The quantities σ_x, σ_y, and τ_{xy} are initially known.

To find the normal stress on the face perpendicular to the wedge face dA (i.e., $\sigma_{y'}$) we replace θ by $\theta + 90°$ in the equation for $\sigma_{x'}$ and obtain

$$\sigma_{y'} = \frac{\sigma_x + \sigma_y}{2} - \frac{\sigma_x - \sigma_y}{2}\cos 2\theta - \tau_{xy}\sin 2\theta. \tag{4.42}$$

If we add this to the equation for $\sigma_{x'}$ we see that $\sigma_{x'} + \sigma_{y'} = \sigma_x + \sigma_y$, meaning that the sum of the normal stresses remains invariant, regardless of orientation.

In *plane strain* problems where $\varepsilon_z = \gamma_{zx} = \gamma_{zy} = 0$, a normal stress σ_z can develop. This stress is given as $\sigma_z = \nu(\sigma_x + \sigma_y)$, where ν is Poisson's ratio. However, the forces resulting from this stress do not enter into the relevant equilibrium equations used to derive stress transformation relations. These equations for $\sigma_{x'}$ and $\sigma_{y'}$ are applicable for *plane stress* and *plane strain*.

There are two angles by which our axes can be rotated to achieve the limiting stress cases of extreme normal and extreme shear stress.

4.3.2 Principal and Maximum Stresses

As we know, we are often interested in determining the maximum stresses induced in members, so that these limiting cases may inform our designs. Now that we have expressions for the stresses at any orientation θ, we can determine the location of maximum values by setting the derivatives of these expressions to zero, for example:

$$\frac{d\sigma_{x'}}{d\theta} = -\frac{\sigma_x - \sigma_y}{2} 2\sin 2\theta + 2\tau_{xy} \cos 2\theta = 0, \tag{4.43}$$

which requires that to maximize the stress $\sigma_{x'}$,

$$\tan 2\theta_N = \frac{\tau_{xy}}{(\sigma_x - \sigma_y)/2}. \tag{4.44}$$

The N subscript on theta is used to signify its status as the angle defining the plane of maximum or minimum normal stress. The equation for θ_N has two roots [since $\tan 2\beta = \tan(2\beta + 180^\circ)$], 90° apart. One of these roots locates the plane on which the maximum normal stress acts, and the other locates the minimum normal stress.

On these planes corresponding to maximum or minimum normal stresses, there are no shear stresses. These planes are called the *principal planes* of stress, and the (purely normal) stresses acting on them are the *principal stresses.*[2]

If we substitute θ_N into the equations for normal stresses, we obtain expressions for these extreme stress values. We denote maximum normal stress by σ_1 and minimum normal stress by σ_2 and find that

$$\left(\sigma_{x'}\right)_{\substack{\max \\ \min}} = \sigma_{1 \text{ or } 2} = \frac{\sigma_x + \sigma_y}{2} \pm \sqrt{\left(\frac{\sigma_x - \sigma_y}{2}\right)^2 + \tau_{xy}^2}. \tag{4.45}$$

These principal stresses are experienced by an element oriented at θ_N to the original axes.

Turning our attention to the shear stress, we note again that shear stress is zero on the plane defined by θ_N. However, there is a plane on which *shear* stress may be maximized or minimized, which we obtain in the same way that we obtained θ_N. We find that the extreme shear stresses act on planes defined by θ_S, where

$$\tan 2\theta_S = -\frac{\left(\sigma_x - \sigma_y\right)/2}{\tau_{xy}}. \tag{4.46}$$

Once again this equation has two roots, 90° apart. Also, the roots of this equation, θ_S, are 45° away from the planes defined by θ_N. Substituting θ_S into our equation for shear stress, we get an expression for the maximum and minimum shear stresses:

$$\tau_{\substack{\max \\ \min}} = \pm\sqrt{\left(\frac{\sigma_x - \sigma_y}{2}\right)^2 + \tau_{xy}^2}. \tag{4.47}$$

The maximum and minimum values differ only by sign. Physically, this sign has no meaning (except that if it is negative, the shear has the opposite sense from that assumed in Figure 4.18), so this shear stress regardless of sign is simply called the *maximum shear stress*. (Please see Chapter 3, Figure 3.7 for a reminder of the shear stress sign convention.) The *magnitude* of the shear is what we need to create robust designs; while a material can have different properties in tension and compression, making the sign of the normal stress important, it responds the same way to shear in either direction.

On the principal axes, the principal stresses were purely normal, with zero shear stress. But the planes where maximum shear stress acts are not necessarily free of normal stresses. If we substitute θ_S into our equation for normal stress, we find that the normal stresses acting on planes of maximum shear stress are

$$\sigma_{\theta_S} = \frac{\sigma_x + \sigma_y}{2}. \tag{4.48}$$

4.3.3 Mohr's Circle for Plane Stress

If we look back at our equations for transformed $\sigma_{x'}$ and $\tau_{x'y'}$, we notice that these are the parametric equations of a circle. If we choose a set of rectangular axes and plot points with coordinates ($\sigma_{x'}$, $\tau_{x'y'}$) for all possible values of θ, all the points will lie on a circle. We can see this more clearly if we rewrite the equations

$$\sigma_{x'} - \frac{\sigma_x + \sigma_y}{2} = \frac{\sigma_x - \sigma_y}{2}\cos 2\theta + \tau_{xy}\sin 2\theta, \tag{4.49}$$

$$\tau_{x'y'} = -\frac{\sigma_x - \sigma_y}{2}\sin 2\theta + \tau_{xy}\cos 2\theta, \tag{4.50}$$

and then square both equations, add them, and simplify to get

$$\left(\sigma_{x'} - \frac{\sigma_x + \sigma_y}{2}\right)^2 + \tau_{x'y'}^2 = \left(\frac{\sigma_x - \sigma_y}{2}\right)^2 + \tau_{xy}^2. \tag{4.51}$$

This equation has the form $(\sigma_{x'} - a)^2 + \tau_{x'y'}{}^2 = b^2$, where the quantities $a = (\sigma_x + \sigma_y)/2$ and $b^2 = [(\sigma_x - \sigma_y)/2]^2 + \tau_{xy}{}^2$ are constants. We remember that $(x - a)^2 + y^2$

$= b^2$ is the equation of a circle of radius b with its center at $(+a, 0)$. Hence, we may plot all points $(\sigma_{x'}, \tau_{x'y'})$ on a circle. The resulting circle is called Mohr's circle of stress, named for Otto Mohr, who first proposed its use in 1882.

From equation (4.51), we can see that the center of this circle will be at $(a, 0)$, or at

$$\left(\frac{\sigma_x + \sigma_y}{2}, 0\right),$$

and that the circle's radius, b, is given by

$$R = \sqrt{\left(\frac{\sigma_x - \sigma_y}{2}\right)^2 + \tau_{xy}^2}.$$

Using Mohr's circle to graphically display stress transformations will offer a big-picture view of a problem and will make certain relationships visually clear. Mohr's circle gives us a way to see all possible stress states at a certain point (i.e., the stress states for all possible axes with their origins at that certain point) at once, as in Figure 4.25.

Certain observations can be made based on Figure 4.25:

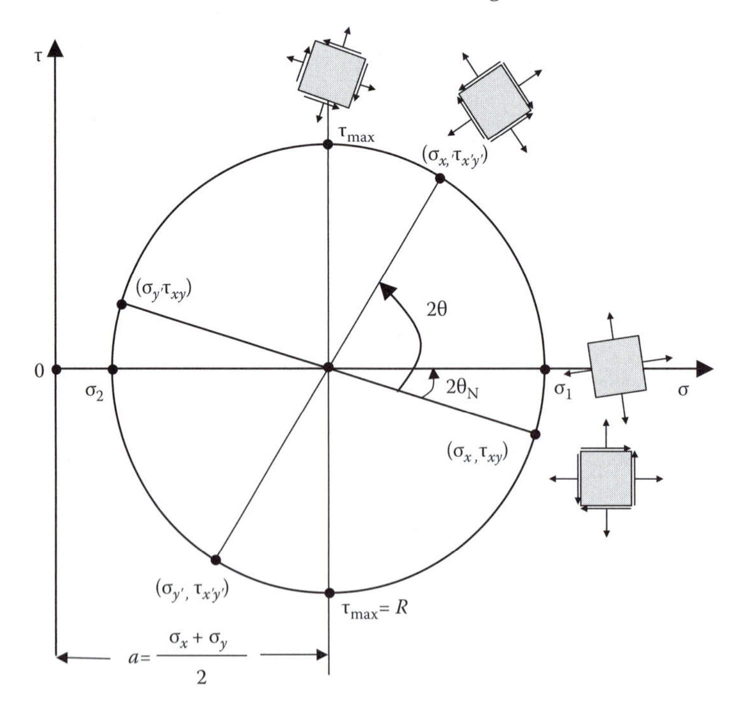

FIGURE 4.25
Mohr's circle.

- The largest possible normal stress is σ_1, and the smallest is σ_2. No shear stresses exist together with either one of these principal stresses.
- The largest shear stress τ_{max} is equal to the radius of the circle, R. A normal stress equal to $(\sigma_1 + \sigma_2)/2$ acts on each of the planes of maximum shear stress.
- If $\sigma_x + \sigma_y = 0$, the center of Mohr's circle coincides with the $\sigma\tau$ coordinate origin, and the state of pure shear exists.
- The sum of the normal stresses on any two mutually perpendicular planes is invariant. That is,

$$\sigma_x + \sigma_y = \sigma_1 + \sigma_2 = \sigma_{x'} + \sigma_{y'} = \text{constant}. \tag{4.52}$$

One tricky aspect of constructing Mohr's circle is whether to plot a given shear stress "above" or "below" the σ axis. There are a variety of conventions used in the literature; the choice of convention is not as critical as is consistency in applying it. The best way to get comfortable with any convention is to work examples, such as Example 4.8.

Mohr's circle also gives us a way to check our earlier results for axial loading and torsional loading. In the case of axial loading, shown in Figure 4.26a, we have already shown that $\sigma_x = P/A$, $\sigma_y = 0$, and $\tau_{xy} = 0$. The corresponding points X and Y define a circle with radius $R = P/2A$, as in Figure 4.26b. Points D and E yield the orientation of the planes of maximum shearing stress (Figure 4.26c, at $\theta = 45^\circ$, as we already knew; these lines are separated by 2θ on Mohr's circle), and the values of maximum shear stress and corresponding normal stress:

$$\tau_{\max} = \sigma' = R = \frac{P}{2A}. \tag{4.53}$$

In torsional loading, we have $\sigma_x = 0$, $\sigma_y = 0$, and $\tau_{xy} = \tau_{max} = Tc/J$ (Figure 4.27a). Points X and Y are on the τ axis, and Mohr's circle (Figure 4.27b) has radius

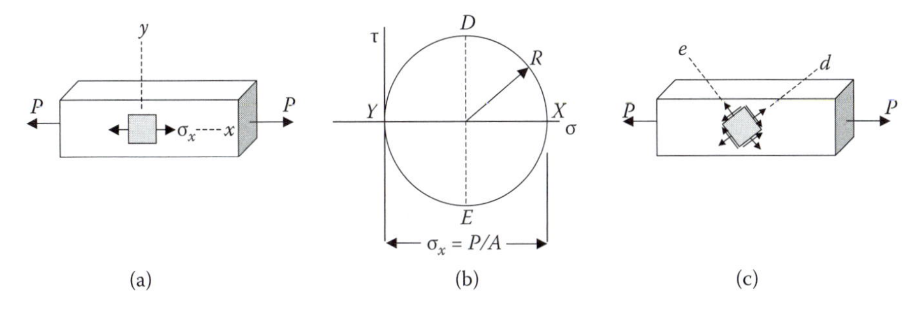

FIGURE 4.26
Mohr's circle for axial loading.

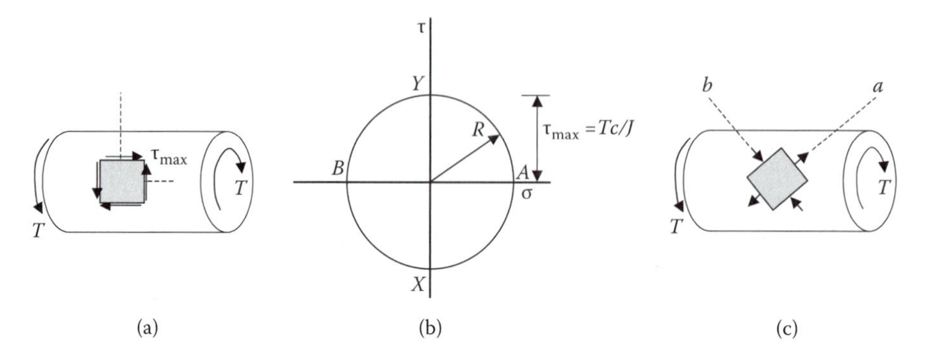

FIGURE 4.27
Mohr's circle for torsional loading.

$R = Tc/J$ and is centered on the origin. Points A and B define the principal planes and the principal stresses:

$$\sigma_{1,2} = \pm R = \pm \frac{Tc}{J}. \tag{4.54}$$

4.3.4 Transformation of Plane Strain

We recall the existence and form of the *strain tensor*, which like the stress tensor gives us six independent components at each location within a member. We may find ourselves in a situation where the initial xy axes are rotated through some angle θ, and we may need to transform strains associated with xy to an equivalent set of strains on the rotated axes. One way to do this is described here.

We consider first an arbitrary point A at point (x, y) in the initial coordinate system. After the rotation through θ, as shown in Figure 4.28, this point A is at (x', y'). From the figure we see that

$$x' = x \cos \theta + y \sin \theta, \tag{4.55a}$$

$$y' = -x \sin \theta + y \cos \theta. \tag{4.55b}$$

These equations may be written in matrix form:

$$\begin{pmatrix} x' \\ y' \end{pmatrix} = \begin{pmatrix} \cos\theta & \sin\theta \\ -\sin\theta & \cos\theta \end{pmatrix} \begin{pmatrix} x \\ y \end{pmatrix}. \tag{4.55}$$

If we want to rearrange this expression, we look at the 2 × 2 matrix and see that its determinant is unity; hence, its transpose is equal to its inverse. So,

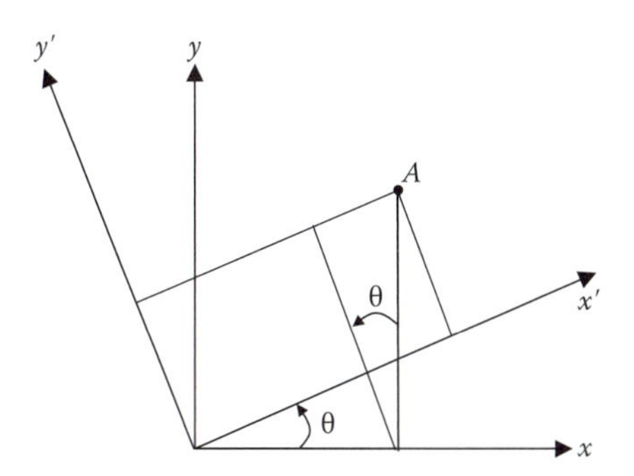

FIGURE 4.28
Coordinate transformation.

$$\begin{pmatrix} x \\ y \end{pmatrix} = \begin{pmatrix} \cos\theta & -\sin\theta \\ \sin\theta & \cos\theta \end{pmatrix} \begin{pmatrix} x' \\ y' \end{pmatrix}. \tag{4.56}$$

The same rules of transformation will apply to the small linear displacements u and v:

$$u' = u \cos\theta + v \sin\theta, \tag{4.57a}$$

$$v' = -u \sin\theta + v \cos\theta. \tag{4.57b}$$

Next, we recall the definition of normal strain from the previous section. Applying the chain rule, we have the normal strain in the x' direction:

$$\varepsilon_{x'} = \frac{\partial u'}{\partial x'} = \frac{\partial u'}{\partial x}\frac{\partial x}{\partial x'} + \frac{\partial u'}{\partial y}\frac{\partial y}{\partial x'}. \tag{4.58}$$

If we differentiate the previous expressions for x and y in terms of the rotated x' and y', we obtain

$$\varepsilon_{x'} = \varepsilon_x \cos^2\theta + \varepsilon_y \sin^2\theta + \gamma_{xy} \sin\theta\cos\theta, \tag{4.59a}$$

$$\varepsilon_{y'} = \frac{\varepsilon_x + \varepsilon_y}{2} + \frac{\varepsilon_x - \varepsilon_y}{2}\cos 2\theta + \gamma_{xy} \sin 2\theta. \tag{4.59b}$$

We may also transform the shear strain, by first writing it in the rotating coordinates as

$$\gamma_{x'y'} = \frac{\partial v'}{\partial x'} + \frac{\partial u'}{\partial y'} \tag{4.60}$$

and then differentiating and simplifying to get

$$\gamma_{x'y'} = -(\varepsilon_x - \varepsilon_y)\sin 2\theta + \gamma_{xy}\cos 2\theta. \tag{4.61}$$

We may also construct Mohr's circle of strain to determine the principal strains and maximum shear strains. In doing this, we must remember that the shear components of the strain tensor are truly "gammas divided by 2"; that is, the shear strain component $\varepsilon_{x'y'}$ is given by

$$\varepsilon_{x'y'} = \frac{\gamma_{x'y'}}{2} = -\frac{(\varepsilon_x - \varepsilon_y)}{2}\sin 2\theta + \frac{\gamma_{xy}}{2}\cos 2\theta. \tag{4.62}$$

So we plot Mohr's circle of strain as the set of all points (ε, $\gamma/2$). The principal (normal) strains are written as

$$\left(\varepsilon_{x'}\right)_{\substack{\max\\ \min}} = \varepsilon_{1,2} = \frac{\varepsilon_x + \varepsilon_y}{2} \pm \sqrt{\left(\frac{\varepsilon_x - \varepsilon_y}{2}\right) + \left(\frac{\gamma_{xy}}{2}\right)^2} \tag{4.63}$$

and they occur on the plane defined by

$$\tan 2\theta_{N\varepsilon} = \frac{\gamma_{xy}}{\varepsilon_x - \varepsilon_y}. \tag{4.64}$$

4.3.5 Three-Dimensional State of Stress

The normal and shear stresses acting on a plane through a material depend on the orientation of that plane. We have found both equations and a graphical technique (Mohr's circle) to determine the normal and shear stresses on a variety of planes, but only for the special case of plane stress, and for a few selected planes. If we keep things more general, we can make a first pass at obtaining the principal stresses in three dimensions.

We remember that our stress tensor is *symmetric*. From linear algebra, we recall that a symmetric matrix may be diagonalized. This suggests that the symmetric stress tensor may also be diagonalized—that there is a certain coordinate system (x', y', z') for which

$$\begin{pmatrix} \sigma_{x'} & \tau_{x'y'} & \tau_{x'z'} \\ \tau_{y'x'} & \sigma_{y'} & \tau_{y'z'} \\ \tau_{z'x'} & \tau_{z'y'} & \sigma_{z'} \end{pmatrix} = \begin{pmatrix} \sigma_1 & 0 & 0 \\ 0 & \sigma_2 & 0 \\ 0 & 0 & \sigma_3 \end{pmatrix} \tag{4.65}$$

The axes x', y', z' are called the principal axes for this state of stress, and σ_1, σ_2, and σ_3 are the principal stresses. It can be shown[3] that the principal stresses are the roots of the cubic equation

$$\sigma^3 - I_1\sigma^2 + I_2\sigma = 0 \tag{4.66}$$

where

$$\begin{aligned} I_1 &= \sigma_x + \sigma_y + \sigma_z \\ I_2 &= \sigma_x\sigma_y + \sigma_y\sigma_z + \sigma_z\sigma_x - \tau_{xy}^2 - \tau_{yz}^2 - \tau_{zx}^2 \\ I_3 &= \sigma_x\sigma_y\sigma_z - \sigma_x\tau_{yz}^2 - \sigma_y\tau_{xz}^2 - \sigma_z\tau_{xy}^2 + 2\tau_{xy}\tau_{yz}\tau_{zx} \end{aligned} \tag{4.67}$$

These quantities I_i are invariants, independent of the orientation of the coordinate system. [I_1 is the extension of the invariant ($\sigma_x + \sigma_y$) in plane strain.] We determine the principal stresses by evaluating the coefficients I_i and solving for σ.

We determine the maximum shear stress in the same way we did for plane stress and find that the absolute maximum shear stress is the largest of the three values:

$$\max \left|\frac{\sigma_1 - \sigma_2}{2}\right|, \left|\frac{\sigma_1 - \sigma_3}{2}\right|, \left|\frac{\sigma_2 - \sigma_3}{2}\right|. \tag{4.68}$$

This absolute maximum shear stress may be visualized by superimposing the Mohr's circles obtained from the three orientations shown in Figure 4.29a. Notice from Figure 4.29b that if $\sigma_1 > \sigma_2 > \sigma_3$, the absolute maximum shear stress is $(\sigma_1 - \sigma_3)/2$.

4.4 Failure Prediction Criteria

In Chapter 3, Section 3.6, we discussed the need for techniques to predict failure for various materials under various loading. In designing structures, it may be necessary to make compromises and trade-offs, based on material

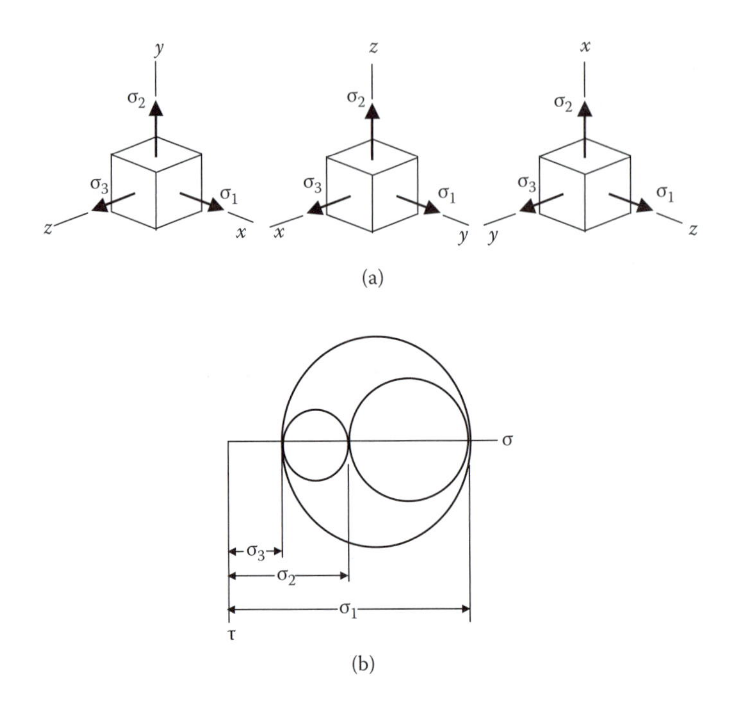

FIGURE 4.29
(a) Different orientations of the coordinate system relative to the element on which the principal stresses act; (b) superimposing the Mohr's circles demonstrates the absolute maximum shear stress.

availability, manufacturability, cost, weight, and aesthetic issues, but avoiding failure is not negotiable. For complex structures subject to general states of stress, various criteria have been proposed for predicting (and so preventing) failure. The applicability of these criteria depends on the nature of the materials and the loading involved.

4.4.1 Failure Criteria for Brittle Materials

We recall from Chapter 3, Figure 3.10c that a brittle material subjected to uniaxial tension fails without necking, on a plane normal to the material's long axis. When such an element is under uniaxial tensile stress, the normal stress that causes it to fail is the ultimate tensile strength of the material. However, when a structural element is in a state of plane stress or a three-dimensional stress state, it is useful to determine the principal stresses at any given point and to use one of the following criteria.

4.4.1.1 Maximum Normal Stress Criterion

According to this criterion, a given structural element fails when the maximum normal stress in that component reaches the material's ultimate tensile strength. This criterion is appropriate for brittle materials, which do not yield or undergo much plastic deformation, since the implied mechanism for the failure is one of separation rather than sliding or shear. Mathematically, we can represent this criterion as saying that failure will occur when

$$\text{Max}\{|\sigma_1|, |\sigma_2|, |\sigma_3|\} = \sigma_U. \tag{4.69}$$

In the case of plane stress ($\sigma_3 = 0$), this criterion can be applied in a straightforward graphical manner (Figure 4.30). The safe values of σ_1 and σ_2 for which failure will not occur are bounded by the square shaded region. For a three-dimensional stress state, the safe stress region is enclosed in a cube.

This criterion suffers from a significant shortcoming: It assumes that the ultimate strength of the material is the same in both tension and compression. As we saw in Table 3.1, this is rarely the case. This criterion also makes no allowance for effects other than normal stresses on the material's failure.

4.4.1.2 Mohr's Criterion

The fact that many materials have different ultimate strengths in tension and compression necessitates a modified version of Figure 4.30. This criterion allows the use of different values of ultimate tensile strength, σ_{UT}, and ultimate compressive strength, σ_{UC}. With these two values, we could construct Mohr's circles corresponding to each value, as in Figure 4.31a, and be assured that a state of stress represented by a circle that was fully contained by either of

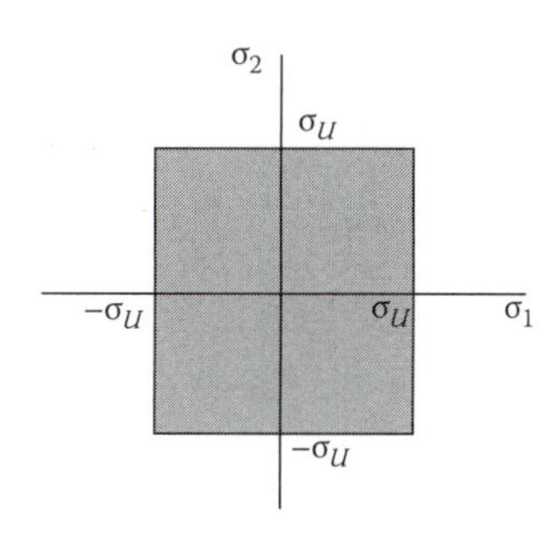

FIGURE 4.30
Graphical use of maximum normal stress criterion.

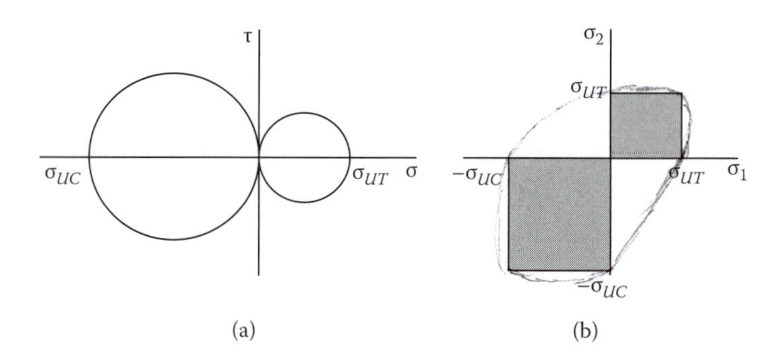

FIGURE 4.31
(a) Mohr's circles corresponding to σ_{UT} and σ_{UC}; (b) graphical use of Mohr's failure criterion for plane stress.

these circles would be safe. The version of Figure 4.30 corresponding to this statement is shown in Figure 4.31b.

If we also know the material's ultimate shear strength, τ_U, we can draw three circles, as in Figure 4.32a. Mohr's criterion states that a state of stress is safe if it can be represented by a circle located entirely within the area bounded by the *envelope* of the circles corresponding to the available data, as in Figure 4.32a and Figure 4.32b.

If the torsional test data (τ_U) that provide the curvature of the bounding curves in Figure 4.32b are unavailable, then it is possible to simplify Mohr's criterion by using the tangents to the circles for σ_{UT} and σ_{UC} and drawing a "safe" shaded region for plane stress, as shown in Figure 4.33a and Figure 4.33b.

To determine whether a structural component will be safe under a given load, we should calculate the stress state at all critical points of the component and particularly at all points where stress concentrations are likely to occur.

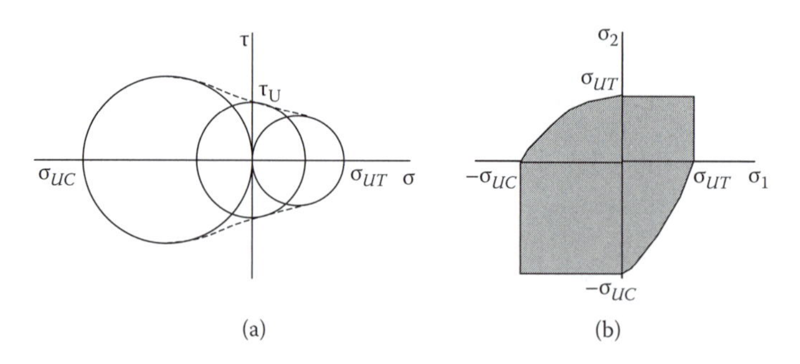

FIGURE 4.32
(a) Mohr's circles corresponding to σ_{UT}, σ_{UC}, andτ_U; (b) graphical use of Mohr's failure criterion for plane stress.

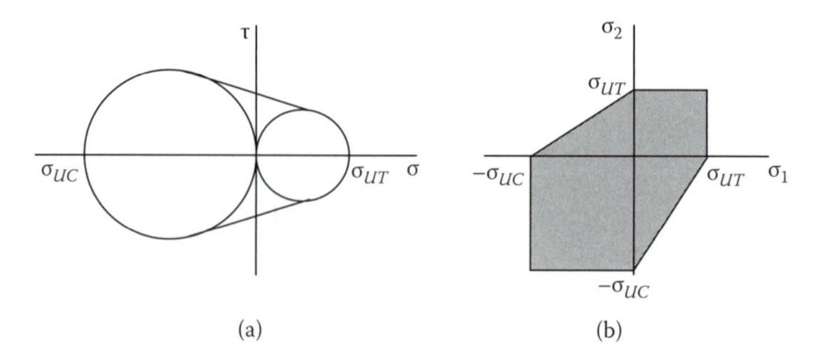

FIGURE 4.33
(a) Mohr's circles corresponding to σ_{UT}, and σ_{UC}; (b) graphical use of Mohr's failure criterion for plane stress.

4.4.2 Yield Criteria for Ductile Materials

We observed in Chapter 3, Figure 3.10b that a ductile material subjected to uniaxial tension yields and fails by slippage along oblique surfaces and is due primarily to shear stresses. A uniaxial tension test produces values of tensile and compressive yield strengths for a ductile material. Although ductile materials do not fracture when they reach their yield strengths, at this point permanent deformation can occur and proper function of a structural element may be lost. We therefore cast our criteria in terms of *yield* and not of fracture.

4.4.2.1 *Maximum Shearing Stress (Tresca) Criterion*

Because the plastic deformation initiated at the yield strength takes place through shear deformation, it is natural to expect failure criteria to be expressed in terms of shear stress. Based on this logic, the Tresca criterion says that a given structural component is safe as long as the maximum shear stress value in that component does not exceed the yield shear strength, τ_Y, of the material. Since for axial loading, as we saw in Chapter 2, Section 2.5, the maximum shear stress is equal to half the value of the corresponding normal, axial stress, we conclude that the maximum shear stress experienced by a uniaxial tensile test specimen is $\tau_Y = \sigma_Y/2$.

We extend this criterion to an arbitrary state of stress by assuming that yielding occurs when the absolute maximum shear stress is equal to τ_Y:

$$\text{Max}\left(\left|\frac{\sigma_1-\sigma_2}{2}\right|,\left|\frac{\sigma_2-\sigma_3}{2}\right|,\left|\frac{\sigma_1-\sigma_3}{2}\right|\right)=\tau_Y. \quad (4.70)$$

Or, using $\tau_Y = \sigma_Y/2$ to recast the criterion in terms of a Tresca equivalent normal stress σ_T:

$$\text{Max}\left(|\sigma_1-\sigma_2|,|\sigma_2-\sigma_3|,|\sigma_1-\sigma_3|\right)\equiv\sigma_T=\sigma_Y. \quad (4.71)$$

So, for plane stress our safe region is bounded by the lines $\sigma_1 - \sigma_2 = \pm\sigma_Y$, $\sigma_1 = \pm\sigma_Y$, and $\sigma_2 = \pm\sigma_Y$, which form the hexagon shown in Figure 4.34.

We can also compare the Tresca equivalent stress with the material's yield stress to determine how close the material is to failure. We obtain a *Tresca safety factor*:

$$S_T=\frac{\sigma_Y}{\sigma_T}. \quad (4.72)$$

Failure will occur when $S_T = 1$, and a safe design will ensure that $S_T >> 1$.

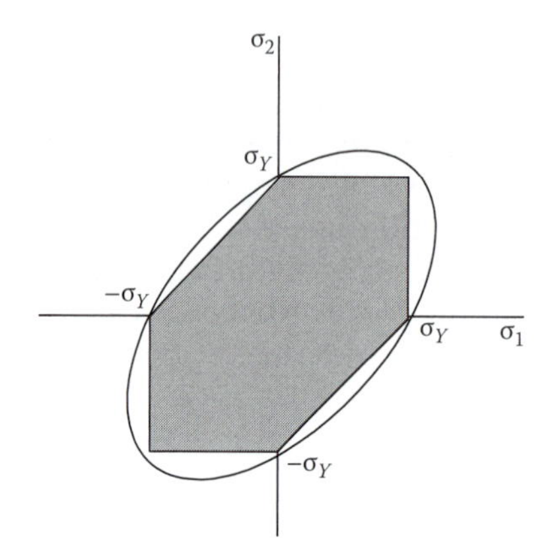

FIGURE 4.34
Failure boundaries for the Tresca (hexagon) and von Mises (ellipse) failure criteria under plane stress.

4.4.2.2 *Von Mises Criterion*

This criterion for failure of ductile materials is derived from strain energy considerations and states that yielding occurs when

$$\tfrac{1}{2}\left[(\sigma_1-\sigma_2)^2+(\sigma_2-\sigma_3)^2+(\sigma_1-\sigma_3)^2\right]=\sigma_Y^2. \tag{4.73}$$

In plane stress the safe region is bounded by the curve, which describes the ellipse in Figure 4.34. We can define a von Mises equivalent stress as

$$\sigma_M \equiv \tfrac{1}{\sqrt{2}}\sqrt{(\sigma_1-\sigma_2)^2+(\sigma_2-\sigma_3)^2+(\sigma_1-\sigma_3)^2} \tag{4.74}$$

and can compare this value with the material's yield stress to determine how close the material is to failure. We obtain a *von Mises safety factor*:

$$\sigma_{yy}=0. \tag{4.75}$$

4.5 Examples

Example 4.1

A 100-mm-diameter core is bored out from a 200-mm-diameter solid circular shaft (Figure 4.35). What percentage of the shaft's torsional strength is lost due to this operation?

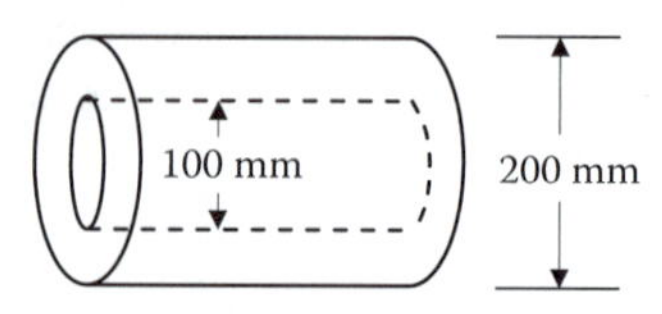

FIGURE 4.35

Given: Dimensions of a shaft to be loaded in torsion.

Find: Percentage of original shaft's strength lost when a hole is bored out.

Assume: Hooke's law applies.

Solution

Torsional strength of a shaft is reflected by how much torque it can withstand without exceeding its allowable τ_{max}. This τ_{max} limit is a material property and does not change due to geometrical alterations. We know how the maximum shear stress induced in the shaft is calculated: $\tau_{max} = Tc/J$.

When the hole specified is bored out of the original shaft, the maximum radius c does not change; however, the polar second moment of area J, which is a property of the cross section, does change.

So,

$$\left(\frac{\tau}{c}\right)_{old} = \left(\frac{\tau}{c}\right)_{new} = \left(\frac{T}{J}\right)_{old} = \left(\frac{T}{J}\right)_{new},$$

and we find the ratio of the allowable torques to be

$$\frac{T_{new}}{T_{old}} = \frac{J_{new}}{J_{old}} = \frac{\pi\,(r_{old}^{\;4} - r_{new}^{\;4})/2}{\pi\, r_{old}^4}.$$

And, solving for T_{new}, we have

$$T_{new} = T_{old}\left(\frac{r_{old}^{\;4} - r_{new}^{\;4}}{r_{old}^4}\right) = T_{old}\left(\frac{(0.100\text{ m})^4 - (0.050\text{ m})^4}{(0.100\text{ m})^4}\right);$$

that is, $T_{new} = 0.938\ T_{old}$.

The bored-out shaft can withstand only 93.8% of the maximum torque withstood by the original shaft. We have gone from 100% to 93.8%; that is, we have lost 6.2% of the torsional strength of the shaft.

Example 4.2

Design a hollow steel shaft to transmit 300 hp at 75 rpm without exceeding a shear stress of 8000 psi. Use 1.2:1 as the ratio of the outside diameter to the inside diameter. What solid shaft could be used instead?

Given: Desired performance of hollow shaft.

Find: Inner and outer diameters of shaft; dimensions of equivalent solid shaft.

Assume: Hooke's law applies.

Solution

We can easily obtain the torque required for 300 hp of power at a rotational frequency of 75 rpm:

$$T = \frac{P}{\omega},$$

and the conversions necessary to find a torque in U.S. units of in. · lb are built in to the following version of this relationship:

$$T\ [\text{in.-lb}] = \frac{63{,}000 \times [\text{hp}]}{\text{N}\ [\text{rpm}]}.$$

We have

$$T = \frac{63{,}000 \times 300\ \text{hp}}{75\ \text{rpm}} = 252{,}000\ \text{in.-lb}.$$

Since the maximum shear stress induced by this torque is given by

$$\tau_{\max} = \frac{Tc}{J},$$

we obtain the value of J/c needed to transmit 600 hp without exceeding the stated stress limit:

$$\frac{J}{c} = \frac{T}{\tau_{\max}} = \frac{252{,}000\ \text{in.-lb}}{8000\ \text{psi}} = 31.5\ \text{in.}^3$$

$$\frac{J}{c} = \frac{\pi}{2}\frac{\left(c^4 - (c/1.2)^4\right)}{c} = 0.813\ c^3 = 31.5\ \text{in.}^3$$

This has the solution $c = 3.38$ in., so the outer diameter necessary is $D_o = 2c = 6.77$ in., and the inner diameter is $D_i = D_i/1.2 = 5.64$ in.

For a solid shaft, J/c has a simpler form, and we require only

$$\frac{J}{c} = \frac{\pi}{2}c^3 = 31.5\ \text{in.}^3$$

Solving for c, the radius of a solid shaft capable of transmitting 600 hp without exceeding a shear stress of 8000 psi, we have $D = 2c = 5.44$ in.

Example 4.3

What must be the length of a 6-mm-diameter aluminum (G = 27 GPa) wire so that it could be twisted through one complete revolution without exceeding a shear stress of 42 MPa?

Given: Cross section of wire, desired deformation, limit on shear stress.

Find: Required length of wire.

Assume: Hooke's law applies.

Solution

The angle of twist of the wire after "one complete revolution" is 2π. The angle of twist is defined as

$$\phi = \frac{TL}{JG},$$

and, using the definition of maximum shear stress, this can also be written as

$$\phi = \frac{\tau_{max} \not{J}}{c} \frac{L}{\not{J}G} = \frac{\tau_{max} L}{cG},$$

which we rearrange to solve for the wire length L:

$$L = \frac{\phi\, cG}{\tau_{max}} = \frac{2\pi \cdot (0.003 \text{ m})(27 \times 10^9 \text{ Pa})}{42 \times 10^6 \text{ Pa}}$$

$$L = 12.12 \text{ m}.$$

Example 4.4

Find the required fillet radius for the juncture of a 6-in.-diameter shaft with a 4-in.-diameter segment if the shaft transmits 110 hp at 100 rpm and the maximum shear stress is limited to 8000 psi.

Given: Dimensions of and requirements for shaft performance.

Find: Fillet radius for connecting two segments of shaft.

Assume: Hooke's law applies. Transition between segments is only stress concentration.

Solution

We make use of the relationship between applied torque, power output, and rotational frequency (see also Example 4.2) to find the applied torque:

$$T = \frac{63{,}000 \cdot 110 \text{ hp}}{100 \text{ rpm}} = 69{,}300 \text{ in.-lb}$$

The shear stress in the shaft cannot exceed 8000 psi. We obtain the maximum allowable stress concentration factor, using the smaller segment's radius for c and in J:

$$K = \frac{\tau_{\max} J}{Tc} = \frac{8000 \text{ psi}\left[\frac{\pi}{2}(2 \text{ in.})^4\right]}{69{,}300 \text{ in.} \cdot \text{lb } (2 \text{ in.})} = 1.45,$$

and

$$\frac{D}{d} = \frac{\text{big shaft diameter}}{\text{small shaft diameter}} = \frac{6}{4} = 1.5.$$

From Figure 4.5, we find that this K and this D/d correspond to an r/d ratio of 0.085. This means that the allowable fillet radius is

$$r = (0.085)(4 \text{ in.}) = 0.340 \text{ in.}$$

Example 4.5

A solid aluminum alloy shaft 60 mm in diameter and 1000 mm long is to be replaced by a tubular steel shaft of the same outer diameter such that the new shaft will exceed neither (1) twice the maximum shear stress nor (2) the angle of twist of the aluminum shaft. What should be the inner radius of the tubular steel shaft? Which of the two criteria (1) strength or (2) stiffness governs?

Given: Dimensions of aluminum shaft.

Find: Dimensions of steel shaft (same length, same outer diameter) that will meet strength and stiffness requirements.

Assume: Hooke's law applies.

Solution

We design first for strength and then for stiffness. From Table 2.1 or another source, we find the appropriate material properties: G_{Al} = 28 GPa and G_{St} = 84 GPa.

1. Designing for Strength

$$\underset{Steel}{\tau_{\max}} \le 2\underset{Al}{\tau_{\max}}.$$

$$\left(\frac{Tc}{J}\right)_{St} \leq 2\left(\frac{Tc}{J}\right)_{Al}.$$

Since we are told that the outer diameters of both shafts are equal, and since the applied torque T does not change, we are simply requiring that

$$J_{St} \geq \frac{1}{2} J_{Al}.$$

$$\frac{\pi}{2}\left[(0.03\text{ m})^4 - r_i^4\right] \geq \frac{\pi}{4}\left[(0.03\text{ m})^4\right]w.$$

$$r_i^4 \leq \frac{2}{\pi}\left[\frac{\pi}{2}(0.03\text{ m})^4 - \frac{\pi}{4}(0.03\text{ m})^4\right].$$

$$r_i^4 \leq \frac{2}{\cancel{\pi}}\frac{\cancel{\pi}}{4}(0.03\text{ m})^4 = 405 \times 10^{-9}\text{ m}^4.$$

$$r_i \leq 25.2\text{ mm}.$$

2. *Designing for Stiffness*

$$\phi_{St} \leq \phi_{Al}.$$

$$\frac{TL}{J_{St}G_{St}} \leq \frac{TL}{J_{Al}G_{Al}}.$$

$$J_{St}G_{St} \geq J_{Al}G_{Al}.$$

$$\frac{\pi}{2}\left[(0.03\text{ m})^4 - r_i^4\right](84 \times 10^9\text{ Pa}) \geq \frac{\pi}{2}\left[(0.03\text{ m})^4\right](28 \times 10^9\text{ Pa}).$$

Solve for $r_i \leq 27.1$ mm. The inner radius of the steel shaft must be $r_i \leq 25.2$ mm, as strength governs.

Example 4.6

Two shafts (G = 28 GPa) A and B are joined and subjected to the torques shown in Figure 4.36. Section A has a solid circular cross section with diameter 40 mm and is 160 mm long; B has a solid circular cross section with diameter 20 mm and is 120 mm long. Find (a) the maximum shear stress in sections A and B; and (b) the angle of twist of the right-most end of B relative to the wall.

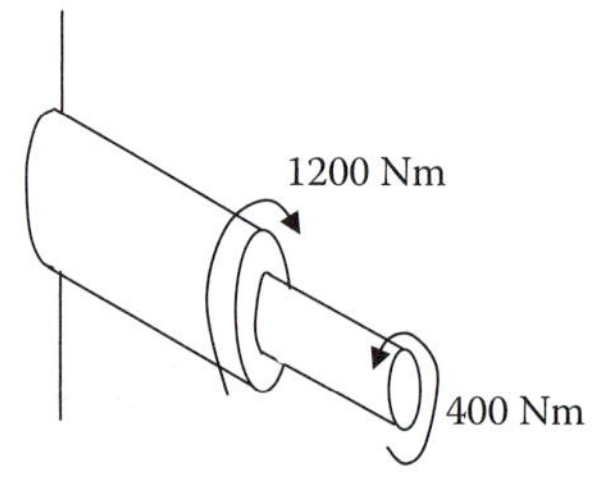

FIGURE 4.36

Given: Dimensions and properties of composite shaft in torsion.

Find: Shear stresses, angle of twist of free end.

Assume: Hooke's law applies.

Solution

Our strategy is to use the method of sections to find the internal torque in each portion of the composite shaft and then find the shear stress and angle of twist induced by this torque. First, we construct an FBD (Figure 4.37):

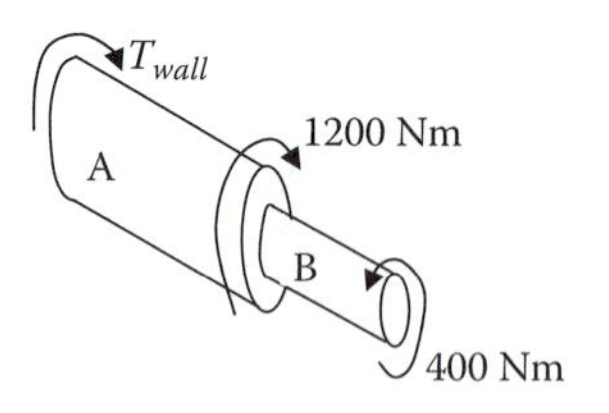

FIGURE 4.37

Equilibrium requires that 400 Nm – 1200 Nm – T_{wall} = 0.

Therefore, T_{wall} = –800 Nm (T_{wall} is clockwise, opposite from what is drawn at left.)

Now use the method of sections on segments A and B (Figure 4.38):

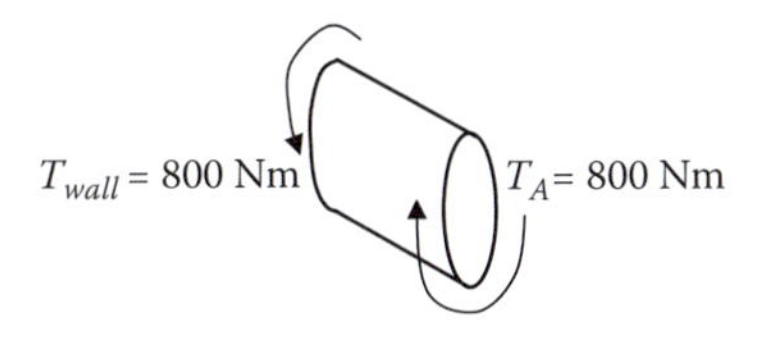

FIGURE 4.38

Internal torque T_A = 800 Nm, and maximum shear stress occurs at c_A = 0.02 m. So,

$$\tau_{\underset{A}{\max}} = \frac{T_A c_A}{J_A} = \frac{T_A c_A}{\frac{\pi}{2} c_A^4} = 63.7 \text{ MPa}.$$

To find the internal resisting torque in section B, we must look at the whole shaft from the wall to our imaginary section cut (Figure 4.39):

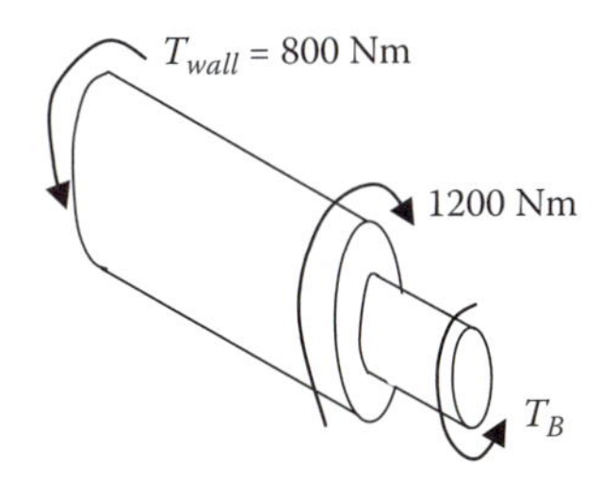

FIGURE 4.39

Equilibrium of this section requires that the internal torque T_B = 400 Nm. Maximum shear stress occurs at c_B = 0.01 m, and

$$\tau_{\underset{B}{\max}} = \frac{T_B c_B}{J_B} = \frac{T_B c_B}{\frac{\pi}{2} c_B^4} = 255 \text{ MPa}.$$

Next, we will calculate the angles of twist of both A and B and then find the resultant twist of the free end with respect to the wall:

$$\phi_A + \phi_B = \phi.$$

Taking counterclockwise twists to be positive, as we take counterclockwise torques to be,

$$\phi_A = \frac{T_A L_A}{J_A G_A} = \frac{(-800\ \text{Nm})(0.16\ \text{m})}{(2.51\times10^{-7}\ \text{m}^4)(28\times10^9\ \text{Pa})} = -0.0182\ \text{rad}\ (-1.04°)$$

$$\phi_B = \frac{T_B L_B}{J_B G_B} = \frac{(400\ \text{Nm})(0.12\ \text{m})}{(1.57\times10^{-8}\ \text{m}^4)(28\times10^9\ \text{Pa})} = 0.1091\ \text{rad}\ (6.25°).$$

The total angle of twist of the free end relative to the wall is then

$$\varphi = \varphi_A + \varphi_B = -1.04° + 6.25° = 5.21°\ \text{(CCW)}.$$

Example 4.7

Calculate the torsional stiffness k_t of the rubber bushing shown in Figure 4.40. Assume that the rubber is bonded both to the steel shaft and to the outer steel tube, which is in turn attached to a machine housing. Assume that the metal parts do not deform, and that the shear modulus of rubber is G.

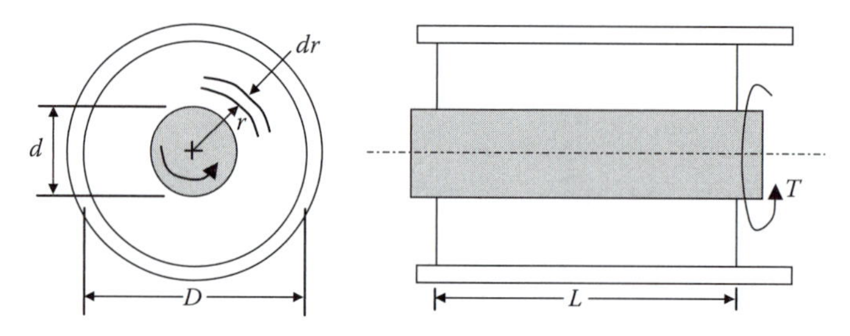

FIGURE 4.40

Given: Rubber bushing of known dimension and shear modulus.

Find: Torsional stiffness k_t.

Assume: Hooke's law applies.

Solution

Axisymmetric torque is resisted by constant shear stresses, $T = \tau(2\pi rL)r$.
From Hooke's law,

$$\gamma = \frac{\tau}{G} = \frac{T}{(2\pi r^2 L)G}.$$

The incremental shaft rotation is $rd\phi \cong \gamma dr$, the total shaft rotation is

$$\phi = \int d\phi = \frac{T}{2\pi LG} \int_{d/2}^{D/2} \frac{dr}{r^3} = \frac{T}{\pi LG}\left(\frac{1}{d^2} - \frac{1}{D^2}\right),$$

and the stiffness is

$$k_t = \frac{T}{\phi} = \frac{\pi LG}{1/d^2 - 1/D^2}.$$

Example 4.8

Calculate the tensile stresses (circumferential and longitudinal) developed in the walls of a cylindrical pressure vessel with inside diameter 18 in. and wall thickness 1/4 in. The vessel is subjected to an internal gage pressure of 300 psi and a simultaneous external axial tensile load of 50,000 lb.

Given: Dimensions of and loading on cylindrical pressure vessel.

Find: Hoop and longitudinal normal stresses.

Assume: We will test whether thin-walled theory may be applied to this vessel.

Solution

Does thin-walled theory apply? Is the thickness $t \le 0.1r_i$?

$$(t = 0.25 \text{ in.}) \le 0.1 \cdot (r_i = 9 \text{ in.}) = 0.90.$$

We can use thin-walled theory.

The circumferential, or hoop stress, is calculated as

$$\sigma_1 = \frac{pr_i}{t} = \frac{(300 \text{ psi})(9 \text{ in.})}{0.25 \text{ in.}} = 10.8 \text{ ksi}.$$

The longitudinal stress due to the internal pressure may be combined with the normal stress due to the axial load by straightforward superposition, as these stresses are in the same direction and act normal to areas with the same orientation:

$$\sigma_2 = \frac{pr_i}{2t} + \frac{P}{A} = \frac{(300 \text{ psi})(9 \text{ in.})}{0.5 \text{ in.}} + \frac{50{,}000 \text{ lb}}{\pi(2 \cdot r_i t)}.$$

$$\sigma_2 = 5.4 \text{ ksi} + 3.5 \text{ ksi} = 8.9 \text{ ksi}.$$

Note: The area on which P acts can also be calculated as $\pi\, r_o^2 - \pi\, r_i^2$; this result is $P/A = 3.49$ ksi and results in a longitudinal stress of 8.89 ksi.

Example 4.9

For the given state of plane stress, construct Mohr's circle, determine the principal stresses, and determine the maximum shearing stress and the corresponding normal stress (Figure 4.41).

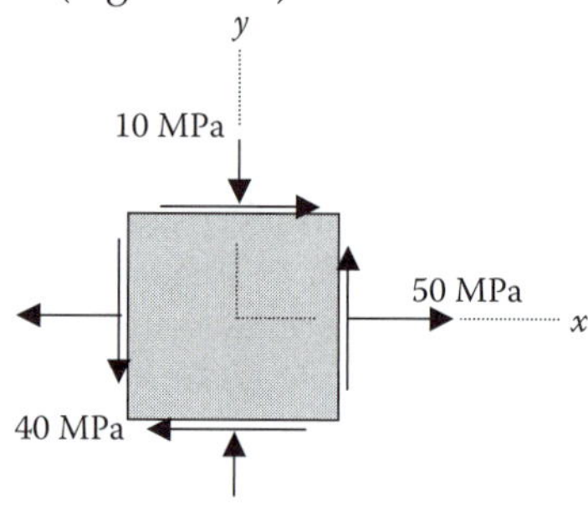

FIGURE 4.41

Given: $\sigma x = 50$ MPa, $\sigma y = -10$ MPa, $\tau xy = 40$ MPa.

Find: Extreme stress states.

Assume: Plane stress: $\sigma z = \tau xz = \tau zy = 0$.

Solution

We outline the steps used to construct Mohr's circle and make the necessary calculations. The steps are as follows:

Plot point X: (σ_x, τ_{xy}).
Plot point Y: (σ_y, τ_{xy}).
Draw line XY, which passes through the circle center: $(\sigma_{ave}, 0)$.
Find radius R and draw in the circle.

1. Plot Point X: (σ_x, τ_{xy})

We note straight off that the shear stress given is "positive," according to Chapter 2, Figure 2.5, but we are not sure how to plot the point (σ_x, τ_{xy}) on Mohr's circle. Finding σ_x on the σ axis is straightforward—the normal stress sign convention simply says that tensile stresses are positive and compressive are negative, but does τ_{xy} lie above or below that axis? We know that for Mohr's circle to work, we must have points X and Y on opposite sides of the σ axis so that their connecting line XY passes through the center of the circle. Our sign convention must ensure this. We therefore make use of a system based on the positive x (and y) faces of our unrotated element.

Looking at the positive x face of our initial element (right-hand face), we see that the component of shear stress on this face is tending to rotate the element counterclockwise. This tells us to plot point X below the σ axis. Our convention is that when this component tends to rotate clockwise, X is above the axis and when counterclockwise, it is below. (This somewhat awkward rule can be remembered by the equally strange mnemonic: "In the kitchen, the clock is above and the counter is below.") We formalize this rule in Figure 4.42. Remember that we'll apply this sign convention to points X and Y separately—for Mohr's circle to work, we must have points X and Y on opposite sides of the σ axis.

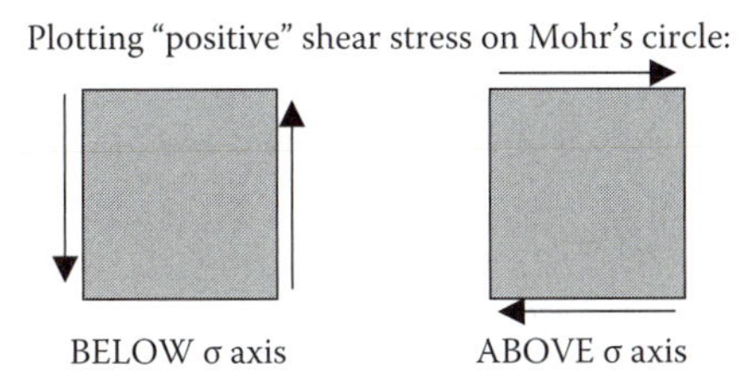

FIGURE 4.42

2. Plot Point Y: (σ_y, τ_{xy})

Following the same reasoning as for point X, we plot point Y to the left of the τ axis, as σ_y is compressive, and above the σ axis, as the shear stress on the positive y (top) face of the element tends to rotate clockwise.

These two points may now be plotted on the $\sigma\tau$ axes (Figure 4.43).

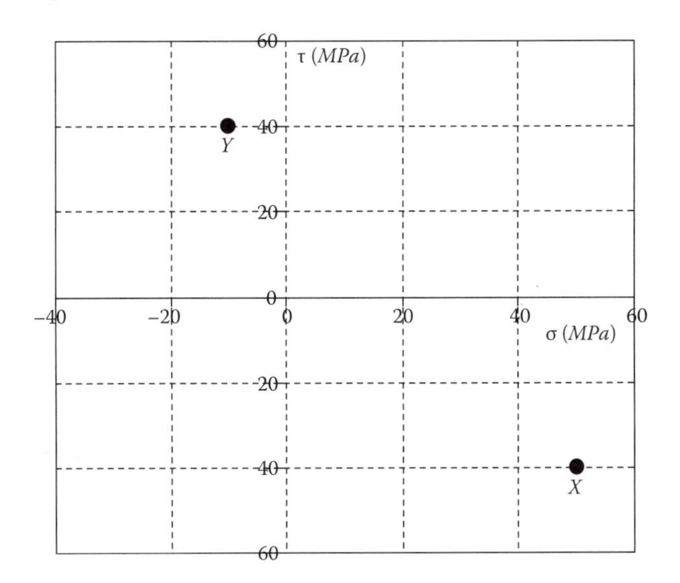

FIGURE 4.43

3. Draw Line XY

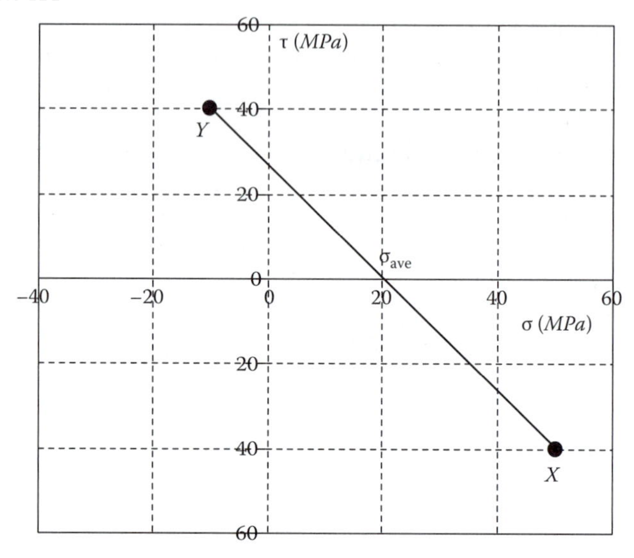

FIGURE 4.44

This line (Figure 4.44) passes through the σ (horizontal) axis at the center of the circle:

$$\left(\sigma_{ave},0\right)=\left(\frac{\sigma_x+\sigma_y}{2},0\right)=(20\ \text{MPa},\,0)$$

4. Find the Radius R and Draw the Circle

We may use the geometry of the first three steps, or the formulas derived in the notes, to calculate the radius of the circle. Graphically, we see that R is the hypotenuse of a right triangle whose other legs have length 40 and 50 – 20 = 30. Thus, $R = ((40)^2 + (30)^2)^{1/2} = 50$ MPa. Alternatively,

$$R=\sqrt{\left(\frac{\sigma_x-\sigma_y}{2}\right)^2+\left(\tau_{xy}\right)^2}=50\ \text{MPa}.$$

We can now sketch Mohr's circle by hand, by using a compass, or by using a software package. The circle contains all the information we need about all possible axes and thus all possible stress states for the given element (Figure 4.45).

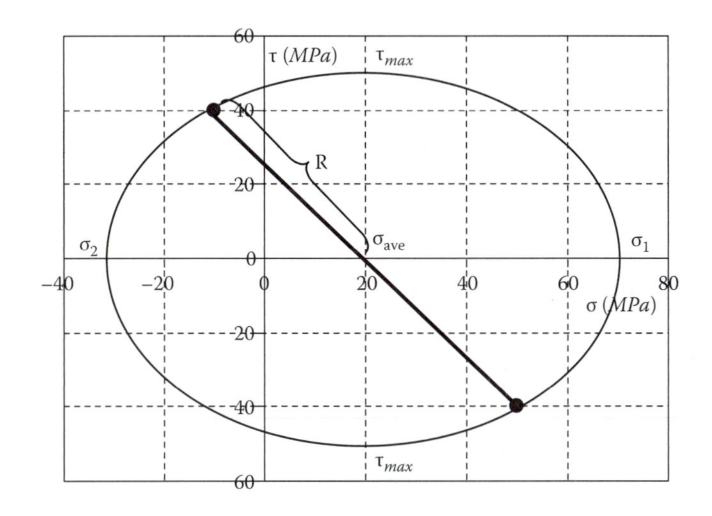

FIGURE 4.45

We can find, and label, the principal stresses:

$$\sigma_1 = \sigma_{ave} + R = 20\text{ MPa} + 50\text{ MPa} = 70\text{ MPa}.$$

$$\sigma_2 = \sigma_{ave} - R = 20\text{ MPa} - 50\text{ MPa} = -30\text{ MPa}.$$

This principal stress state (extreme normal stress, no shear stress) occurs when the axes are rotated by θ_N. (Or, when line XY is rotated around Mohr's circle by $2\theta_N$.) We can find $2\theta_N$ using a protractor, or we can use our formulas: $\theta_N = ½ \tan^{-1}(2\tau_{xy}/(\sigma_x - \sigma_y))$. At this θ_N we can calculate that the value of $\sigma_{x'}$ (rather than $\sigma_{y'}$) is 70 MPa, so we draw our properly oriented element that experiences this principal stress state (Figure 4.46).

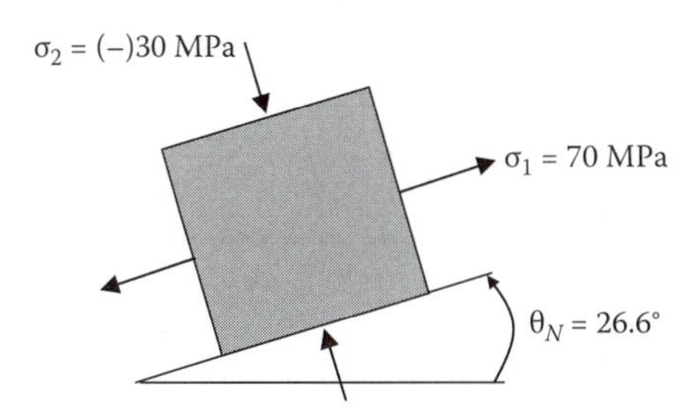

FIGURE 4.46

Next, we calculate the maximum shear stress and the corresponding normal stress, which we can see from Mohr's circle is the average normal stress, σ_{ave}:

$$\tau_{max} = R = 50\text{ MPa}.$$

$$\sigma_{ave} = 20 \text{ MPa}.$$

From the principal stress state, we can see on Mohr's circle that it will take $2\theta = 90°$ to get to this stress state (σ_{ave}, τ_{max}). We need, then, to rotate our element's axes 45° counterclockwise past the principal stress orientation. From our initial orientation, this rotation is given by

$$\theta_S = \theta_N + 45° = 26.6° + 45° = 71.6°$$

We can again draw a properly oriented element experiencing the maximum shear stress, having been rotated by 71.6° counterclockwise from its initial orientation:

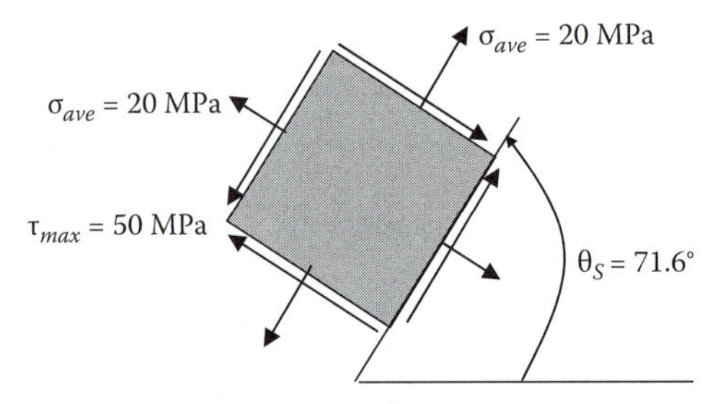

FIGURE 4.47

We obtain the proper sense of the shear stress from Mohr's circle. By rotating line XY counterclockwise by $2\theta_S$, we get to a point above the σ axis. Thus, on the rotated positive x face, we must have a shear stress that tends to rotate the element *clockwise*.

Finally, we can visualize these rotations on Mohr's circle (Figure 4.48).

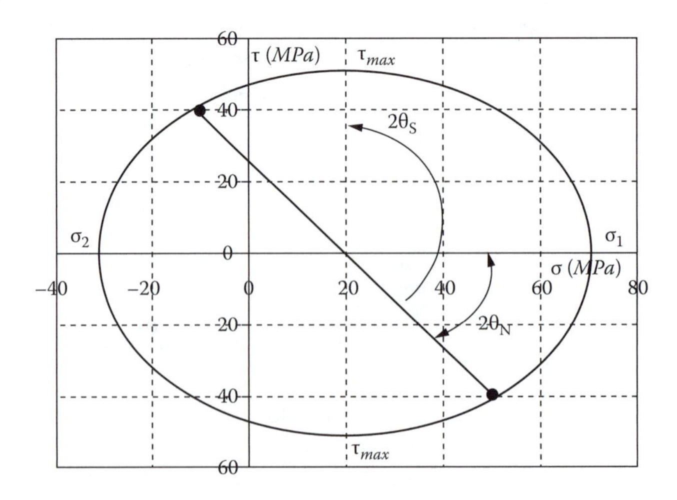

FIGURE 4.48

Note: A "positive" rotation (i.e., a positive value in degrees or radians) is counterclockwise, both in physical space and on Mohr's circle.

Example 4.10

A state of plane stress consists of a tensile stress $\sigma_o = 8$ ksi exerted on vertical surfaces and unknown shear stresses τ_o (Figure 4.49). Determine (a) the magnitude of the shear stress τ_o for which the maximum normal stress is 10 ksi, and (b) the corresponding maximum shear stress.

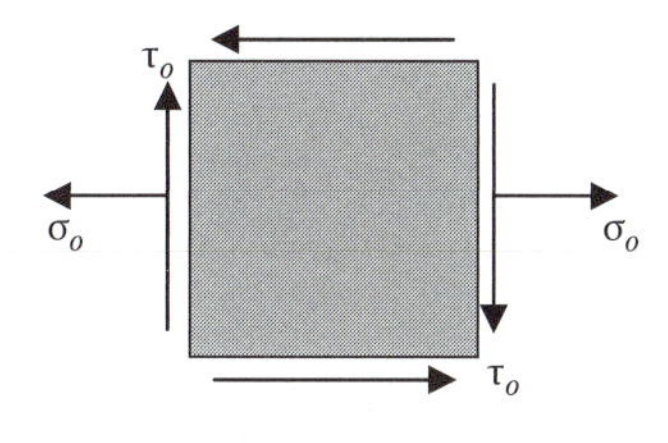

FIGURE 4.49

Given: Partial plane stress state.

Find: Shear stress τ_o; maximum shear stress.

Assume: Plane stress.

Solution

We assume a sense (sign) for the unknown shear stress and can construct Mohr's circle. The shearing stress τ_o on faces normal to the x axis tends to rotate the element clockwise, so we plot point X, whose coordinates are (σ_o, τ_o), above the σ axis. We see that in our initial state σ_y is zero and that on faces normal to the y axis τ_o tends to rotate the element counterclockwise; thus, we plot point Y $(0, \tau_o)$, below the σ axis (Figure 4.50).

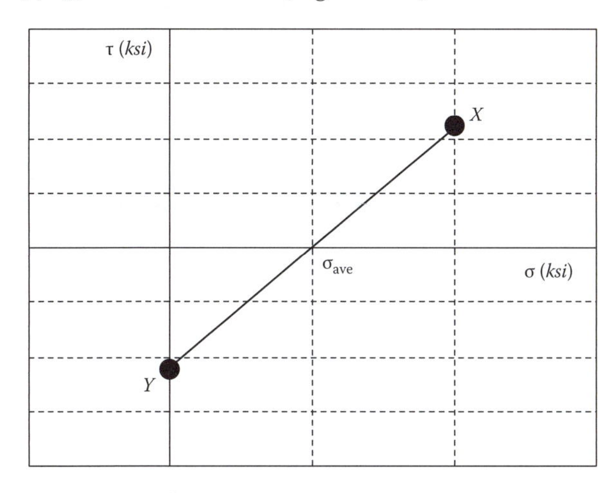

FIGURE 4.50

Line XY passes through the center of our circle, at

$$\sigma_{ave} = ½ (\sigma_x + \sigma_y) = ½ (8 + 0) = 4 \text{ ksi}.$$

We determine the radius R of the circle by observing that the maximum normal stress, given as 10 ksi, appears a distance R to the right of the circle's center (Figure 4.51):

$$\sigma_1 = \sigma_{ave} + R.$$

$$R = \sigma_1 - \sigma_{ave}.$$

$$R = 10 \text{ ksi} - 4 \text{ ksi} = 6 \text{ ksi}.$$

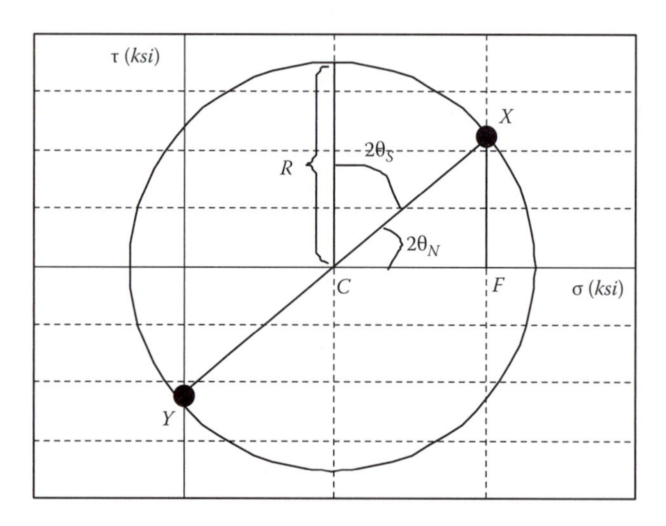

FIGURE 4.51

Now we have Mohr's circle to work with. We see that the rotation required to get from our initial stress state (point X) to the principal stress state at (10 ksi, 0) is either clockwise $2\theta_N$ as shown or counterclockwise $360 - 2\theta_N$. We choose to work with the more manageable clockwise rotation, and consider the right triangle CFX.

$$\cos 2\theta_N = \frac{CF}{CX} = \frac{CF}{R} = \frac{4 \text{ ksi}}{6 \text{ ksi}}$$

$$\theta_N = -24.1°$$

This rotation, again, is clockwise, as reflected by the negative sign. The right triangle *CFX* also allows us to compute the unknown shear stress, τ_o, which is experienced at point *X*:

$$\tau_o = FX = R \sin 2\theta_N = (6\text{ ksi})\cdot\sin 48.2^\circ = 4.47\text{ ksi}.$$

The maximum shear stress is also apparent from Mohr's circle. It is simply the radius of the circle, *R*:

$$\tau_{max} = R = 6\text{ ksi}.$$

The corresponding normal stress at this stress state is $\sigma_{ave} = 4$ ksi. Mohr's circle indicates that to get from the initial stress state to the state of maximum shear stress, we must rotate the circle diameter *XY* counterclockwise by $2\theta_S$ or to rotate the element itself by θ_S. It is clear from the circle that $2\theta_S + |2\theta_N| = 90^\circ$. Hence,

$$2\theta_S = 90^\circ - |2\theta_N| = 90^\circ - 48.2^\circ = 41.8^\circ.$$

With all this information in hand, we can draw properly oriented elements in each of the identified stress states (Figure 4.52).

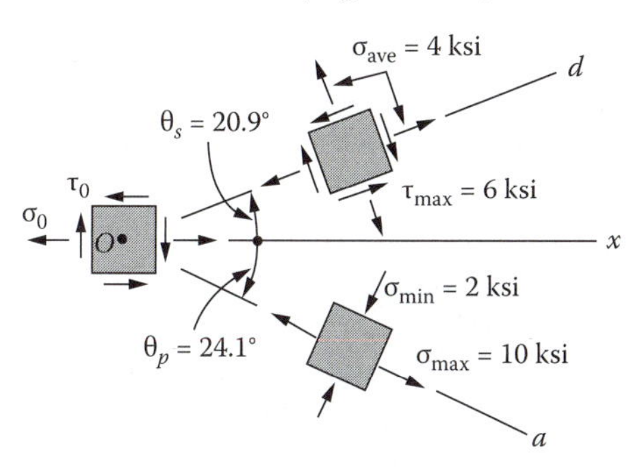

FIGURE 4.52

Note: If we had originally assumed the opposite sense of the unknown τ_o, we would have obtained the same numerical answers, but the orientation of the elements would be as shown in Figure 4.53.

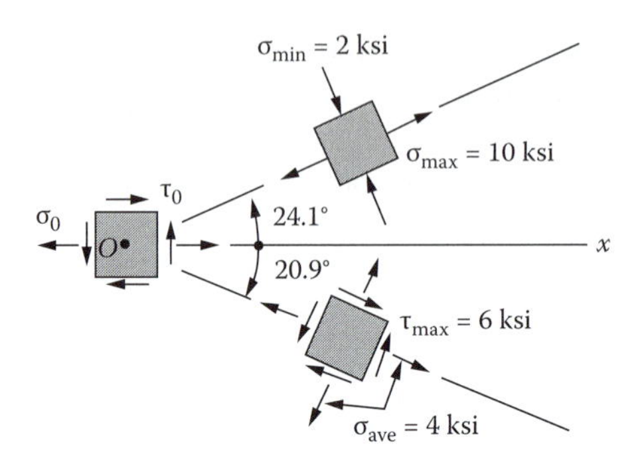

FIGURE 4.53

Example 4.11

A compressed-air tank is supported by two cradles as shown in Figure 4.54. Relative to the effects of the air pressure inside the tank, the effects of the cradle supports are negligible. The cylindrical body of the tank has a 30 in. outer diameter and is fabricated from a 3/8-in. steel plate by welding along a helix that forms an angle of 25° with a transverse (vertical) plane. The end caps are spherical and have a uniform wall thickness of 5/16 in. For an internal gage pressure of 180 psi, determine (a) the normal stresses and maximum shear stresses in the spherical caps; and (b) the stresses in directions perpendicular and parallel to the helical weld.

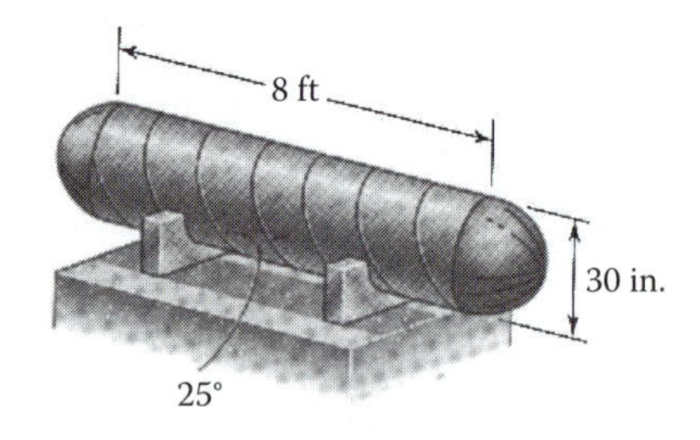

FIGURE 4.54

Given: Dimensions of and pressure on compressed-air tank.

Find: Stress states in spherical end caps and along welds in cylindrical body.

Assume: Thin-walled pressure vessel theory applies.

Solution

First, we validate our assumption that thin-walled theory will apply in both the spherical end caps and the cylindrical body. We must have $t \leq 0.1r$ in both sections. So,

In spherical cap, $t = 5/16$ in. and $r = 15 - (5/16) = 14.688$ in. So, $t = 0.0212r$.
In cylindrical body, $t = 3/8$ in. and inner radius $r = 14.625$ in. So, $t = 0.0256r$.

In a spherical pressure vessel, we have equal hoop and longitudinal stresses (Figure 4.55):

$$\sigma_1 = \sigma_2 = \frac{pr}{2t} = \frac{(180 \text{ psi})(14.688 \text{ in.})}{2(0.3125 \text{ in.})} = 4230 \text{ psi.}$$

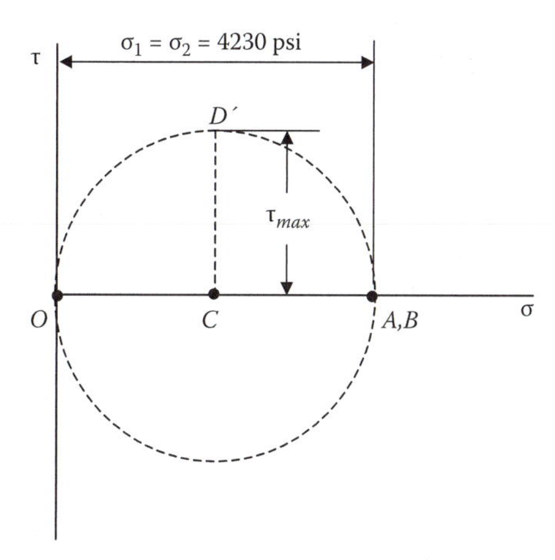

FIGURE 4.55

So, in a plane tangent to the cap, Mohr's circle reduces to a point *(A, B)* on the horizontal (σ) axis, and all in-plane shear stresses are zero. On the surface of the cap, the third principal stress is zero, corresponding to point *O*. On a Mohr's circle of diameter *AO*, point *D′* represents the maximum shear stress; it occurs on planes inclined at 45° to the plane tangent to the cap. (This is as we would expect for purely normal loading in the reference axes, as for an axially loaded bar that experiences maximum normal stress on planes inclined at 45° to the bar axis.) Hence,

$$\tau_{\max} = \tfrac{1}{2}(4230 \text{ psi}) = 2115 \text{ psi.}$$

In the cylindrical body of the tank, we have hoop and longitudinal normal stresses:

$$\sigma_1 = \frac{pr}{t} = \frac{(180 \text{ psi})(14.625 \text{ in.})}{0.375 \text{ in.}} = 7020 \text{ psi}$$

$$\sigma_2 = \frac{pr}{2t} = \frac{(180 \text{ psi})(14.625 \text{ in.})}{2(0.375 \text{ in.})} = 3510 \text{ psi.}$$

Here, the average normal stress is

$$\sigma_{ave} = \tfrac{1}{2}(\sigma_1 + \sigma_2) = 5265 \text{ psi}$$

and the radius of Mohr's circle is

$$R = \tfrac{1}{2}(\sigma_1 - \sigma_2) = 1755 \text{ psi.}$$

We want to rotate our axes from their initial configuration, shown at left in Figure 4.56, so that our element has a face parallel to the weld, as shown at right; the transformed σ_x' and τ_{xy}'', or σ_w and τ_w, will be the requested stresses.

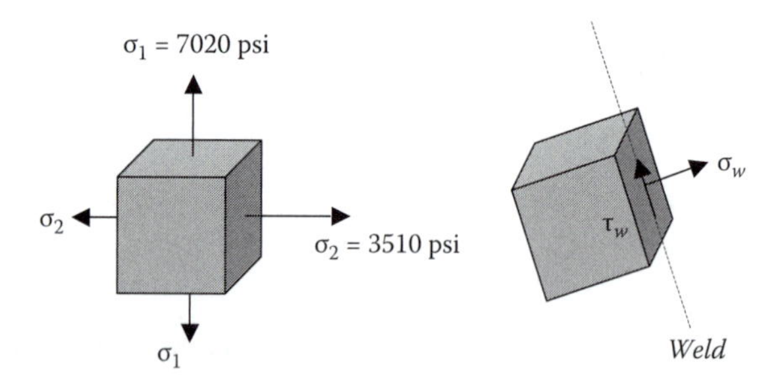

FIGURE 4.56

Using the average stress (center) and radius R just found, we construct Mohr's circle and find these transformed stress components.

Since we want to rotate the element by $\theta = 25°$, we rotate around Mohr's circle by $2\theta = 50°$, to arrive at point X'. This point has the following coordinates:

$$\sigma_w = \sigma_{ave} - R \cos 50°$$

$$= 5265 - 1755 \cos 50°$$

$$= 4140 \text{ psi (tensile)}$$

$$\tau_w = R \sin 50° = 1755 \sin 50°$$

$$= 1344 \text{ psi.}$$

Since point X' is below the horizontal axis, τ_w tends to rotate the element counterclockwise, as assumed in Figure 4.57.

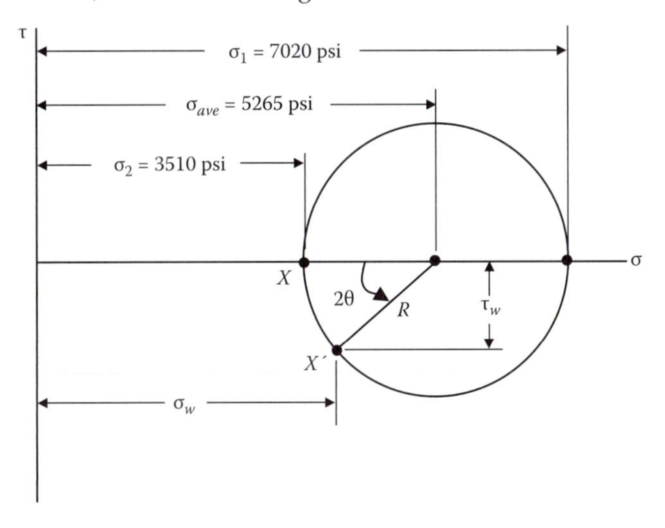

FIGURE 4.57

4.6 Problems

4.1 For the state of stress shown in Figure 4.58, determine (a) the principal planes; (b) the principal stresses; (c) the orientation of the planes of maximum shear stress; and (d) the extreme shear stresses and (any) associated normal stresses.

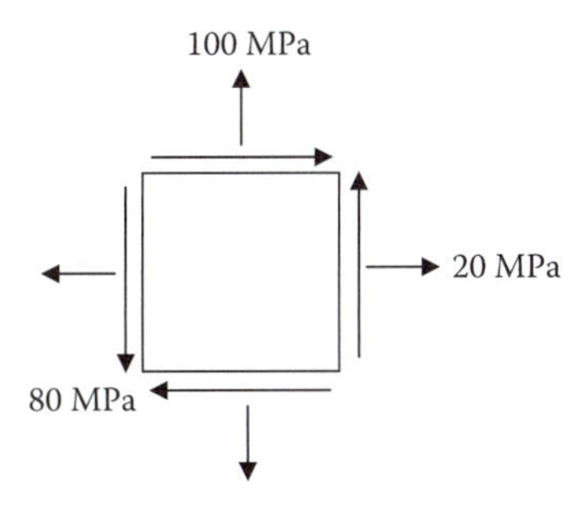

FIGURE 4.58

4.2 Consider the torsion of a thin-walled tube. Determine an approximate expression for the torque if the shear stress must be less than a given working stress τw. Express this result in terms of the tube's mean radius R and its thickness t. (Hint: The binomial theorem is useful here.) Also, derive an approximate expression for the strength-to-weight ratio of the tube in terms of the working stress, its radius and length L, and its specific weight ρg. This result is widely used in aircraft design.

4.3 Determine the reaction torques at the fixed end of the circular shaft shown in Figure 4.59.

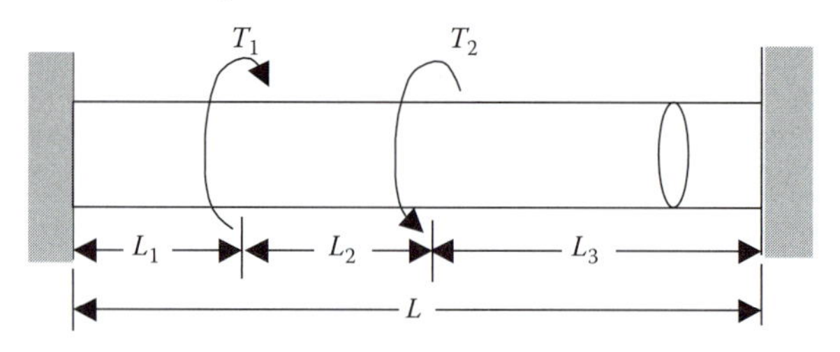

FIGURE 4.59

4.4 A solid circular shaft has a slight uniform taper (Figure 4.60). Find the error committed if the angle of twist for a given length is calculated using the mean radius of the shaft when $b/a = 1.2$.

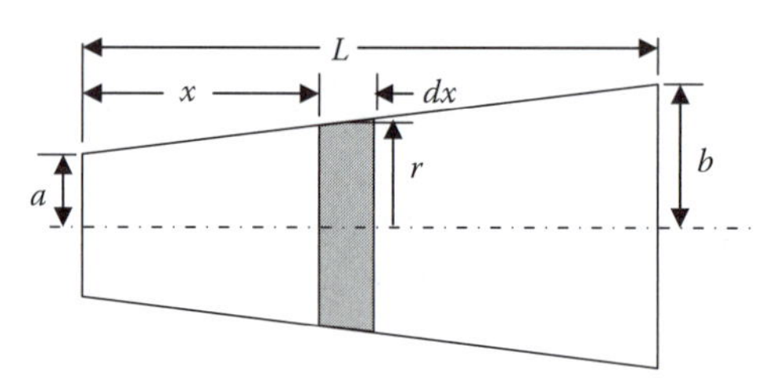

FIGURE 4.60

4.5 A solid circular shaft of 40 mm diameter is to be replaced by a hollow circular tube. If the outside diameter of the tube is limited to 60 mm, what must be the thickness of the tube for the same linearly elastic material working at the same maximum stress? Determine the ratio of weights for the two shafts.

4.6 The propeller of a wind generator is supported by a hollow circular shaft with 0.4-m outer radius and 0.3-m inner radius (Figure 4.61). The shear modulus of the material is G = 80 GPa. (a) If the propeller exerts an 840 kN-m torque on the shaft, what is the resulting maximum shear stress? (b) What is the angle of twist of the propeller shaft per meter of length?

FIGURE 4.61

4.7 A solid aluminum-alloy shaft 60 mm in diameter and 1000 mm long is to be replaced by a tubular steel shaft of the same outer diameter such that the new shaft would exceed neither twice the maximum shear stress nor the angle of twist of the aluminum shaft. (a) What should be the inner radius of the tubular steel shaft? (b) Which of the two criteria governs the design?

4.8 A cylindrical pressure vessel of 120 in. *outside* diameter, used for processing rubber, is 36 ft long. If the cylindrical portion of the vessel is made from 1-in.-thick steel ($E = 29 \times 10^6$ psi, $v = 0.25$) plate and the vessel operates at 120 psi internal pressure, determine the total elongation of the circumference and the increase in the length caused by the operating pressure.

4.9 You are asked to design a scuba tank with a radius $R = 16$ cm to a pressure of $p = 12.0$ MPa at a factor of safety of 2.0 with respect to the yield stress. The relevant tabulated yield values for the steel of which the tank is intended to be made are 290 MPa in tension and 124 MPa in shear. What wall thickness t would you recommend?

4.10 A closed cylindrical tank of length L, radius R, and wall thickness t contains a liquid at pressure p. If a hole is suddenly made in the cylinder, determine (a) how much the tank radius R changes; and (b) how much the tank length L changes.

4.11 An inflatable cylindrical Quonset hut of length L, radius $R = 30$ ft from material with thickness $t = 2.5$ mm, $E = 30.7$ GPa, and $v = 0.24$ has a longitudinal seam that runs the entire length of the hut at its highest point (Figure 4.62). The hut is closed at each end by a quarter of a sphere.

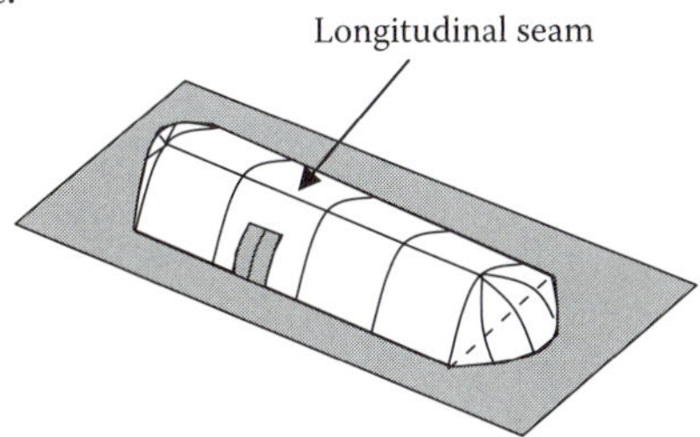

FIGURE 4.62

If the hut is inflated to a pressure $p = 3.44$ kPa, determine (a) the maximum tension (a force per length along the seam) that the longitudinal seam must withstand with a factor of safety of 2; (b) how much higher the peak of the roof gets just before tearing occurs (use the result of part (a), including the safety factor); and (c) the maximum tension that a seam in the quasi-spherical end cap must withstand to maintain the same safety factor of 2.

FIGURE 4.63

Note: Quonset huts (Figure 4.63) were lightweight, prefabricated structures developed to be used as military barracks and offices during WWII. The Quonset hut skeleton was a row of semicircular steel ribs covered with corrugated sheet metal. The ribs sat on a low steel-frame foundation with a plywood floor. The basic model was 20 ft wide and 48 ft long with 720 sq. ft of usable floor space. A larger model was 40 × 100 ft. Approximately 170,000 Quonset huts were produced during the war. After the war, the military sold the huts to civilians for about $1,000 each.

4.12 A strain gage is installed in the longitudinal direction on the surface of an aluminum beverage can (Figure 4.64). The radius-to-thickness ratio of the can is 200. When the lid of can is popped open, the magnitude of the strain reading changes by 180 μstrain. (a) What was the internal pressure p in the can? (b) When the can was pressurized, what was the factor of safety with respect to yielding in the cylindrical wall?

FIGURE 4.64

4.13 For the state of stress given {$\sigma_x = -8$ ksi, $\sigma_y = 6$ ksi, and $\tau_{xy} = -6$ ksi}, determine (a) the principal planes; (b) the principal stresses; and (c) the orientation of the planes of maximum shear stress.

4.14 For the (same) state of stress given in Problem 4.8, determine (a) the maximum shear stress; (b) the normal stresses on the plane of maximum shear stress; and (c) the normal and shear stresses after the element has been rotated through an angle of 30° clockwise.

4.15 Using the equations for stress transformation: (a) confirm the angles that define the planes of maximum and minimum shear stress; and (b) determine the maximum and minimum values of the shear stress.

4.16 For the results of Problem 4.8, (a) what is the normal stress that acts on the plane of maximum shear? (b) How does this result differ from the plane of maximum normal stress? (c) How do the planes of maximum shear stress relate to the principal stress planes?

4.17 A cylindrical pressure vessel with hemispherical endcaps has radius $r = 2$ m, wall thickness $t = 10$ mm, and is made of steel with yield stress $\sigma y = 1800$ MPa. It is internally pressurized at $p = 2$ MPa. Compare the Tresca and von Mises safety factors.

4.18 A solid aluminum-alloy shaft 60 mm in diameter and 1000 mm long is to be replaced by a tubular steel shaft of the same outer diameter such that the new shaft would exceed neither twice the maximum shear stress nor the angle of twist of the aluminum shaft. (a) What should be the inner radius of the tubular steel shaft? (b) Which of the two criteria governs the design?

4.19 A cylindrical pressure vessel of 120 in *outside* diameter, used for processing rubber, is 36 ft long. If the cylindrical portion of the vessel is made from 1-in.-thick steel ($E = 29 \times 10^6$ psi, $v = 0.25$) plate and the vessel operates at 120 psi internal pressure, determine the total elongation of the circumference and the increase in the length caused by the operating pressure.

Case Study 2: Pressure Vessel Safety

Pressure vessels are structures that are designed to contain or preclude a significant pressure—that is, a force distributed over the entire surface of the vessel in question. Pressure vessels show up in a variety of settings and typically are in one of two shapes. Some are spherical: balloons of all sorts, gas storage tanks (Figure CS2.1a), and basketballs. Many are cylindrical: pressurized cabins in aircraft, rocket motors, scuba tanks, oil storage tanks, aerosol spray cans, and fire extinguishers. Some of the cylindrical tanks have flat ends or caps, as in spray cans and home heating oil storage tanks. Often, though, cylindrical tanks have slightly rounded caps (Figure CS2.1b) or spherical caps, as do submarines (Figure CS2.1c). Nuclear reactor containment vessels are often cylinders with spherical caps, although newer nuclear plants tend to have spherical containment tanks.

Pressure vessels have given way or exploded in some rather dramatic fashions. Among the most notorious are the explosion of a molasses storage tank in Boston in 1919 that resulted in 21 deaths and more than 150 injured as 2 million gallons of thick, brown molasses swept through Boston's North End (Figure CS2.2, Problem CS2.1); the burning of the Hindenburg blimp in Lakehurst, New Jersey in 1937; the rupture of the Apollo 13 oxygen tank in 1970; and the implosion of several submarines, including the USS *Thresher* in 1963, the USS *Scorpion* in 1968, and the Russian submarine *Kursk* in 2001. Though the causes of these and other catastrophes varied, serious pressure build-ups and the failures of connections or joints or seams were involved in most. Thus, the design and construction of a pressure vessel is at least as important as its shape. In fact, a major piece of regulatory code is the American Society of Mechanical Engineers (ASME) International Boiler and Pressure Vessel Code (IBPVC) that governs the design and manufacture of pressure vessels.

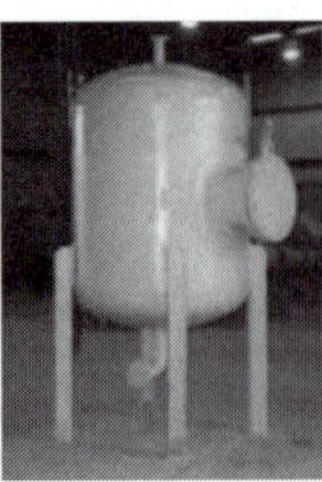

FIGURE CS2.1
(a) A spherical pressure vessel; (b) a cylindrical pressure vessel with a slightly rounded top; (c) a submarine: a cylindrical pressure vessel with a rounded cap.

FIGURE CS2.2
A glimpse of the aftermath of the 1919 failure of five-story-high tank that unleashed 12,000 tons of molasses on Boston's North End. (Photograph by Leslie Jones, *Boston Herald*. With permission.)

Why Are Pressure Vessels Spheres and Cylinders?

Why are pressure vessels curved rather than flat? Two important reasons become evident when we review the physics of pressure vessels. The first has to do with material properties. As we noted in Chapter 2, Section 2.10 and Chapter 3, Section 3.7, cracks propagate in metals, even in ductile metals. Further, crack propagation is especially likely to propagate from the *stress concentrations* that typically form at corners that subsume an angle less than 90°, termed *re-entrant corners* (Figure CS2.3). That is why, for example, airplane windows have rounded corners. A similar situation occurs when we bend a piece of metal to make a corner; a similar stress concentration is created, and there is an increased likelihood of crack propagation when we have, for example, rectangular tubes that include four corners.

The second reason that pressure vessels are spheres or cylinders is that when such shapes are pressurized, they respond with a set of normal stresses that are distributed uniformly through the thickness and always directed along tangents to the surface enclosed (see Section 4.2). These stress states are called *membrane stresses*. These shapes and their membrane stress states produce much *stiffer* structural forms than their beam counterparts, and thus they deflect or deform much less. Consider the pressurized cylinder originally depicted in Figure 4.16. We have already seen in Section 4.2 that

FIGURE CS2.3
Cracks emanating (a) from a re-entrant corner inside a groove; (b) from a filleted corner. (J. A. King.)

a thin-walled pressure vessel experiences a circumferential stress, the *hoop stress* σ_h, of magnitude:

$$\sigma_h = p\frac{R}{t}. \tag{CS2.1}$$

Remember, too, that the wall thickness of a thin-walled vessel is always significantly smaller than the radius (i.e., $t << R$) so that the stresses induced by the pressure are significantly larger than the pressure itself.

Now consider what happens to the geometry of a cylinder when that cylinder is subjected to this pressure. We would expect it to expand symmetrically, meaning that its radius will become larger (Figure CS2.4). Thus, if we denote w as the radial motion or deflection of the cylinder, the circle that originally was of mean radius R will become a circle of radius $(R + w)$. So the circumference of the cylinder increases from $2\pi R$ to $2\pi\,(R + w)$, and the resulting hoop strain ε_h is given by

$$\varepsilon_h = \frac{2\pi(R+w) - 2\pi R}{2\pi R} = \frac{w}{R}. \tag{CS2.2}$$

For simplicity's sake, let us assume a one-dimensional stress–strain law, $\sigma = E\varepsilon$, which means that we can find (see Problem CS2.2 and Problem CS2.3) that the radial expansion or deflection is

$$\frac{w}{t} = \frac{p}{E}\left(\frac{R}{t}\right)^2. \tag{CS2.3}$$

Now imagine that instead of a cylinder of radius R and thickness t, we were using a square tube of dimensions $H \times H$ (and wall thickness t) to contain a gas at the same pressure. We assume that the tube's side lengths H are comparable in magnitude to the cylinder radius R (see Problem CS2.3) and that $t \ll H$. How does the shape of this square tube change when the gas is subjected to pressure? As we look at Figure CS2.5 we can envision that a given side—say, the one marked BC—will move upward or outward due to two effects: (1) the upward movement of points B and C as sides AB and CD are stretched; and (2) the vertical or *transverse* motion of the side BC due to the pressure acting on that surface. The side stretching is just like the extension of a one-dimensional bar, so the equal vertical movement of points B (from the stretching of AB) and C (from the stretching of CD) can be shown (see Problem CS2.4) to be

$$\frac{u_{B,C}}{t} = \frac{p}{2E}\left(\frac{H}{t}\right)^2. \tag{CS2.4}$$

On the other hand, the transverse motion of the side BC is actually due to the *bending* of that side under the pressure load. As we will see in Chapter 6 when we analyze the deflections of bent beams, the maximum deflection of such a beam occurs at its center and can modeled as

$$\frac{w_{\text{beam}}}{t} = \frac{5p}{32E}\left(\frac{H}{t}\right)^4. \tag{CS2.5}$$

Note that this beam deflection is proportional to the ratio (H/t) raised to the fourth power. Compare that with the deflection due to the stretching of the sides (equation CS2.4), which is proportional to the same ratio squared—which means that the bending deflection of a tube face is going to be much, much larger than movement due to the stretching of the sides (again, see Figure CS2.5). Further, compare the beam bending deflection (equation CS2.5) to the radial expansion of a circular cylinder (equation CS2.3) of the same sheet material and under the same pressure:

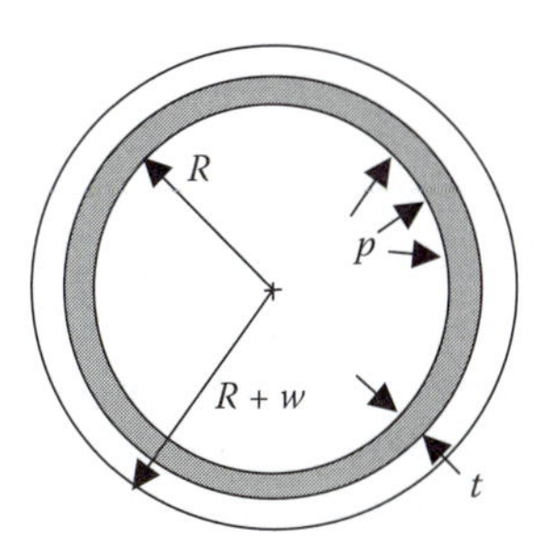

FIGURE CS2.4
The development of hoop strain in a pressurized cylinder due to the (greatly exaggerated) axisymmetric radial deflection w.

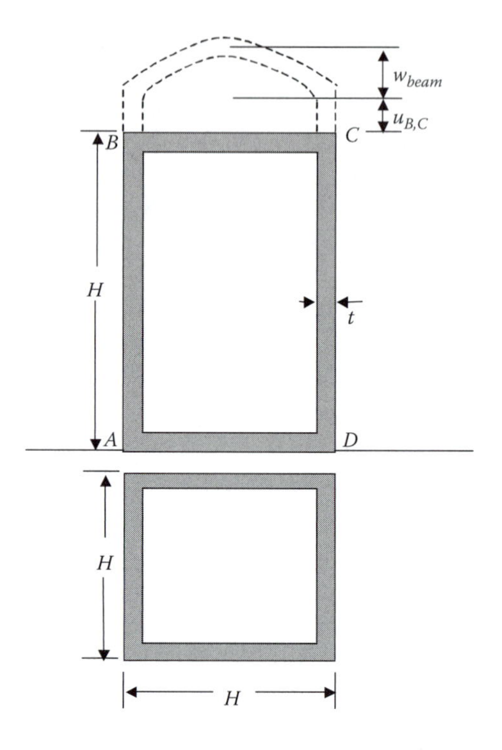

FIGURE CS2.5
A pressurized square tube with cross-section side dimensions $H \times H$ and wall thickness t. Note that the movement of side BC is due in part to the upward movement of points B and C as sides AB and CD are stretched and in part due to the bending of the side BC due to the pressure acting on that surface. Deformations are not to scale; $w_{beam} >> u_{B,C}$.

$$\frac{w_{\text{beam}}/t}{w_{\text{cyl}}/t} = \frac{5p}{32E}\left(\frac{H}{t}\right)^4 \frac{E}{p}\left(\frac{t}{R}\right)^2 = \frac{5}{32}\left(\frac{H}{t}\right)^2\left(\frac{H}{R}\right)^2, \tag{CS2.6}$$

or, since $H \sim R$ (see Problem CS2.4 again!),

$$\frac{w_{\text{beam}}}{w_{\text{cyl}}} \sim \left(\frac{H}{t}\right)^2 >> 1. \tag{CS2.7}$$

Equation (CS2.7) clearly shows that the deflections due to the bending of the sides of a rectangular tube are two orders of magnitude larger than the radial motion due to the extension of the walls of a circular cylinder. Thus, a cylinder, like a sphere, responds to pressure as a *stiff* structural form characterized by large membrane forces and stresses and relatively small (compared with the corresponding bending of thin-walled cylinders that are not pressurized) deflections (see Problem CS2.6). We will say more about this when we describe beam bending in Chapters 5 and 6.

We have noted that some cylindrical tanks have spherical caps. While we're on the subject of radial expansion of such tanks, it is interesting to examine another aspect of pressure vessel behavior: Can we put a hemispherical cap on the end of a cylinder of the same radius R? For a cylinder of finite length, we noted in Section 4.2 that both axial and hoop stresses result from an internal pressure p:

$$\sigma_h = p\frac{R}{t} \quad \text{and} \quad \sigma_a = p\frac{R}{2t}. \tag{CS2.8}$$

The hoop strain for a (two-dimensional) state of plane stress within the cylinder surface would follow from Chapter 3, equation (3.43) rather than the one-dimensional version just used. Thus,

$$\varepsilon_h = \frac{\sigma_h - \nu\sigma_a}{E}, \tag{CS2.9}$$

so that we can eliminate the hoop strain between equation (CS2.2) and equation (CS2.9) and the hoop stresses from the equations in (CS2.8) to find the radial expansion to be

$$\frac{w_{\text{cyl}}}{t} = \frac{(2-\nu)p}{2E}\left(\frac{R}{t}\right). \tag{CS2.10}$$

A comparable analysis for the sphere would look much like the cylinder's, with the obvious exception that the stresses in a hemispherical cap were found in Section 4.2 to be

$$\sigma_\phi = \sigma_\theta = p\frac{R}{2t}, \tag{CS2.11}$$

so that here the analysis of the sphere's hoop strain yields the following radial deflection for the sphere:

$$\frac{w_{\text{sph}}}{t} = \frac{(1-\nu)p}{2E}\left(\frac{R}{t}\right). \tag{CS2.12}$$

Clearly both p and R must be the same for the mated cylinder and sphere, and the radial deflections are compatible only if they are equal. If we set the right-hand side of equation (CS2.10) to equal that of equation (CS2.12), we find that the radial deflections are equal when

$$\frac{(Eh)_{\text{cyl}}}{(Eh)_{\text{sph}}} = \frac{(2-\nu_{\text{cyl}})}{(1-\nu_{\text{sph}})}, \tag{CS2.13}$$

where we have added appropriate subscripts to distinguish the thicknesses and materials. If the materials are the same, which seems a reasonable assumption, then

$$\frac{t_{\text{cyl}}}{t_{\text{sph}}} = \frac{2-\nu}{1-\nu}, \tag{CS2.14}$$

which suggests that the thickness of the cylinder should be much larger than that of the cap. For typical materials for which $\nu = 0.30$, the ratio (t_{cyl}/t_{sph}) is about 2.43! Thus, the cylinder should be thicker by a factor of almost 2.5.

What happens in "real life" is, of course, more complicated. The mismatch caused by the mating of spheres and cylinders produces some modest bending effects that are superposed on the basic membrane states caused by the pressure. The bending stresses add a modest amount (~30%) to the membrane stresses, and they decay fairly rapidly as we move away from the joint or intersection of cap and circular tube. As a result, cylinders are tapered near the joints, with locally increased thickness designed to accommodate the added bending stresses. The complete analysis of these *edge effects* allows us to carefully and safely design such intersections—and thus to avoid a catastrophic failure due to a bad joint!

Why Do Pressure Vessels Fail?

Gas pressure vessels typically contain a large volume of gas that has been compressed to fit into the vessel's much smaller volume, which thus produces the constant, unremitting pressure that acts on the container's inner wall. When such vessels fail, they explode because the pent-up gas wants to return to its initial volume as quickly as it can (see Problem CS2.9, Problem CS2.10, Problem CS2.11, and Problem CS2.12). However, pressure vessels containing incompressible liquids also fail, as did the Boston molasses tank mentioned earlier. The common link is that tank failures typically arise because their designers either failed to properly anticipate possible sources of crack propagation or failed to adequately analyze the stresses at connections. For example, the owners of the Boston molasses tank, the Purity Distilling Company, claimed that it failed because of (variously) an explosion, vibration from an adjacent elevated train track, and a runaway trolley car colliding with the tank. However, forensic analysis of the tank ruins showed that its joints were inadequately designed and, further, that the tank was fabricated with even thinner materials than required by the (already inadequate) design.[4]

The detailed design of joints, and of the connections of pipes and gauges and doors, is beyond our scope. However, understanding the nature of the stress fields in pressure vessels is not. The estimates of the membrane stresses given for the cylinder (equation CS2.8) and sphere (equation CS2.11) are correct as far as they go, but they are incomplete—and sufficiently incomplete that by themselves they do not form an adequate basis for a comprehensive design. Thus, we explore the stress states in greater depth to show that *shear* is present in pressure vessels and that Mohr's circle can be used to advantage in the design of joints.

Consider first the sphere. If we plunk down an x, y coordinate system tangent to the sphere's surface at any point on the sphere, the equations for stress transformation (4.38, 4.39, and 4.40) quickly confirm that the stresses in the plane of the sphere are always normal stresses, that is,

$$\sigma_{x'x'} = \sigma_{y'y'} = \frac{pR}{2t} = \sigma_1 = \sigma_2. \tag{CS2.15}$$

But what happens through the thickness, or in the z direction? Consider the element shown in Figure CS2.6. We see there a stress state

$$\sigma_{yy} = \frac{pR}{2t}, \quad \sigma_{zz} = -p. \tag{CS2.16}$$

If we then apply the stress transformation equations (4.38, 4.39, and 4.40) to this stress state (see Problem CS2.13), we would find that in the z–y plane the stresses vary as

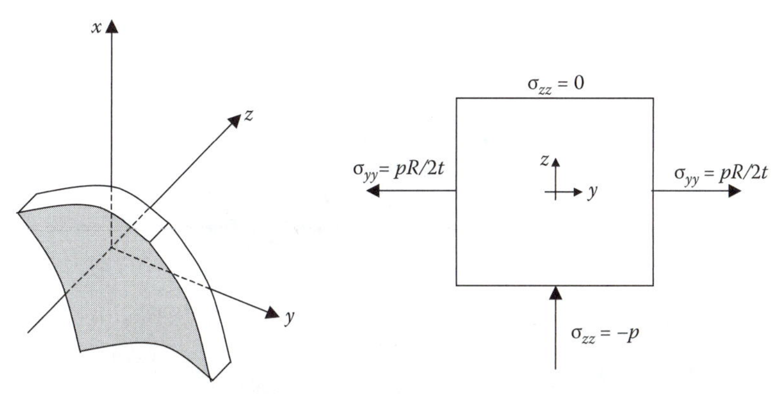

FIGURE CS2.6
An element in the skin of a spherical pressure vessel and a "blow-up" of the y–z plane showing the stress $_{zz} = -p$ due to the internal pressure p acting on the sphere's inner wall and the stress $_{yy} = pR/2t$, acting as shown.

$$\sigma_{z'z'} = \frac{1}{2}\left(-p + \frac{pR}{2t}\right) - \frac{1}{2}\left(p + \frac{pR}{2t}\right)\cos 2\theta, \tag{CS2.17a}$$

$$\tau_{z'y'} = +\frac{1}{2}\left(p + \frac{pR}{2t}\right)\sin 2\theta, \tag{CS2.17b}$$

$$\sigma_{y'y'} = \frac{1}{2}\left(-p + \frac{pR}{2t}\right) + \frac{1}{2}\left(p + \frac{pR}{2t}\right)\cos 2\theta. \tag{CS2.17c}$$

Equation (CS2.17b) shows that there are shear stresses in the wall of a spherical pressure vessels and that the maximum shear stress occurs at $\theta = \pi/4$ and has the value

$$\tau_{\max} = \frac{p}{2}\left(1 + \frac{R}{2t}\right). \tag{CS2.18}$$

Thus, there is a shear stress that acts through the thickness: Its magnitude is the same as the principal membrane stresses, and it must be accounted for in the design of spherical tanks (see Problem CS2.14).

A similar situation occurs in the case of cylindrical tanks, with an interesting twist that arises because of the different ways that cylindrical tanks are actually made. While hollow reeds and bamboo tubes occur quite naturally, we have to manufacture cylindrical tanks. Typically that means forming flat, rectangular sheets around a rigid form, termed a *mandrel*, and welded together along seams that can be longitudinal, transversely circumferential, or even helically wound around the cylinder's axis (Figure CS2.7). This

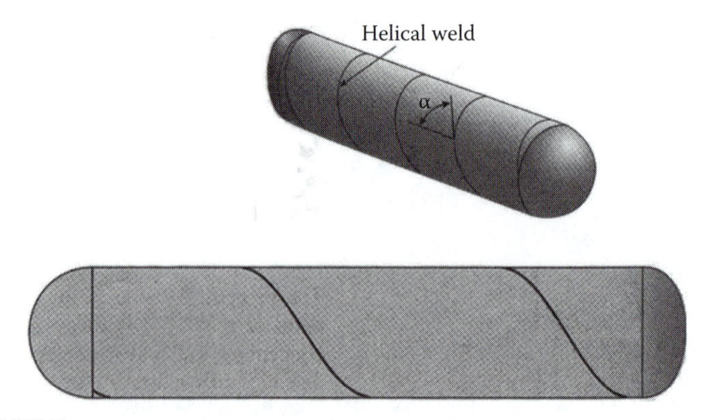

FIGURE CS2.7
A cylindrical tank with helical seams. (From Gere, J. M. and Timoshenko, S. P., *Mechanics of Materials*, 4th ed., WS Publishing, 1997. With permission.)

means that the stresses along the seams are especially of special interest and, thus, that the transformation of stresses needs to be considered. In fact, if we identify x as the axial coordinate in a cylinder and y as the circumferential coordinate, the membrane stress state of a pressurized cylinder is

$$\sigma_{xx} = \frac{pR}{2t}, \quad \sigma_{yy} = \frac{pR}{t}. \tag{CS2.19}$$

Then this stress state substituted into the stress transformation equations (4.38, 4.39, and 4.40) yields the following stresses in the xy plane (see Problem CS2.15):

$$\sigma_{x'x'} = \frac{3pR}{4t} - \frac{pR}{4t}\cos 2\theta, \tag{CS2.20a}$$

$$\tau_{x'y'} = -\frac{pR}{4t}\sin 2\theta, \tag{CS2.20b}$$

$$\sigma_{y'y'} = \frac{3pR}{4t} + \frac{pR}{4t}\cos 2\theta. \tag{CS2.20c}$$

Equation (CS2.20a), equation (CS2.20b), and equation (CS2.20c) allow us to determine the variation of the stresses with the angle θ. We can thus calculate the in-plane or membrane stresses along any intended seams (see Problem CS2.16 and Problem CS2.17). Of course, in addition to the membrane stresses just analyzed, cylindrical tanks also have shear stress components that are directed in the thickness or z direction.

Cylindrical pressure vessels that are made from welded steel sheets are very common; this is the least expensive way to manufacture cylindrical tubing. It is also common to see cylinders that have been extruded from a solid piece of steel or aluminum, with welding only necessary at the end caps. This extrusion method is preferred for applications requiring higher safety factors than welded tubing, for example, in scuba cylinders. A third manufacturing technique of interest is *spin casting*, which helps to reduce the weight of the pressure vessel and also requires no welds along the length of the vessel. The tank material, generally aluminum, is melted and poured into a rotating cylindrical mold, where it solidifies in the desired vessel shape.

Problems

CS2.1 Determine in both customary and metric units the volume that 12,000 tons of molasses occupies.

CS2.2 Derive equation (CS2.3) by substituting equation (CS2.1) and equation (CS2.2) into the appropriate one-dimensional stress–strain law.

CS2.3 Is equation (CS2.3) dimensionally correct? Explain your answer.

CS2.4 Determine how much the sides *AB* and *CD* of the square tube in Figure CS2.7 are stretched due to an upward pressure p acting on the bottom surface of side *BC*. How does this answer compare with equation (CS2.4)?

CS2.5 Develop three scenarios for comparable circular cylinders of radius R and square tubes of side H that allow one to say $R \sim H$. (Hint: What geometric attributes of the cylinder and tube might be made equal?)

CS2.6 If a structural stiffness parameter was defined in terms of the pressure/radial deflection ratio (i.e., p/w), compare the stiffness parameters for a circular cylinder of radius R with that of a square tube of side H. What are the physical dimensions of these stiffness parameters and of their ratio? Assume that $R \sim H$. (Hint: Recall equation CS2.3 and equation CS2.5.)

CS2.7 Given that the hoop strain is likely to be a very small number, estimate the pressure-to-modulus ratio, p/E.

CS2.8 Given the result of Problem CS2.7, estimate the magnitude of the radial deflection of a pressurized cylinder as a fraction of its thickness. (Hint: Equation CS2.3 might be handy.)

CS2.9 The adiabatic compression of an ideal gas obeys the following law: $p^{V}{}_{-}$ = constant, where p is the pressure, V the volume, and $\gamma = 1.4$. Assuming that the ideal law provides a reasonable rough estimate of the gas's behavior, determine the pressure in a tank of 1 ft^3 volume when it stores 100 ft^3 of standard atmospheric air.

CS2.10 For the two scuba cylinders shown in Figure CS2.8, of radii $R = 4$ in. and length $L = 25$ in., estimate the pressure reading if 80 ft^3 of air was compressed into them.

FIGURE CS2.8

CS2.11 For the assumptions stated in Problem CS2.9, show that the work done in adiabatically compressing an ideal gas is

$$W_{1\text{-}2} = -\int_1^2 p dV = \frac{p_1 V_1}{\gamma - 1}\left[\left(\frac{V_1}{V_2}\right)^{\gamma-1} - 1\right].$$

CS2.12 Determine how much work was required to undertake the compression specified in Problem CS2.9. Is that a lot of work (or energy)? Explain your answer, perhaps by providing a suitable comparison.

CS2.13 Verify that equation (CS2.17a), equation (CS2.17b), and equation (CS2.17c) are correct by substituting the stress state of equation (CS2.16) into the stress transformation equations (4.49 and 4.50).

CS2.14 Determine an appropriate approximation to equation (CS2.18) for thin-walled pressure vessels (i.e., $t/R \ll 1$). From which of the original components of stress does the dominant, surviving term originate?

CS2.15 Verify that equation (CS2.20a), equation (CS2.20b), and equation (CS2.20c) are correct by substituting the stress state of equation (CS2.19) into the stress transformation equations (4.38, 4.39, and 4.40).

CS2.16 Determine the maximum in-plane stresses for a cylindrical tank made of steel (E = 205 GPa, ν = 0.30), having a mean radius of 2 m and a thickness of 20 mm and subjected to an internal pressure $p = 1$ MPa.

CS2.17 Determine the in-plane normal and shear stresses along the seam of the tank of Problem CS2.16 if it is helically wound at angle $\theta = 60^{\circ}$.

Notes

1. J is commonly called the polar moment of inertia, though it is an area moment of inertia and is more correctly referred to as polar second moment of area. Moments of area are geometric properties of certain areas, reflecting how effectively those areas resist deformation. A large J indicates a cross section that will resist torsion. Please see Appendix A for a table of values for common areas.
2. We recognize that the principal stress state, in which an element experiences only normal stresses, signifies that we have in essence diagonalized the symmetric stress tensor. We also note that the subscript convention for pressure vessels, where stress components $\sigma_{\theta\theta}$ and σ_{xx} were known as σ_1 and σ_2, was an implicit acknowledgment that the stress state corresponding to conventional cylindrical or spherical coordinates is the principal stress state for a pressure vessel. However, we might still be interested in the stress state under different reference axes to learn the design constraints for a weld used in constructing a pressure vessel from a flat sheet of material. So even when the principal stress state is what we see with our usual coordinates, we will have a motivation to transform the stress state to different axes and orientations.
3. For the details of this analysis, first proposed by French mathematician A. L. Cauchy in the 1820s, see Timoshenko and Goodier (1970, sec. 77).
4. The 50-ft-high, 90-ft-diameter steel molasses tank had been ordered from Hammond Iron Works in 1915 by the Purity Distilling Company on authorization of U.S. Industrial Alcohol. The treasurer of Purity ordered it without consulting an engineer. The only constraint given was that the tank should have a factor of safety of three for the storage of molasses, which is 50% heavier than water, weighing 12 lb per gallon. All the steel sheets used in construction of the tank actually proved less thick than shown on the drawings used to obtain the building permit. For instance, the bottom ring—the most stressed part of the structure—was supposed to be 0.687 in.; as built, it was only 0.667 in. The steel thicknesses for the other six rings were similarly found to be 5% to 10% less than the values indicated on the permit plans. The tank was completed early in 1916 and tested with only 6 in. of water. During the tank's 3 years of service it had on several occasions contained a maximum of around 1.9 million gallons (for periods up to 25 days). At the time of failure the tank had been near maximum capacity (at 2.3 million gallons) for 4 days. Months later, at the legal proceedings, several recalled that the seams of the tank were leaking molasses before the disaster.

5

Beams

The word *beam* is derived from Germanic words meaning *tree* or *structural member.* (One would guess that *tree* came first.) Beams are among the most common structural elements, popping up in the support structures of cars, aircraft, and buildings. They carry loads applied at right angles to the longitudinal axis of the member, which causes the member to bend. In practice, structural members may experience complex loading including axial, torsional, and traditional *beam loading.* As we have already examined axial and torsioal loading, for now we consider the isolated effects of beam loading. The four fundamental elements of continuum mechanics serve us well: We need equilibrium, constitutive laws, and compatibility. This time we start with equilibrium and then develop our definitions of stress and strain, relatable by Hooke's law when deformations are small. We examine first the internal forces and moments in the beams, then the resulting stresses on the beams, and finally (in Chapter 6) the beams' deflections due to this loading.

5.1 Calculation of Reactions

Unsurprisingly, our first step in analyzing a beam is to draw a free-body diagram (FBD) and to determine the reactions at its supports. A beam's behavior when subjected to an external load depends on the type of supports and on the type of loading.

There are three basic types of supports for planar structures such as beams: (1) the roller or link, which is capable of resisting a force in only one specific line of action; (2) the pin, which is capable of resisting a force in any direction of the plane and whose reaction force hence has two components; and (3) the fixed support, which is capable of resisting a force in any direction and is also capable of resisting a moment or a couple. This third type of support is obtained by building a beam into a wall, by casting it into concrete, or by welding the end of a member to the main structure. Figure 5.1 shows physical and idealized diagrams of these three types of supports and the resisting reactions they offer.

In addition to the type of supports, we also take into account the type of loading on the beam. In this book we have considered a number of concentrated "point" loads; we can also see this type of loading on beams. We also consider problems where the loads are distributed, either uniformly or not.

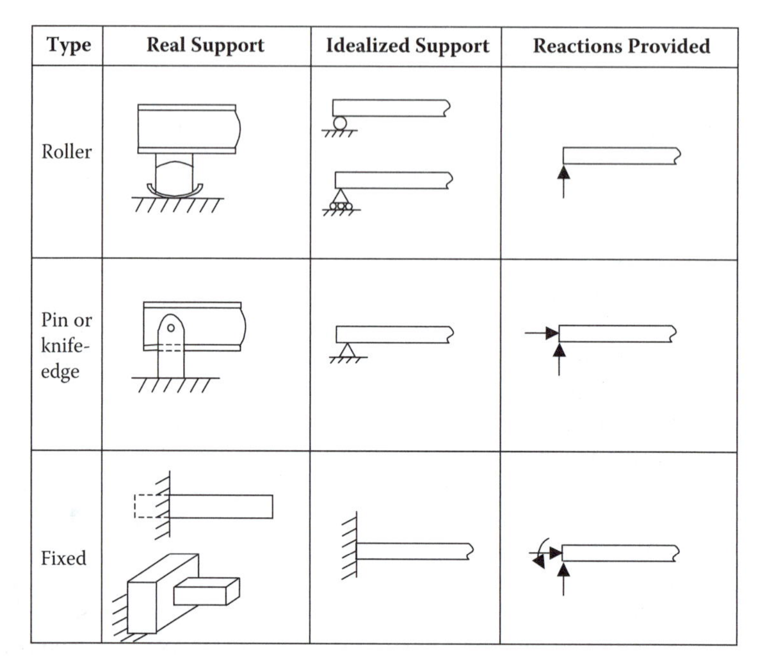

FIGURE 5.1
Beam supports.

Figure 5.2 gives an idea of how these load diagrams will look and what their real-world equivalents might be.

Often, we are able to replace a distributed load by an equivalent concentrated resultant load, acting through the centroid (center of force) of the distributed load. We also are able to classify beam problems, as we have classified axial bar and torsion problems, into statically determinate and statically indeterminate scenarios. And as before, whenever we encounter a statically indeterminate problem, we supplement the equations of static equilibrium with geometry and constitutive laws.

Armed with this information about beam supports and loads, we are prepared to calculate beam reaction forces and moments. In statically determinate cases, the equations of static equilibrium suffice.

5.2 Method of Sections: Axial Force, Shear, Bending Moment

By now we are good friends with the method of sections: the idea that if a whole body is in equilibrium, any part or section of this body is in equilibrium itself. We exploit this method to determine the complete force system of a body, including both external and internal forces and moments. In the particular case of a beam, the externally applied forces and the support reactions keep the *entire body* in equilibrium. When we make "cuts" to apply the

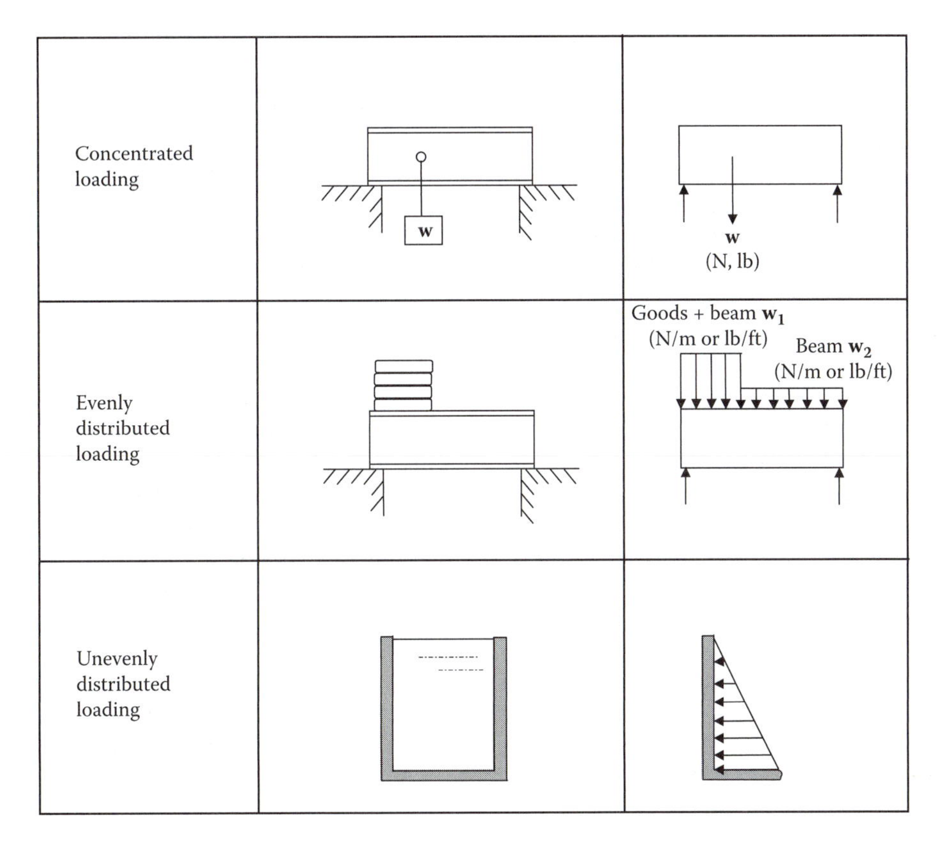

FIGURE 5.2
Types of loading conditions for beams.

method of sections, equilibrium requires the existence of *internal* forces at the cut section. These internal "resisting" forces and moments are what keep the *cut sections* in equilibrium. At each section we may find any or all of a vertical force, a horizontal force, and a moment necessary to maintain equilibrium.

Axial Force in Beams

A horizontal force P may be necessary at a beam section to satisfy equilibrium. The magnitude and sense of this force P are obtained from the solution of $\Sigma F_x = 0$. If the force P acts toward the section, it is sometimes called a *thrust* or *compressive force,* as we have already seen; if it acts away from the section, P is called *axial tension.* Its line of action is always directed through the centroid[1] of the beam's cross-sectional area.

Shear in Beams

Typically, to keep a section in equilibrium, there must be an internal vertical force V at the cut. Because this internal force acts normal to the beam axis and therefore *parallel* to the beam's cross-sectional area, it is called a *shear force*. The shear's magnitude is the sum of the vertical components of all the external forces acting on this cut section, and it is in the opposite direction to balance the external resultant.

If we look at two adjacent sections, the shear on their shared face is defined as in Figure 5.3. The shear on this face should clearly have the same magnitude no matter which way we choose to look at it; the direction of the shear depends on the face. If we are looking at the left-hand section, this shear is upward: The beam provides upward support to keep the section in (vertical) equilibrium. And if we are looking at the right-hand section, the shear is downward. This could get quite confusing unless we establish a consistent *sign convention* for talking about shear. This convention is to say that "positive shear" involves downward V on the left-hand segment of a beam and upward V on the adjacent right-hand segment, as shown in Figure 5.4. This tells us that our previous example, in Figure 5.3, was an example of "negative shear." So in addition to specifying the direction of V, it is important to make sure we have associated it with a particular side of a section.

We can think of this sign convention and nomenclature in terms of the physical action of the forces:

> Shear in a beam is positive if the segment of the beam to the left of a cutting plane tends to move upward relative to the segment on the right (due to external forces). Negative shear reflects the left segment moving downward relative to the right segment.

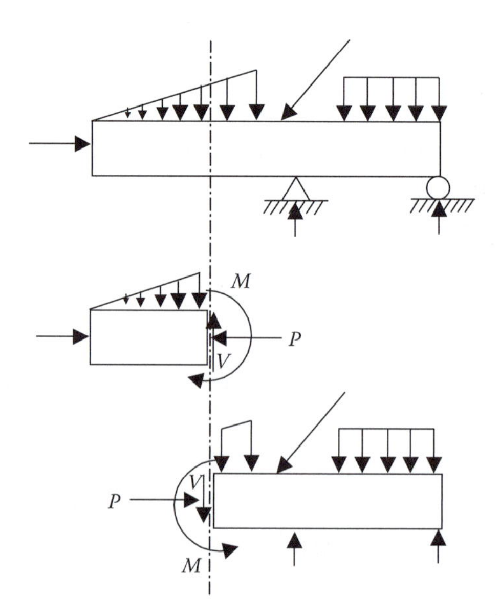

FIGURE 5.3
Application of method of sections. (*Note:* The internal shears and moments shown here turn out to be *negative*, once we define our sign convention!)

Bending Moment in Beams

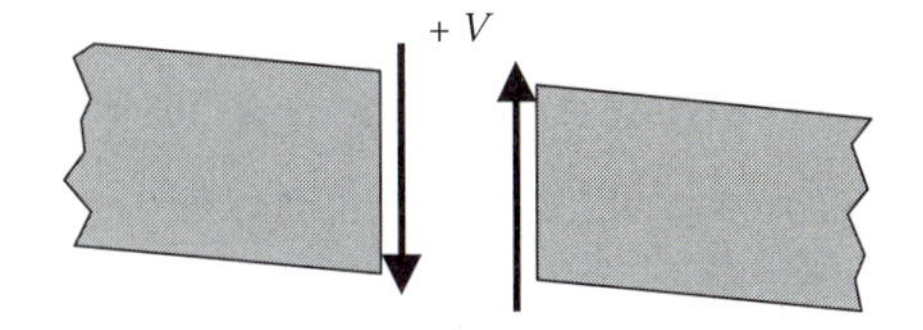

FIGURE 5.4
Definition of positive shear.

We have two internal forces, an axial force P and a shear force V, to assist us in satisfying equilibrium equations for a beam. Clearly, P and V will help out with $\Sigma F_x = 0$ and $\Sigma F_z = 0$. The remaining equilibrium equation for a planar problem is $\Sigma M_y = 0$, and we will generally need an *internal resisting moment* to help us meet this requirement, to balance the moment caused by external loads. This internal moment is developed within the cross-sectional area of the cut, in a direction opposite to the resultant external moment. The magnitude of the internal resisting moment, it should be apparent, equals the external moment. These moments tend to bend a beam in the plane of the loads and are hence called *bending moments.*

In the method of sections, this external moment can be defined as the sum of the moments of all the external forces acting on one side of the cutting plane. We also make use of a sign convention that can be stated:

> The bending moment in a beam is positive when the bottom fibers are in tension and the top fibers are in compression. The bending moment is negative when the bottom fibers are in compression and the top fibers are in tension.

An illustration of this convention is shown in Figure 5.5. As with the shear, it is critical to associate the moment with a particular side of a section, because of this sign convention.

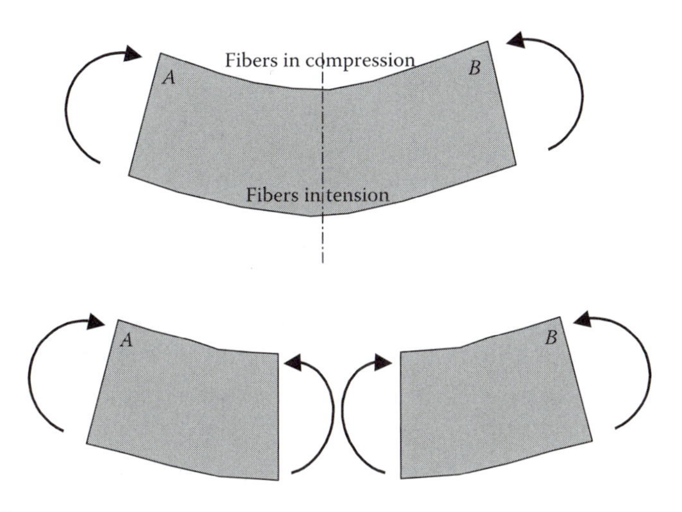

FIGURE 5.5
Definition of positive bending moment.

5.3 Shear and Bending Moment Diagrams

Now that we have sign conventions for the internal forces and moments in a beam, we have the ability to represent the varying values of internal forces and bending moment throughout the length of the beam. We do this by means of separate diagrams for each quantity. These diagrams are called, rather unimaginatively, *axial force, shear,* and *bending moment diagrams.* We rarely use axial force diagrams since most of the beams we investigate (and most beams in practice) are loaded by forces acting perpendicular to the beam axis, and for these cases there are no axial forces at any section.

These diagrams can be quick sketches, typically made just below the free-body diagram of the beam. From a quick glance at such a diagram, a designer can ascertain the type of performance that is required of a beam at every section.

We first construct these diagrams by inspection of the free-body diagram; later, we use integration to evaluate more complex cases.

Rules and Regulations for Shear and Bending Moment Diagrams

Shear Diagrams

Protocol

1. Sketch free-body diagram of beam.
2. Find reactions.
3. Draw V diagram directly below load diagram.
4. By solving $\Sigma F_z = 0$ on sections, find and plot V.
5. Locate points of zero shear.

Fun Facts

- For any part of the beam where there are no external loads, the shear diagram will be a straight horizontal line.
- The shear diagram at the point of application of a concentrated load will be a vertical line—that is, there will be a sudden change in the shear.
- Where there is a uniformly distributed line load, the shear diagram will be a straight line with slope equal to the load intensity.
- For a simply supported beam subjected to vertical loads, the absolute values of the positive and negative areas contained by the shear diagram are equal.

Moment Diagrams

Protocol

1. Draw M diagram directly below shear diagram.
2. Either (a) calculate shear areas between key points[2] and then calculate moments by adding shear areas beginning at the left end of the beam; or (b) use free-body diagrams of sections beginning at the left end of the beam to compute moments at key points and points of zero shear.
3. Plot moment values. Sketch shape between plotted points by referring to the shear diagram.

Fun Facts

- For a simply supported, single span beam, bending moment at both ends is equal to zero.
- For a cantilever beam acted on only by vertical downward loads, bending moment is zero at the free end and maximum at the fixed end. (Shear is also maximum at the fixed end.)
- Bending moment is positive for simply supported beams and negative for a cantilever beam.
- Except for cantilever beams, maximum bending moment occurs at points of zero shear, or where V goes through zero.

5.4 Integration Methods for Shear and Bending Moment

To develop a more elegant method for calculating the internal forces and moments (P, V, and M) within a beam, we derive a few differential relations. To do this, we imagine using the method of sections on an infinitesimally small section of the beam, say, one with length dx. To keep things as general as possible, we say that this beam is acted on by a distributed force with intensity $q(x)$. (Remember that q has units of force per unit length; also note that a positive $q(x)$ is defined as a load in the positive z direction, or *downward*.) A free-body diagram of such a segment is shown in Figure 5.6.

The changes in shear and moment from the left face of element dx to the right are denoted by dV *and* dM, respectively.[3] Our next step is to write the equations of equilibrium for this element:

$$\Sigma F_z = -V + q\,dx + (V + dV) = 0, \tag{5.1}$$

which simplifies to

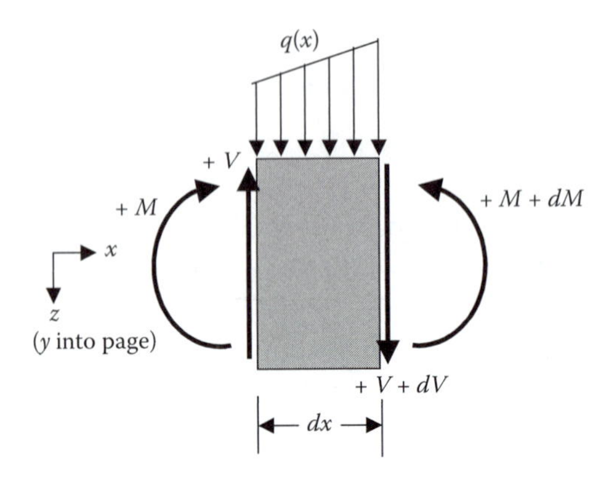

FIGURE 5.6
Differential element.

$$\frac{dV}{dx} = -q\,. \tag{5.2}$$

We also sum the moments about the center of the right face of our element, using the convention that counterclockwise moments are positive:

$$(M + dM) - V\,dx - M + (q\,dx)(dx/2) = 0, \tag{5.3}$$

which gives us

$$\frac{dM}{dx} = V - \frac{qdx}{2}. \tag{5.4}$$

If we take the limit as $dx \to 0$, we see that we have two equations:

$$\begin{aligned} \frac{dV}{dx} &= -q \\ \frac{dM}{dx} &= V \end{aligned} \tag{5.5}$$

And furthermore, by substituting $dM/dx = V$ into $dV/dx = -q$, we get

$$\frac{d^2M}{dx^2} = -q, \tag{5.6}$$

which we will be able to exploit to determine reactions of beams from the boundary conditions. Equation (5.5) and equation (5.6) are very useful in the construction of shear and moment diagrams, as we will see.

Integrating equation (5.2), we obtain an expression for shear at any x:

$$V = -\int q\, dx + C_1. \tag{5.7}$$

From this integral it should be clear that the shear at any section is simply a sum (i.e., an integral) of the vertical forces along the beam from the left end of the beam *to the section of interest*, plus a constant of integration C_1. This constant is equal to the shear on the left-hand end of the beam (at $x = 0$). So, between any two sections of a beam, the shear V changes by the amount of vertical force included *between* these two sections. If no force occurs between any two sections, there is no change in the shear (i.e., the shear diagram is a horizontal line). If a concentrated force occurs, a discontinuity or jump in the value of V occurs at the point of application. The *slope* of the shear diagram comes from the load intensity q. If, for example, the applied distributed load is downward and uniformly distributed ($q = q_o$ = constant), then the slope of $V(x)$ is negative and also constant. For nonuniform distributed loads, the slope of the shear diagram is determined from the trend of q. (Similarly, for a V diagram with positive slope, the corresponding M diagram is concave up, and for V with negative slope, M is concave down. This follows nicely from the differential relations we have just derived: If V goes as $+x$, M goes as $+x^2$, and if V goes as $-x$, M goes as $-x^2$.)

Once again, to determine a shear diagram in this way we must first find the reactions. Then we can start summing vertical forces to calculate the shear at any point.

Integrating the dM/dx equation, we obtain a relation for bending moment at any x:

$$M = \int V\, dx + C_2, \tag{5.8}$$

where, again, C_2 is a constant of integration, determined from boundary conditions at $x = 0$. If the ends of the beam are on rollers, pins, or free, the moments at these ends are zero. If an end of the beam is fixed, the moment at this end is known from the reactions. For cantilever beams, the maximum moment occurs at the fixed end, and zero moment is felt at the free end.

The meaning of the term $V\, dx$ represents the area beneath the V diagram over a length dx. The sum of these areas over a length x, according to equation (5.8), give us the bending moment $M(x)$. By proceeding from the leftmost ($x = 0$) end of a given beam to the right, we can construct a moment diagram.

5.5 Normal Stresses in Beams

We know now that a system of internal forces may occur in a beam subject to external loads. We have already considered the stresses due to internal axial forces such as P. Now, we want to develop a way to talk about the stresses due to the shear force V and bending moment M. For simplicity, we begin our discussion of these stresses by focusing on beams with *symmetric* cross sections, and we first consider a load state known as *pure bending* or *flexure*. In pure bending, only bending moments are applied to the beam.

We use a similar approach to the one we used to consider the effects of torsion. First, we make a plausible assumption about the deformation to ensure that we are able to deal with the problem analytically. Figure 5.7 should help you visualize this assumption for pure bending. In Figure 5.7a, a horizontal beam with a vertical axis of symmetry is shown. The horizontal line through the cross-section centroid is called the *axis* of the beam. If we look at a segment of this beam when it is subjected to a bending moment, as in Figure 5.7b, the beam bends in the plane of symmetry. Although the planes initially perpendicular to the beam axis slightly tilt, the lines defining their boundaries remain straight, that is:

> Plane sections through a beam taken normal to its axis remain plane after the beam is subjected to bending.[4]

This assumption is completely valid for elastic, rectangular members in pure bending. If shears are also introduced, we will make some small corrections to the theory. But in practice, this theory is remarkably robust and cabable of supporting the stress analysis of all beams.

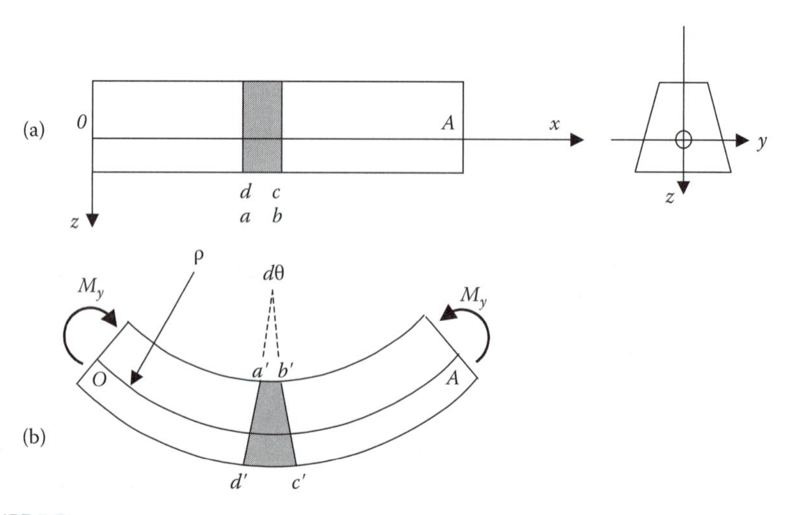

FIGURE 5.7
Behavior of elastic beam in bending.

Looking again at Figure 5.7b, we see that the beam axis has deformed into a portion of a circle of radius ρ. For an element (shaded) defined by an infinitesimal element $d\theta$, the so-called fiber length ds of the beam axis (an arc-length on this circle with radius ρ containing angle $d\theta$) is given by $ds = \rho\, d\theta$. Rearranging,

$$\frac{d\theta}{ds} = \frac{1}{\rho} \equiv \kappa, \tag{5.9}$$

where the reciprocal of ρ is defined as the axis *curvature* κ. For bending of prismatic beams, both ρ and κ are constant. In the course of solving the bending problem, we hope to find a way of determining κ.

If we imagine another curve, parallel to the beam axis, at some radius $\rho + z$, we can find the arc length contained in our shaded segment. We call this arc length ds', and $ds' = (\rho + z)\, d\theta$, as shown in Figure 5.8. We write the difference between our two arc lengths:

$$ds' - ds = (\rho + z)\, d\theta - \rho\, d\theta = +\, z\, d\theta. \tag{5.10}$$

We then divide this difference by the first arc length ds, the initial length of the segment of interest. The axis through the centroid, also called the *neutral axis*, does not change length under pure bending. In Figure 5.7b, the horizontal lines above the neutral axis have been compressed, and those below it have been lengthened, but the neutral axis has not changed length. The first arc length ds was in fact the length of all horizontal lines in the shaded segment before the bending moments were applied. Hence, dividing this new change in arc lengths, $ds' - ds$, by the old one, ds, we should get an expression for *strain*—and we do:

$$\varepsilon_x = \frac{ds' - ds}{ds} = z\frac{d\theta}{ds} = \kappa z\,. \tag{5.11}$$

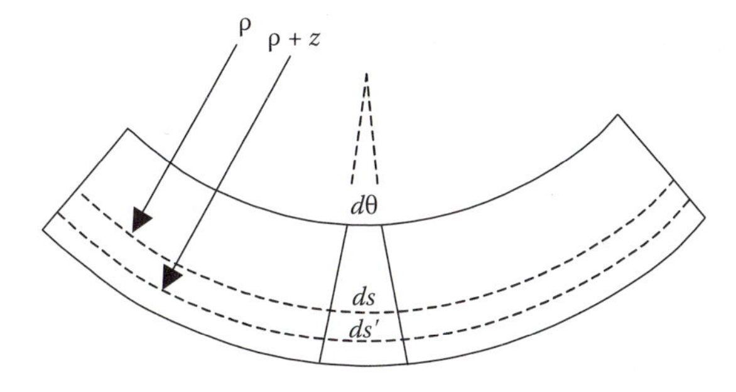

FIGURE 5.8
Nomenclature for deformation of beam in bending.

This is a normal strain, a measure of how much dimensions in the x direction have changed under this bending moment. We see that it depends linearly on z.

By using Hooke's law, we obtain a relation for normal longitudinal stress in the beam:

$$\sigma_x = E\varepsilon_x = E\kappa z. \tag{5.12}$$

Note that due to the position of the origin of the z axis in the beam, z can have either positive or negative values. Also remember that z has been defined positive downward. We need to develop a way of determining the location of the origin. How can we find out where the neutral axis is?

To answer this question, we turn to the equations of equilibrium. In pure bending, the sum of all forces at a section in the x direction must vanish, so

$$\sum F_x = 0 \qquad \int_A \sigma_x dA = 0, \tag{5.13}$$

where the integration over A represents summation over the entire cross-sectional area A of the beam. Using Hooke's law, we can rewrite this integral as

$$\int_A E\kappa z dA = E\kappa \int_A z dA = 0. \tag{5.14}$$

(Since E and κ are constant, we have taken them outside the integral.) By definition, the remaining integral is

$$\int z dA = z_c A,$$

where z_c is the distance from the origin to the centroid of the area. Since the integral must equal zero, this distance z_c must equal zero, and hence the origin must coincide with the centroid: The x axis must pass through the centroid of the cross section. Along this x axis, equations (5.11 and 5.12) tell us that both normal strain ε_x and normal stress σ_x equal zero. In bending theory, this axis is called the *neutral axis* of the beam, as we have already discussed.

To finish up our solution of the bending problem, we use the second relative equation of equilibrium. The sum of the externally applied and the internal resisting moments must vanish, so, using the convention of counterclockwise moments being positive,

$$\sum M_o = 0 \qquad M_y - \int_A \underbrace{\underbrace{E\kappa z}_{\text{stress}}\underbrace{dA}_{\text{area}}}_{force} \underbrace{y}_{arm} = 0 \tag{5.15}$$

Recognizing that E and κ are constants, we can write

$$M_y = E\kappa \int_A z^2 dA, \tag{5.16}$$

and we are now reacquainted with the *second moment of the (rectangular) area*, typically and erroneously called the *moment of inertia*, defined with respect to the cross section's neutral (centroidal), or y axis:

$$I_y = \int_A z^2 dA \,. \tag{5.17}$$

If we replace the integral in the previous expression by I_y, we have

$$\kappa = \frac{M_y}{EI_y}, \tag{5.18}$$

and when we substitute this expression for κ back into our equation for normal stress, we get the elastic *flexure formula* for pure bending of beams:

$$\sigma_x = \frac{M_y}{I_y} z, \tag{5.19}$$

which, to demonstrate the dependence of normal stress on both x and z position along and in the beam, we can also write as

$$\sigma_x(x,z) = \frac{M_y(x)}{I_y} z \,. \tag{5.20}$$

We must note that this derivation has been carried out for the coordinate axes as shown in Figure 5.7. Since z is negative at the top of the beam and positive at its bottom edge, under a positive bending moment, the normal stress will be negative (reflecting compression) at the top of the beam and positive (reflecting tension) at the bottom.

Since we are interested in the limiting behavior of beams, it is valuable to have an expression for the maximum stress. For beams with symmetric cross sections, bent in the plane of symmetry, we designate z_{max} as c and obtain

$$\sigma_{max} = \frac{Mc}{I} \,. \tag{5.21}$$

The convention is to dispense with the sign in this expression, because the sense of the normal stresses can be determined by inspection of the beam in question, and with the subscripts.

Normal stress in the direction of the beam's long axis (here the x axis) is the *only* stress resulting from pure bending.[5] The stress tensor's matrix representation is therefore

$$\underline{\underline{T}} = \begin{pmatrix} \sigma_x & 0 & 0 \\ 0 & 0 & 0 \\ 0 & 0 & 0 \end{pmatrix}. \tag{5.22}$$

Remembering Poisson's ratio, we will have strains in the y and z directions: $\varepsilon_y = \varepsilon_z = -\nu\varepsilon_x$, where ε_x is given by σ_x/E or Mz/EI.

5.6 Shear Stresses in Beams

We now consider shear stresses in beams caused by transverse shear. (Remember, *transverse* here means normal to the beam's long axis.) We also give some thought to the attachment of separate parts of a beam by bolts, gluing, or welding.

For problems of torsion and pure bending, we began by assuming a strain distribution across the cross section. (In both cases, this distribution followed from the assumption made about "plane sections remaining plane.") We cannot make any analogous assumption about the strain distribution due to shear force. However, we are able to use the expressions for normal stress that we have developed in the previous section.

By examining the equilibrium of an infinitesimal beam element, we saw that $dM = Vdx$, that is, the shear force V is linked with a *change* in bending moment. So, if a shear and a bending moment are present at one section of a beam, the adjoining section will have a different bending moment, even if the shear remains constant. This variation in moment establishes shear stresses on the conceptual parallel longitudinal planes of the beam. (As when we first defined shear stress, we can imagine the beam to be composed of thin planes that are allowed to slide with respect to each other.) Even when we seem to be talking about isolated shear forces, we must remember that these forces are linked with a change in the bending moment along the beam's length.

The shear and moment diagrams in Figure 5.9 show this: Bending moment varies over sections with constant shear, while in regions of no shear there is no change in the moment.

The distribution of shear stress over the beam cross section is much different than that of normal stress. The shear stress is zero at those points

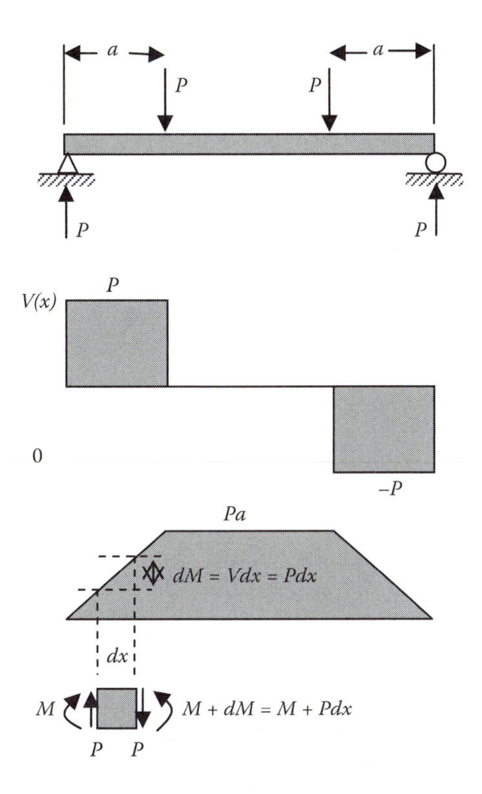

FIGURE 5.9
Shear and bending moment diagrams for the loading shown. (After Popov, E. P., *Engineering Mechanics of Solids,* Prentice Hall, 1998.)

where the bending normal stress is a maximum. And maximum shear stress almost always occurs at the neutral axis (where normal strain and stress are both zero). How do we know that? We know because the theory of elasticity we derived in Chapter 3 tells us so.

Consider the beam as a long, slender body with rectangular cross section ($b \times h$) and length L, as shown in Figure 5.10. *Slender* beams are those for which b and h are both much less than L. The beam of interest is in plane stress in the y direction, that is,

$$\sigma_{yy} = \tau_{xy} = \tau_{yz} = 0\,. \tag{5.23}$$

For a thin elastic beam in plane stress, the two (remaining) elasticity equations of equilibrium are

$$\frac{\partial \sigma_{xx}}{\partial x} + \frac{\partial \tau_{zx}}{\partial z} = 0. \tag{5.24a}$$

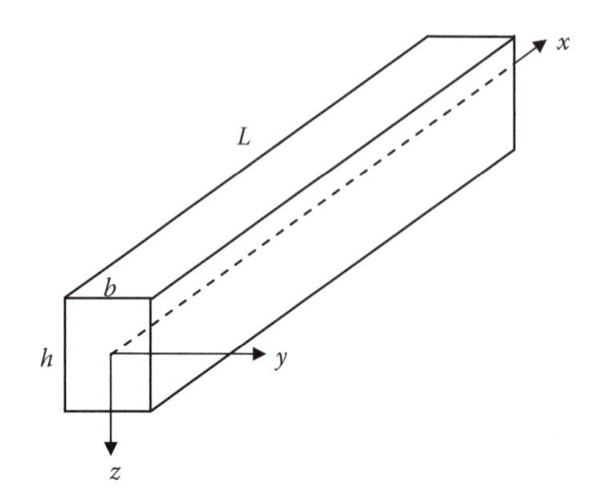

FIGURE 5.10
Rectangular beam.

$$\frac{\partial \tau_{xz}}{\partial x} + \frac{\partial \sigma_{zz}}{\partial z} = 0 \,. \tag{5.24b}$$

Note that we are still taking the z axis to be positive downward, conforming to a long-established notation for elastic beams.

While we have already derived (in Section 5.5) the basic equation for a beam's bending stress, we briefly re-derive that result on our way to an expression for shear stress. We begin assuming what we have already shown, that the normal stress in the axial direction varies linearly through the thickness:

$$\sigma_{xx}(x,z) = f(x) \cdot z, \tag{5.25}$$

where $f(x)$ is an (as yet) unknown function of x. Thus, it follows that

$$\int_{-h/2}^{h/2} \sigma_{xx}(x,z) b dz = f(x) \cdot b \int_{-h/2}^{h/2} z dz = 0 \,. \tag{5.26}$$

This equation simply states that the average axial normal stress across the thickness is identically zero. So, there is no net axial resultant in our beam, because that resultant would be calculated by an integral just like that in equation (5.26).

Now, let's take the moment (about the y axis) of the axial normal stress acting on an element of area bdz, that is,

$$dm_y = (\sigma_{xx} b dz) z \,. \tag{5.27}$$

And if we sum this moment for all elements through the thickness, we obtain a positive internal moment, $M(x)$:

$$M(x) = \int_{-h/2}^{h/2} dm_y = \int_{-h/2}^{h/2} z(\sigma_{xx} b dz) = f(x) \int_{-h/2}^{h/2} bz^2 dz \,. \tag{5.28}$$

We recognize this integral as the *second moment of the beam's cross-sectional area* about a horizontal (here, y) axis through its center, that is,

$$I_{(y)} = \int_{-h/2}^{h/2} bz^2 dz \,. \tag{5.29}$$

For the time being we drop the y subscript on I. It then follows from the previous results that

$$f(x) = \frac{M(x)}{I} \,. \tag{5.30}$$

Consequently, we can now write the equation for the normal stress in the previously derived, very famous form:

$$\sigma_{xx}(x,z) = \frac{M(x)}{I} z \,. \tag{5.31}$$

Having confirmed the axial stress equation, we can now find the shear stress distribution by substituting the bending stress into the first, axial equation of equilibrium:

$$\frac{\partial \tau_{xz}}{\partial z} = -\frac{\partial \sigma_{xx}}{\partial x} = -\frac{z}{I}\frac{dM}{dx} \,. \tag{5.32}$$

We can now integrate through the beam thickness, over the z coordinate, to obtain

$$\tau_{xz} = \frac{1}{2I}\frac{dM}{dx}\left(\frac{h^2}{4} - z^2\right), \tag{5.33}$$

and we can derive a transverse internal shear force, $V(x)$, acting on any cross section as the integral of the shear stress over that area:

$$V(x) = \int_{-h/2}^{h/2} \tau_{xz}(x,z) b dz \,. \tag{5.34}$$

Then, by substituting this result into the previous equation, we find that

$$V(x) = \frac{1}{2I}\frac{dM}{dx}\int_{-h/2}^{h/2}\left(\frac{h^2}{4} - z^2\right)bdz \equiv \frac{dM}{dx}, \tag{5.35}$$

which explicitly equates the resultant shear force to the spatial gradient of the moment. This is consistent with our existing intuition about beams!

By combining the previous results, we can arrive at an explicit expression of the relationship between shear stress and its resultant (integrated) force:

$$\tau_{xz} = \frac{1}{2I}V(x)\left(\frac{h^2}{4} - z^2\right). \tag{5.36}$$

We see that the shear stress is distributed quadratically (and symmetrically) through the thickness, achieving its maximum value at the beam centerline ($z = 0$) and being zero on both the top and bottom surfaces ($z = \pm h/2$). The latter point is consistent with the assumptions we've made about the loading of bent beams. Figure 5.11 allows us to compare the distributions of normal and shear stress along the height of the cross section.

Now we integrate the second equation of equilibrium to learn how the normal stress in the z direction is distributed through the beam thickness. First,

$$\frac{\partial \sigma_{zz}}{\partial z} = -\frac{\partial \tau_{xz}}{\partial x} = -\frac{1}{2I}\frac{dV}{dx}\left(\frac{h^2}{4} - z^2\right), \tag{5.37}$$

from which it follows by integration that

$$\sigma_{zz}(x, h/2) - \sigma_{zz}(x, -h/2) = -\frac{h^3}{12I}\frac{dV}{dx}. \tag{5.38}$$

If we assume that the top surface, which has the beam width b, is loaded with a line load $q(x)$ over its length and that the bottom surface is not loaded at all,

$$\sigma_{zz}(x, -h/2) = -q(x)/b \text{ and } \sigma_{zz}(x, h/2) = 0, \tag{5.39}$$

then it follows that the equation for transverse (vertical) equilibrium is

$$\frac{dV}{dx} + q(x) = 0, \tag{5.40}$$

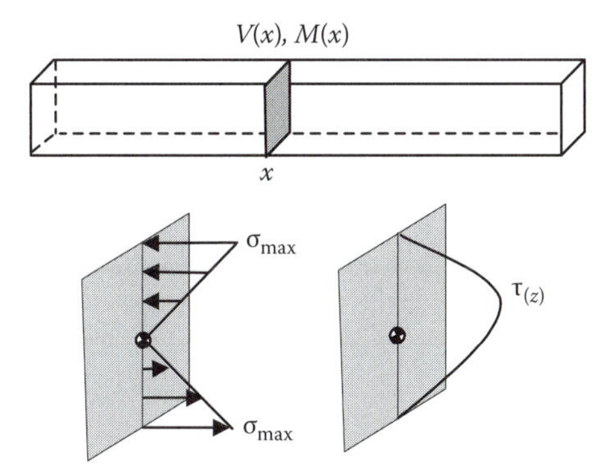

FIGURE 5.11
Stress distributions for rectangular cross section.

which we had already derived in Section 5.4. Therefore, it should not surprise us to learn that we can combine our shear-bending moment relation with equation (5.40) to yield the same equation expressed in terms of the moment (also as seen in Section 5.4):

$$\frac{d^2M}{dx^2} + q(x) = 0 \ . \tag{5.41}$$

Thus, we have now once again derived equilibrium, in a more formal way. We defer discussion of the boundary conditions for this differential equation until we express equilibrium in terms of a beam's deflections (Chapter 6).

It's worth noting that while we've derived all of the previous results assuming a rectangular cross section, $b \times h$, our results are straightforwardly extendable to any cross section. The calculation of the second moment of area, I, is applicable to any cross section. We recognize that the width vary with depth along the thickness, $b = b(z)$, so that

$$I = \int_{-h/2}^{h/2} b(z) z^2 dz \ . \tag{5.42}$$

The effects on shear are slightly more complicated, because those results reflect integration over the z coordinate. However, it turns out that the parabolic variation term that appears in the shear stress equation (5.36) represents the *first moment of area* between a coordinate value z and the cross-sectional area above or below that coordinate, which is called $Q(z)$. We must ask ourselves: When a beam is bent, what tends to *slide*, at a height z? The answer is that what's above z tends to slide with respect to what's below z.

Let's again start with our rectangular beam and then generalize. If we take the first moment of area of the lower area shown in Figure 5.12, we integrate from z to $h/2$.

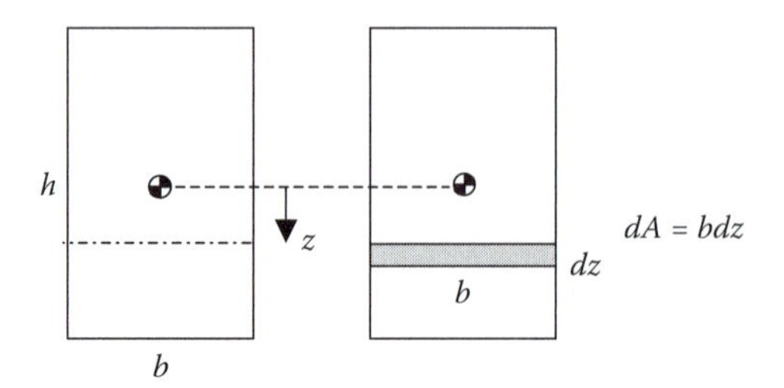

FIGURE 5.12
(a) Rectangular cross section; (b) integration for Q.

$$Q = \int_{\text{from } z \text{ down}} z dA = \int_{z}^{h/2} z' \, b dz' = \frac{z^2}{2} b \Big]_z^{h/2} = \frac{b}{2}\left[\left(\frac{h}{2}\right)^2 - z^2\right]. \tag{5.43}$$

So, the average shear stress at any height y can be written as

$$\tau\,(y) = \frac{VQ}{bI} = \frac{V}{2I}\left[\left(\frac{h}{2}\right)^2 - z^2\right] = \frac{6V}{bh^3}\left[\left(\frac{h}{2}\right)^2 - z^2\right]. \tag{5.44}$$

We see that this is a parabolic distribution of shear stress, with a maximum at $z = 0$, the neutral axis, where

$$\tau_{\max} = \frac{3V}{2bh}.$$

Equation (5.44), remember, is valid only for beams with rectangular cross sections $b \times h$.

For a nonrectangular geometry, as illustrated in Figure 5.13, we can define Q, the first moment of area *fghj* around the neutral axis, by the integral

$$Q = \int_{\substack{\text{area} \\ fghj}} z dA = A_{fghj} z_{c_{fghj}}, \tag{5.45}$$

where $z_{c,fghj}$ is the distance from the neutral axis to the centroid of area A_{fghj}.

We now combine this result with our previous expression for shear stress (from equation 5.34) to provide a general formula for average shear stress at a longitudinal cut, in a plane parallel to the axis where the cross section has thickness b:

$$\tau_{xz} \equiv \frac{VQ}{Ib}. \tag{5.46}$$

This, like the classical bending stress equation, is a well-known and important result. The Q in question is the Q of the portion of the cross section that would tend to slide relative to the plane with thickness b. The shear stress thus varies with height along the cross section, as normal stress does, though the trends are different: Shear stress is maximized at the neutral axis and is zero at the outer surfaces of the beam, as we sketched in Figure 5.11. The details of the shear stress distribution depend on the cross-section geometry.

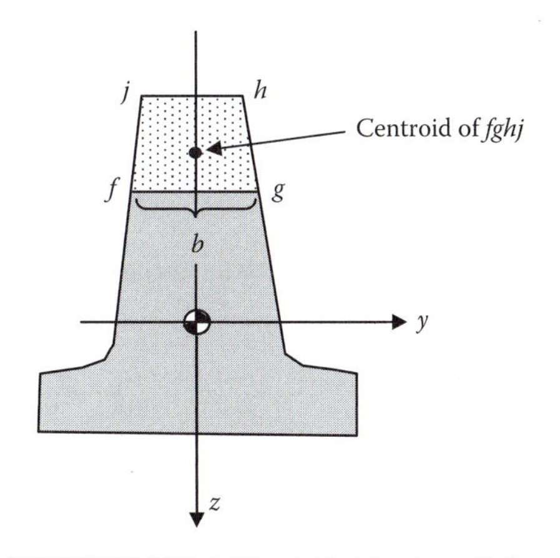

FIGURE 5.13
Elements used in derivation of shear stress in a beam. (Adapted from Popov, E. P., *Engineering Mechanics of Solids*, Prentice Hall, 1998.)

5.7 Examples

Example 5.1

Find the reactions and determine the axial force P, the shear V, and the bending moment M caused by the applied loads at the specified sections in Figure 5.14. Also, draw free-body diagrams indicating the sense (direction) of all forces and moments.

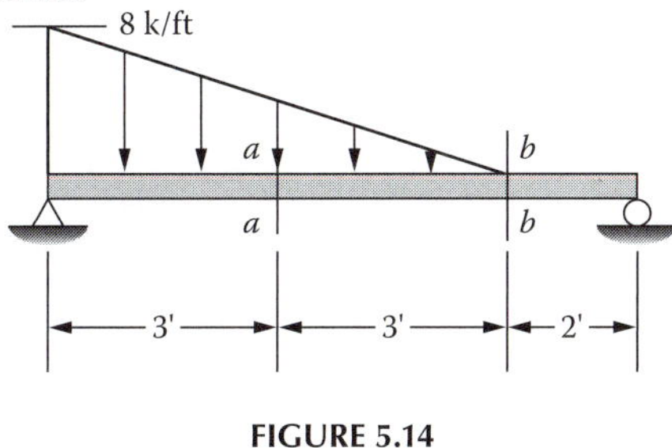

FIGURE 5.14

Given: Dimensions of and loading on beam.

Find: Internal forces and bending moment.

Assume: The only assumptions necessary are implicitly made throughout this textbook: equilibrium and Saint-Venant's Principle.

Solution

Our strategy is to find the reactions at the supports from the whole beam's equilibrium and then to use the method of sections to find the internal forces and moments at the specified locations.

For the purpose of finding reaction forces, we can replace the distributed load by its equivalent concentrated load. The magnitude of this concentrated load is simply the area under the distributed load, in this case $W = ½ (8 \text{ k/ft})(6 \text{ ft}) = 24$ kips. It acts at the centroid of the triangular area under the distributed load: one third of the way from its maximum intensity, or 2 ft from the left end of the beam. We use this load in our FBD (Figure 5.15):

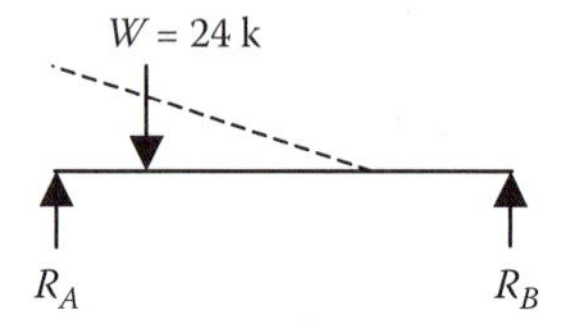

FIGURE 5.15

$$\Sigma M_B = 0 = -R_A(8 \text{ ft}) + (24 \text{ k})(6 \text{ ft}) \rightarrow R_A = 18 \text{ kips}.$$

$$\Sigma M_A = 0 = -24 \text{ k}(2 \text{ ft}) + R_B(8 \text{ ft}) \rightarrow R_B = 6 \text{ kips}.$$

$$\Sigma F_z = 0 \text{ is then used as a check: } R_A + R_B = 24 \text{ kips}.$$

Next, we consider sections *a-a* and *b-b*. We make an imaginary cut at the specified location and realize that considering the loading to the left or to the right of the *a-a* cut yields equivalent results (Figure 5.16):

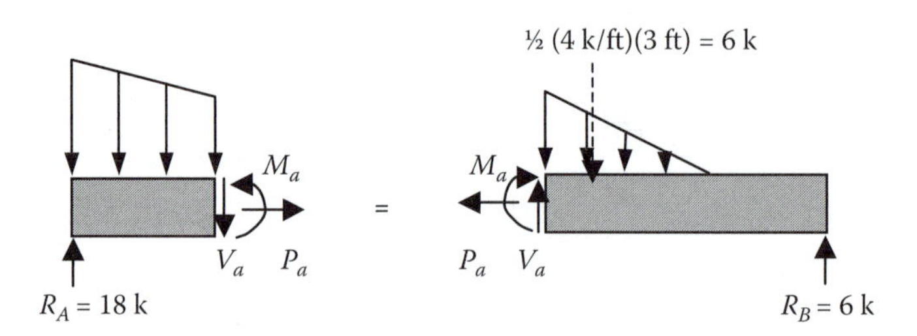

FIGURE 5.16

We choose the simpler side to calculate, in this case the portion of the beam to the right of *a-a*. We simply apply the equilibrium equations to this section of the beam:

$$\Sigma F_x = 0 = P_a.$$

$$\Sigma F_z = 0 = -V_a - R_B + 6 \text{ k} \rightarrow V_a = 0 \text{ kips}.$$

$$\Sigma M_a = 0 = -M_a - (6\text{ k})(1\text{ ft}) + R_B(5\text{ ft}) \rightarrow M_a = 24\text{ ft-kips.}$$

Next comes the cut at *b-b*. It is clear that using the portion of the beam to the right of *b-b* is easier, so we construct an FBD and apply equilibrium (Figure 5.17):

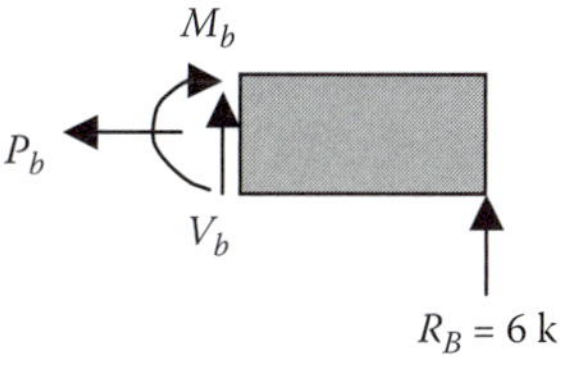

FIGURE 5.17

$$\Sigma F_x = 0 = P_b.$$

$$\Sigma F_z = 0 = -V_b - R_B \rightarrow V_b = -6\text{ kips.}$$

$$\Sigma M_b = 0 = R_B(2\text{ ft}) - M_b \rightarrow M_b = 12\text{ ft-kips.}$$

Note: The negative sign on the shear at cut *b-b* indicates that the shear at this cut is "negative shear," opposite from the way it is drawn in our FBD. It is convenient to assume positive shear when constructing FBDs so that a negative sign always represents negative shear. Please refer to Figure 5.4 for a reminder of the sign convention for shear; Figure 5.5 shows the sign convention for bending moment.

Example 5.2

Plot shear and moment diagrams for the beams shown in Figure 5.18:

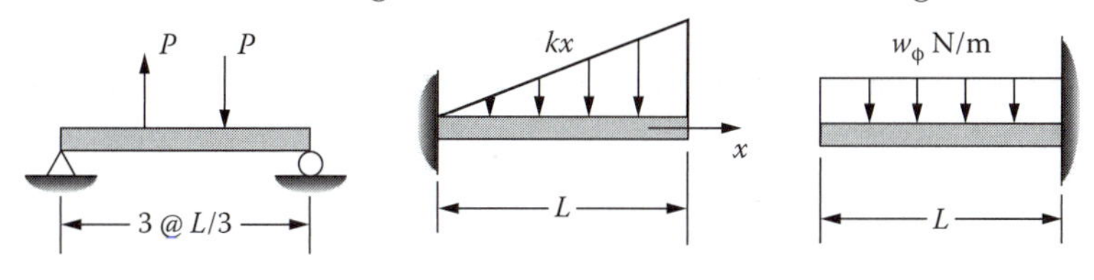

FIGURE 5.18

Given: Loading on three beams.

Find: Internal response to this loading.

Assume: The only assumptions necessary are implicitly made throughout: equilibrium and Saint-Venant's Principle.

Solution

In each case, we first find the external reaction forces or moments (or both) and then use the method of sections at points of interest along the beam to construct the diagrams of $V(x)$ and $M(x)$. We will start with the first case, shown in Figure 5.19.

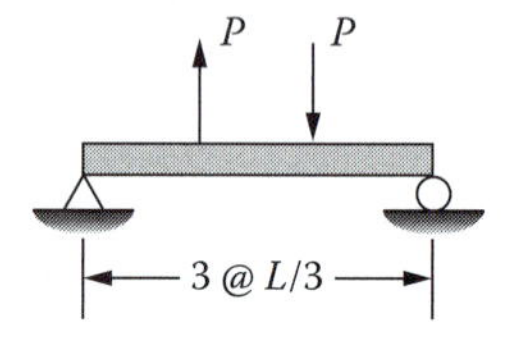

FIGURE 5.19

To find reactions, we need an FBD of the whole beam (Figure 5.20):

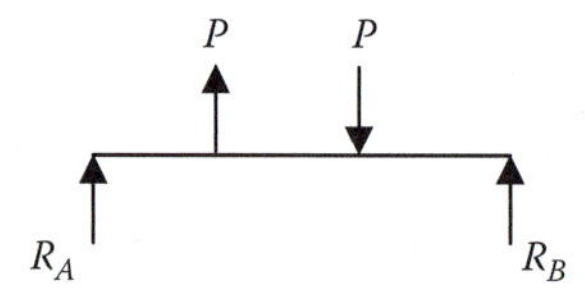

FIGURE 5.20

$$\Sigma M_A = 0 = \frac{PL}{3} - \frac{2PL}{3} + R_B L \rightarrow R_B = \frac{P}{3}(up).$$

$$\Sigma M_B = 0 = \frac{PL}{3} - \frac{2PL}{3} - R_A L \rightarrow R_A = -\frac{P}{3}(down).$$

To construct the V diagram, look only at the points where the loading conditions change:

- At the left-hand end of the beam, V must balance R_A, so $V = -P/3$ (Figure 5.21).

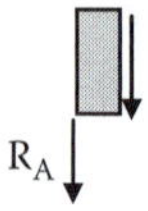

FIGURE 5.21

- At the right-hand end, V must balance R_B, so $V = -P/3$ (Figure 5.22).

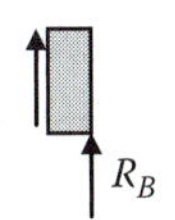

FIGURE 5.22

- Just to the right of the upward applied load *P*, $V = +2P/3$ (Figure 5.23).

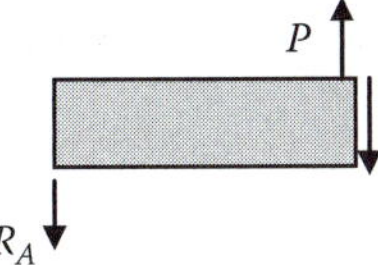

FIGURE 5.23

To construct the *M* diagram:

- $M = 0$ at simply supported ends.
- $M = 0$ at center by symmetry.
- At upward *P*, $M = -\frac{P}{3}\left(\frac{L}{3}\right) = -\frac{PL}{9}$.
- At downward *P*, $M = -\frac{P}{3}\left(\frac{2L}{3}\right) + P\left(\frac{L}{3}\right) = +\frac{PL}{9}$.

Plot the results (Figure 5.24).

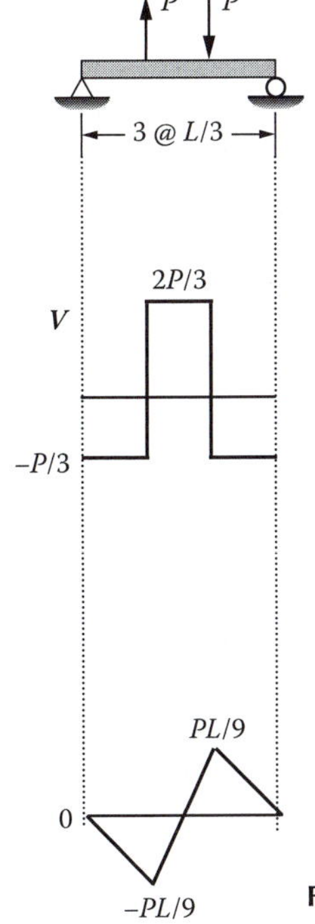

FIGURE 5.24

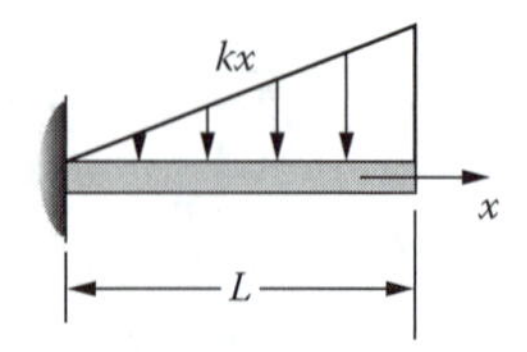

FIGURE 5.25

For Figure 5.25, we start with the external reactions, using the equivalent concentrated load in place of the distributed one (Figure 5.26).

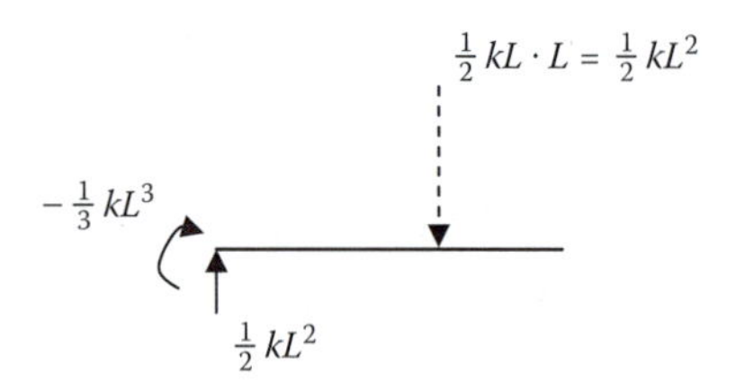

FIGURE 5.26

Since the distributed load is linearly distributed, the shear distribution is parabolic, and the moment distribution is cubic. Or we may proceed either by integrating the distributed load $q(x) = kx$ once for $V(x)$ and twice for $M(x)$ or by making our imaginary cut at some distance x from the end of the beam and finding the internal shear and moment. Both methods provide the same results.

Integration

$$V(x) = -\int q dx = \int kx dx = -\tfrac{1}{2}kx^2 + C.$$

$$V(0) = \tfrac{1}{2}kL^2 = C.$$

$$V(x) = \tfrac{1}{2}kL^2 - \tfrac{1}{2}kx^2.$$

$$M(x) = \int V dx = \tfrac{1}{2}kL^2 x - \tfrac{1}{6}kx^3 + C.$$

$$M(0) = -\tfrac{1}{3}kL^3 = C.$$

$$M(x) = \tfrac{1}{2}kL^2 x - \tfrac{1}{6}kx^3 - \tfrac{1}{3}kL^3.$$

Method of Sections

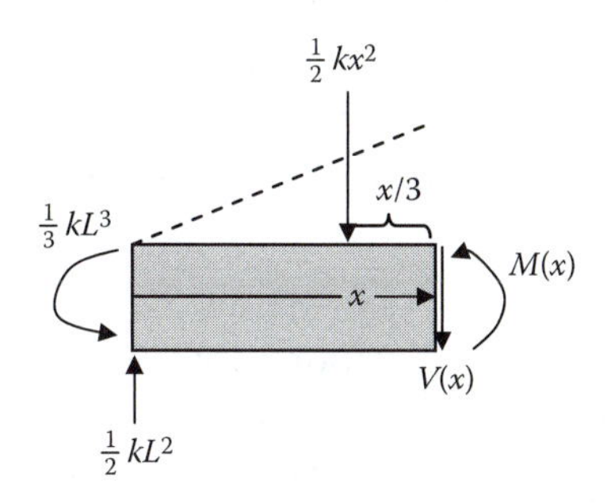

FIGURE 5.27

$$\Sigma F_z = 0 = -\tfrac{1}{2}kL^2 + \tfrac{1}{2}kx^2 + V(x).$$

$$\text{so } V(x) = \tfrac{1}{2}kL^2 - \tfrac{1}{2}kx^2.$$

$$\Sigma M_x = 0 = M(x) + \tfrac{1}{3}kL^3 - \tfrac{1}{2}kL^2x + \tfrac{1}{2}kx^2\left(\tfrac{x}{3}\right).$$

We note that the internal shear $V(x)$ does not cause a moment about the cut at x, so

$$M(x) = -\tfrac{1}{3}kL^3 + \tfrac{1}{2}kL^2x - \tfrac{1}{6}kx^3.$$

The resulting shear and moment diagrams are as shown in Figure 5.28.

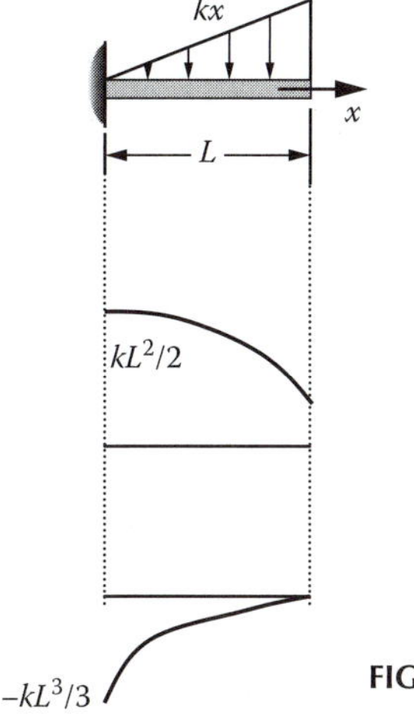

FIGURE 5.28

We now consider the third beam (Figure 5.29).

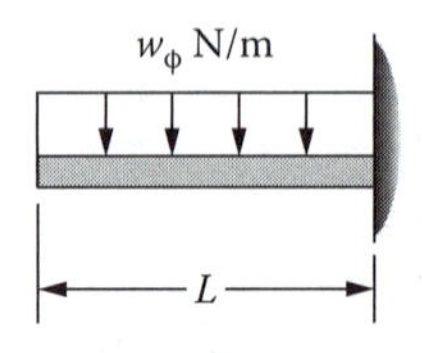

FIGURE 5.29

The fixed support can offer both reaction forces and a moment, which are found using an FBD (Figure 5.30).

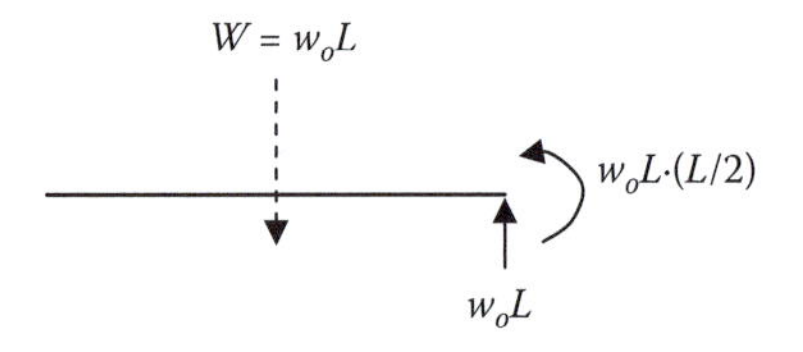

FIGURE 5.30

$$\Sigma F_z = 0.$$

$$\Sigma M_o = 0.$$

The shear V is zero at the free end of this cantilever beam and must balance the upward reaction force w_oL at the fixed end. Since the distributed load is uniformly distributed, the shear distribution is linear.

The moment M is zero at the free end and must balance the reaction moment $-w_oL^2/2$ at the fixed end. Since the shear distribution is linear ($\sim -x$), the bending moment distribution is parabolic ($\sim -x^2$). The shear and moment diagrams are shown in Figure 5.31.

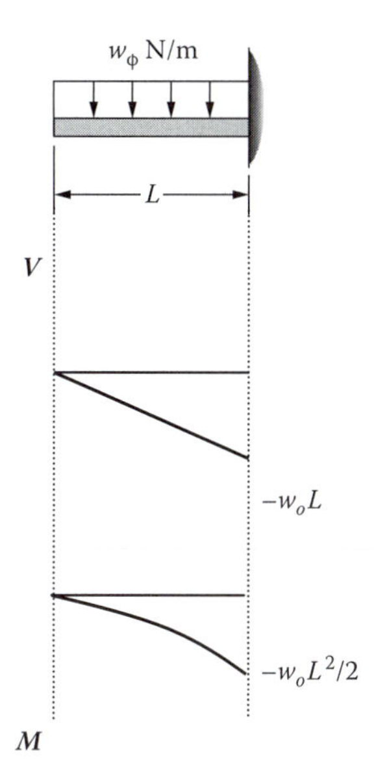

FIGURE 5.31

Example 5.3

Find the centroid and the second moment of area about the horizontal (y) axis of the cross section shown in Figure 5.32. All dimensions are in millimeters. If a beam is constructed with the cross section shown from steel whose maximum allowable tensile stress is 400 MPa, what is the maximum bending moment that may be applied to the beam?

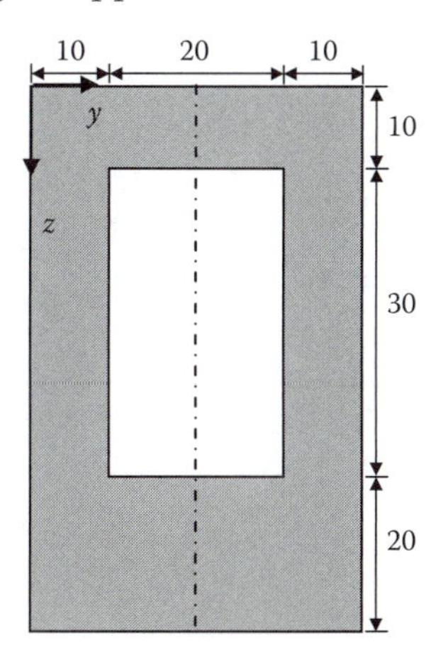

FIGURE 5.32

Given: Dimensions of beam cross section; limiting stress.

Find: Location of centroid; maximum applied moment.

Assume: Hooke's law applies.

Solution

The symmetry of the cross section shown in Figure 5.32 suggests that the horizontal coordinate of the centroid will be on the vertical centerline, as sketched. The value y_c is then 20 cm. We need only locate the vertical coordinate of the centroid (z_c). Several strategies are available to us. We recognize that the cross section is a large rectangle, with an inner rectangular hole. It is thus possible for us to find the area and second moment of area of the large outer rectangle and simply to subtract off the properties of the inner rectangle.

Recall that

$$z_c = \frac{\int z dA}{\int dA} = \frac{\sum z dA}{\sum dA}.$$

For clarity, results are tabulated as follows:

	A (mm²)	z (mm)	$A \cdot z$ (mm³)	
outer	$40 \times 60 = 2400$	30	72,000	**Centroid:** $z_c = \frac{\sum Az}{\sum A} = 31.7$ mm (from top) or 28.3 mm from bottom.
inner	$-20 \times 30 = -600$	25	−15,000	
Σ	**1800**		**57,000**	

	$I_o = bh^3/12$ **(mm⁴)**	d **(mm)**	$A \cdot d^2$ **(mm⁴)**	
outer	720,000	31.7 – 30 = 1.7	6940	**Second Moment of Area:** $I = \sum\left(\frac{bh^3}{12} + Ad^2\right)$
inner	–45,000	31.7 – 25 = 6.7	–26,940	= 655,000 mm⁴
Σ	**675,000**		**–20,000**	

If the maximum allowable normal stress is 400 MPa, we can find the maximum moment that can be applied using the relationship

$$\sigma_{\max} = \frac{Mc}{I}.$$

We have found the second moment of area I, and c is the maximum distance from the centroid attainable on the cross section, in this case 31.7 mm. Solving for M,

$$M = \frac{\sigma_{\max} I}{c} = \frac{(400 \text{ N/mm}^2)(655{,}000 \text{ mm}^4)}{31.7 \text{ mm}} = 8.26 \text{ MN} \cdot \text{mm} = 8.26 \text{ kN} \cdot \text{m} .$$

Example 5.4

A steel T beam is used in an inverted position to span 400 mm. If, due to the application of the three forces shown in Figure 5.33, the longitudinal strain gage at A (3 mm down from top of beam) registers a compressive strain of 50×10^{-5}, how large are the applied forces?

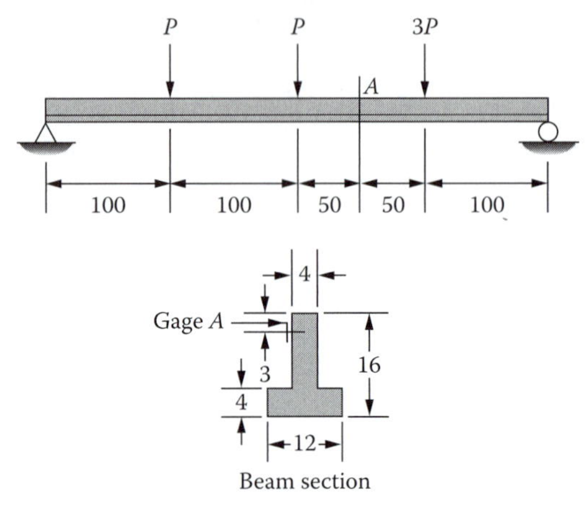

FIGURE 5.33

Given: Dimensions of and strain in T beam.

Find: Magnitude of applied force P.

Assume: Hooke's law applies.

Solution

The gage at A, in the upper portion of the cross section, registers a positive strain. This tells us that the bending moment in the beam at A is positive. Using Hooke's law, we are able to relate this measured strain to a normal stress in the beam at this point, which we can then relate to the local bending moment. To do these calculations, we need to know the location of the centroid and second moment of area of the inverted T cross section.

As in Example 5.3, the centroid is clearly on the vertical line of symmetry. We need z_c (Figure 5.34):

FIGURE 5.34

	A (mm^2)	z(mm)	Az (mm^3)
1	4 (12 = 48	6	288
2	12 (4 = 48	14	672
Σ	96		960

Hence, $z_c = \dfrac{\sum zA}{\sum A}$ = 10 mm from the top, or 6 mm from bottom. Next comes the second moment of area I:

	$bh^3/12$ (mm^4)	d (mm)	$A \cdot d^2$ (mm^4)
1	576	4	768
2	64	4	768
Σ	640		1536

$$I = 2176 \text{ mm}^4.$$

We are now ready to apply Hooke's law and to find the stress corresponding to the strain measured at A. The beam is steel, so its Young's modulus is E = 200 GPa, and

$$\sigma_{x_A} = E\varepsilon_{x_A} = (200 \times 10^9 \text{ Pa})(50 \times 10^{-5}) = 10^8 \text{ Pa}.$$

This is measured at z_A = 3 mm from the top of the T, or 7 mm from the neutral axis.

Next, we relate this stress to the internal bending moment at A:

$$\sigma_{x_A} = \frac{M_A z_A}{I}$$

$$10^8 \text{ Pa} = \frac{M_A (0.007 \text{ m})}{2.176 \times 10^{-9} \text{ m}^4}$$

$$\text{so } M_A = \frac{(10^8 \text{ Pa})(2.176 \times 10^{-9} \text{ m}^4)}{0.007 \text{ m}} = 31.06 \text{ N} \cdot \text{m}$$

We then consider the loading on the beam to relate this local bending moment to the applied loads P. To do this, we must construct an FBD (Figure 5.35).

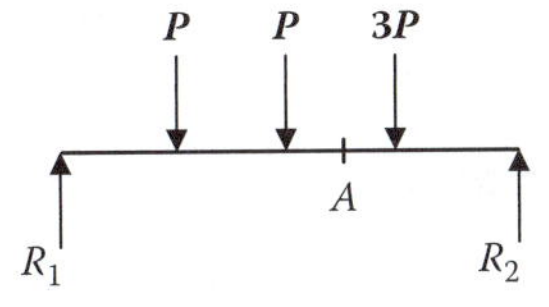

FIGURE 5.35

Equilibrium requires that $\Sigma F_z = 0$, or $R_1 + R_2 = 5P$, and if we also impose $\Sigma M_1 = 0$ we will have $0.1P + 0.2P + 0.3(3P) - 0.4R_2 = 0$, and solving these two equations we have $R_1 = 3P$ and $R_2 = 2P$. We can then use the method of sections to (Figure 5.36) find the bending moment at A:

$$\Sigma M_A = 0 = -M_A - 3P\cdot(0.05 \text{ m}) + 3P\cdot(0.150 \text{ m}).$$

$$M_A = 0.450P - 0.150P = 0.3P.$$

So, knowing that $M_A = 31.06$ N·m and that $M_A = 0.3P$, we find that

$$P = \frac{31.06 \text{ N} \cdot \text{m}}{0.3 \text{ m}} = 103.5 \text{ N} .$$

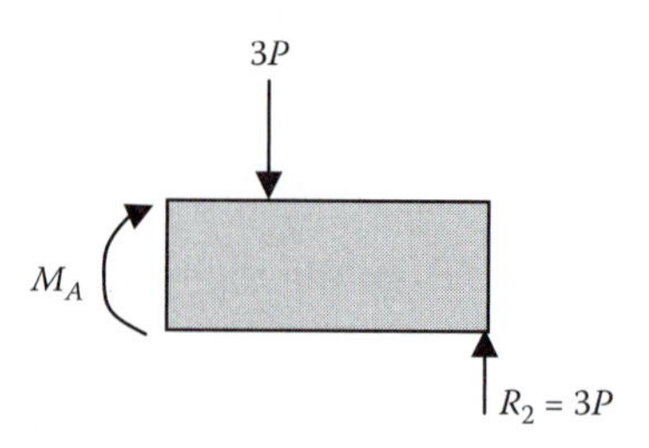

FIGURE 5.36

Example 5.5

An I beam is made by gluing five wood planks together, as shown in Figure 5.37. At a given axial position, the beam is subjected to a shear force V = 6000 lb. (a) What is the average shear stress at the neutral axis $z = 0$? (b) What are the magnitudes of the average shear stresses acting on each glued joint?

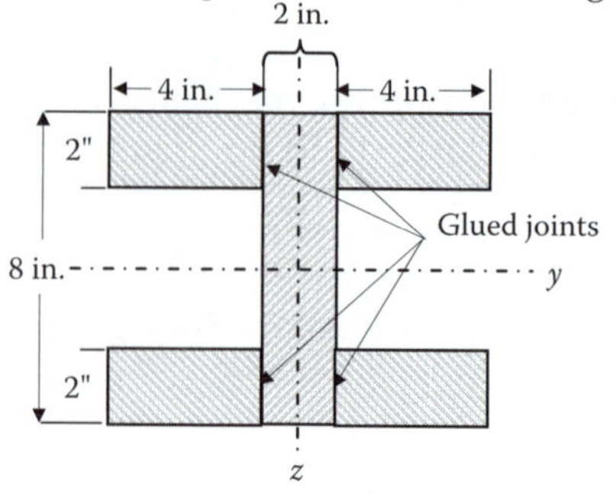

FIGURE 5.37

Given: Cross section, local loading.

Find: Average shear stresses.

Assume: Hooke's law applies.

Solution

We obtained a formula for shear stress at a given height, $V = VQ/Ib$, where Q and b depend on the height in question. I in this relationship is the second moment of area of the entire cross section about the z axis. We have been given V. So we must calculate I once and then calculate the appropriate values of Q and b for both parts of this problem.

By inspection of the cross section's symmetry, we see that the centroid is at the geometric center of the I beam (Figure 5.38). For the central vertical segment, therefore, d, the distance between the centroid of the segment and the centroid of the entire cross section, is zero. The four remaining segments each have the same second moment of area about their own horizontal bisectors and the same areas and lengths d. Thus, we can write

$$I = I_{vertical} + 4I_{smaller} \quad I = \left(\frac{1}{12}bh^3\right)_{vertical} + 4\left[\frac{1}{12}bh^3 + Ad^2\right]_{smaller}$$

$$I = \frac{1}{12}(2\text{ in.})(8\text{ in.})^3 + 4\left[\frac{1}{12}(4\text{ in.})(2\text{ in.})^3 + (2\text{ in.} \times 4\text{ in.})(3\text{ in.})^2\right] = 384\text{ in.}^2$$

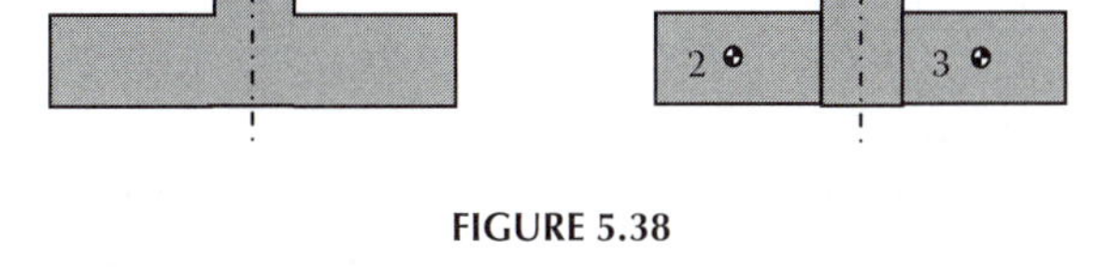

FIGURE 5.38

We can calculate Q at the neutral axis by finding the centroid and area of the shaded area on the left, or by summing the contributions due to the other planks, as shown at right.

$$Q = \int zA = (dA)_1 + (dA)_2 + (dA)_3$$

$$= (2)(2 \cdot 4) + 3(4 \cdot 2) + 3(4 \cdot 2) = 64 \text{ in.}^3$$

So the average shear stress at the neutral axis is

$$\tau = \frac{VQ}{Ib} = \frac{(6000 \text{ lb})(64 \text{ in.}^3)}{(384 \text{ in.}^4)(2 \text{ in.})} = 500 \text{ psi.}$$

As an exercise, verify that each glued joint is subjected to the same average shear stress. We determine only the average shear stress acting on the lower-right glued joint by using the area A and length of contact b as shown in Figure 5.39. The value of Q is $(3)(4 \cdot 2) = 24$ in.^{3}, and the average shear stress is $VQ/Ib = 188$ psi.

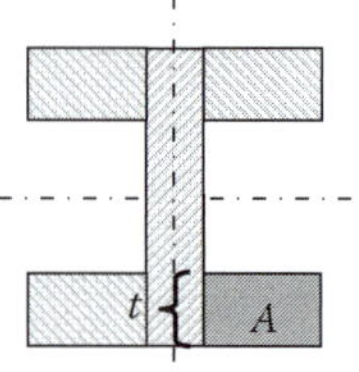

FIGURE 5.39

Example 5.6

The beam shown in Figure 5.40 is subjected to a distributed load. For the cross section at $x = 0.6$ m, determine the average shear stress (a) at the neutral axis, and (b) at $z = 0.02$ m.

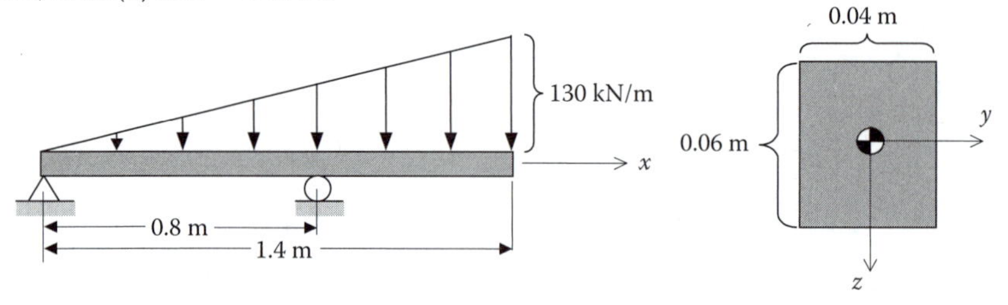

FIGURE 5.40

Given: Dimensions of and loading on simply supported beam.

Find: Shear stress at two locations along height of cross section at $x = 0.6$ m.

Assume: Hooke's law applies.

Solution

We are able to make use of the relationship between average shear stress and height derived for rectangular cross sections in Section 5.6. First, though, we need to consult our FBD and find the reactions at the supports (Figure 5.41).

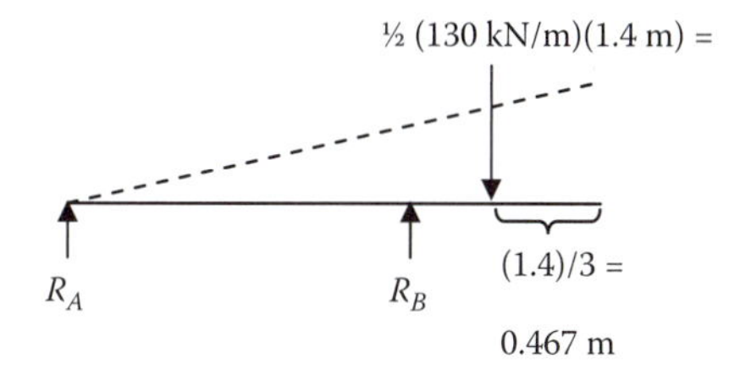

FIGURE 5.41

$$\Sigma F_y = 0 \rightarrow R_A + R_B = 91 \text{ kN}.$$

$$\Sigma M_B = 0 = R_A(0.8 \text{ m}) + 91 \text{ kN}(0.133 \text{ m}).$$

$$\rightarrow R_A = -15{,}130 \text{ N (downward)}.$$

$$\rightarrow R_B = 106{,}130 \text{ N (upward)}.$$

We are interested in the cross section at $x = 0.6$ m. We know that the average shear stress depends on the internal shear force in the beam at the point of interest, so we need to calculate the shear $V(x = 0.6 \text{ m})$. To do this, we make an imaginary cut at $x = 0.6$ m (Figure 5.42).

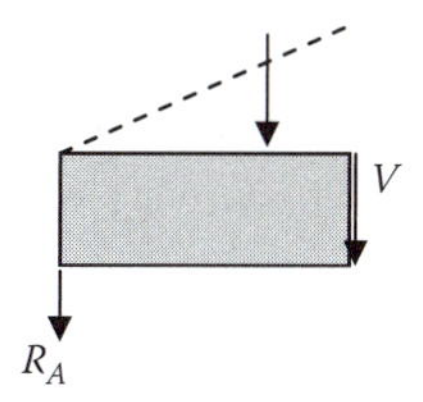

FIGURE 5.42

In this 0.6-m-long span, the distributed load has a maximum intensity of

$$\left(130 \frac{\text{kN}}{\text{m}}\right)\frac{0.6 \text{ m}}{1.4 \text{ m}} = 55.7 \frac{\text{kN}}{\text{m}},$$

so the equivalent concentrated load acting on the 0.6-m-long segment is the area under this load:

$$\frac{1}{2}\left(55.7\frac{\text{kN}}{\text{m}}\right)(0.6\text{ m})=16{,}714\text{ N}.$$

Equilibrium of our 0.6-m segment is

$$\Sigma\,F_y = 0 = -V - R_A - 16{,}714\text{ N}$$

$$\rightarrow V = -(16{,}714\text{ N} + 15{,}130\text{ N}) = -31{,}844\text{ N (negative shear).}$$

We can now use our derived expression for rectangular cross sections, which includes the y dependence of Q:

$$\tau(z)=\frac{6V}{bh^3}\left[\left(\frac{h}{2}\right)^2-z^2\right].$$

At the height of the centroid or neutral axis, $z = 0$, and this becomes

$$\tau(z=0)=\frac{6V}{bh^3}\left[\left(\frac{h}{2}\right)^2\right]=\frac{3V}{2bh}$$

$$=\frac{3(-31{,}844\text{ N})}{2(0.04\text{ m})(0.06\text{ m})}=-19.9\text{ MPa}.$$

At $z = 0.02$ m below the neutral axis, we find that

$$\tau(z)=\frac{6V}{bh^3}\left[\left(\frac{h}{2}\right)^2-z^2\right]=\frac{6(-31{,}844\text{ N})}{(0.04\text{ m})(0.06\text{ m})^3}\left[\left(\frac{0.06\text{ m}}{2}\right)^2-(0.02\text{ m})^2\right]=-11.06\text{ MPa}.$$

5.8 Problems

5.1 Draw a shear diagram for the beam shown in Figure 5.43.

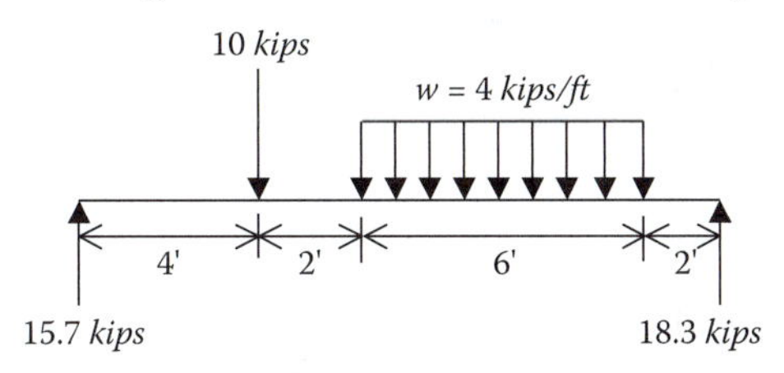

FIGURE 5.43

5.2 Draw shear and bending moment diagrams for the beam shown in Figure 5.44.

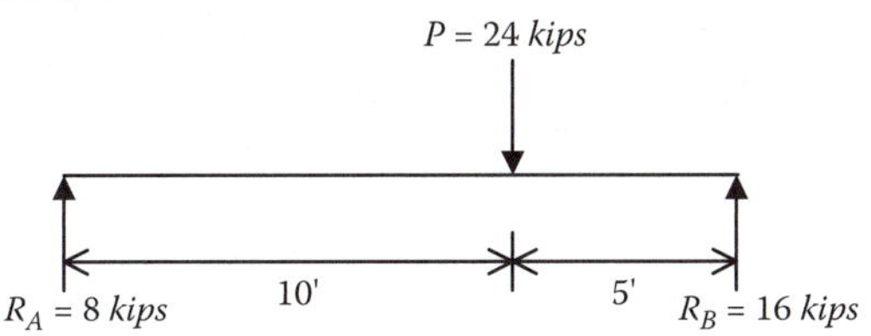

FIGURE 5.44

5.3 Draw shear and bending moment diagrams for the *overhanging* beam in Figure 5.45.

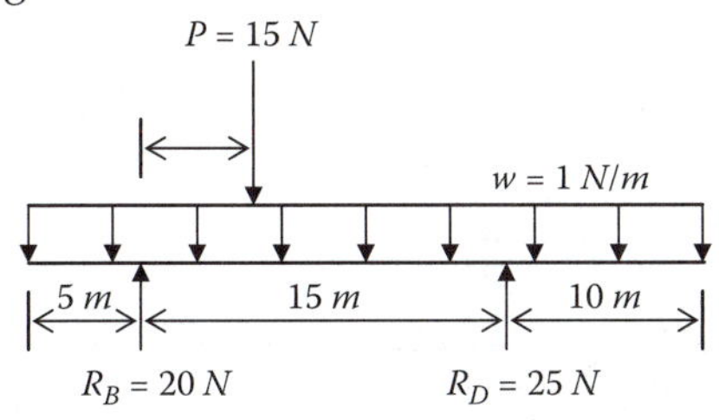

FIGURE 5.45

5.4 Draw shear and bending moment diagrams for the beam shown in Figure 5.46:

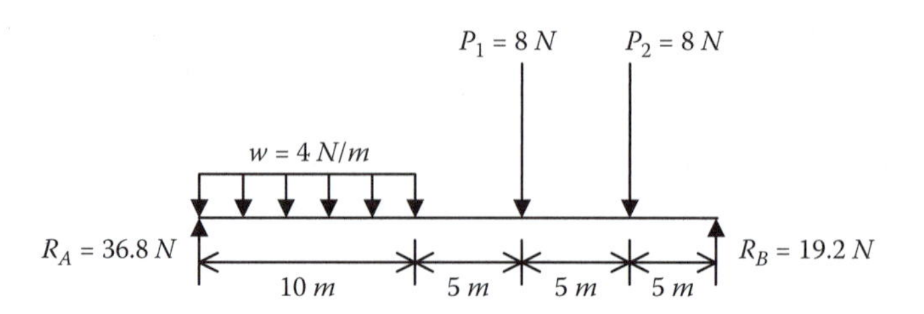

FIGURE 5.46

5.5 Construct axial force, shear, and bending moment diagrams using the integration process for the loaded beam shown in Figure 5.47. *Note*: Drawing is not to scale.

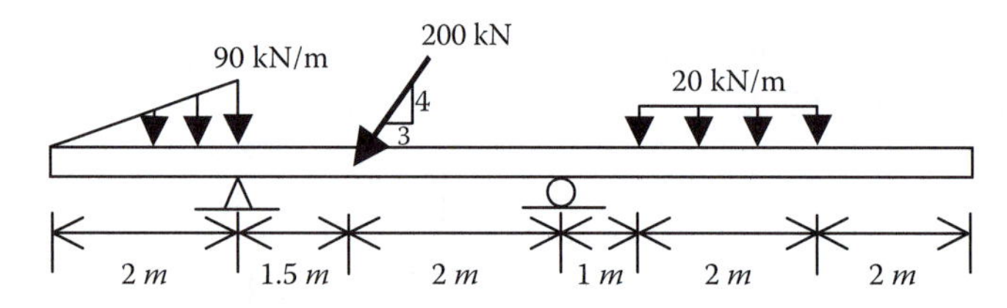

FIGURE 5.47

5.6 Given the shearing force and bending moment diagrams shown in Figure 5.48, what are the x and z coordinates of the points at which you would expect to find the maximum shear stress if the cross section has the shape in the figure? The location of the centroid is indicated.

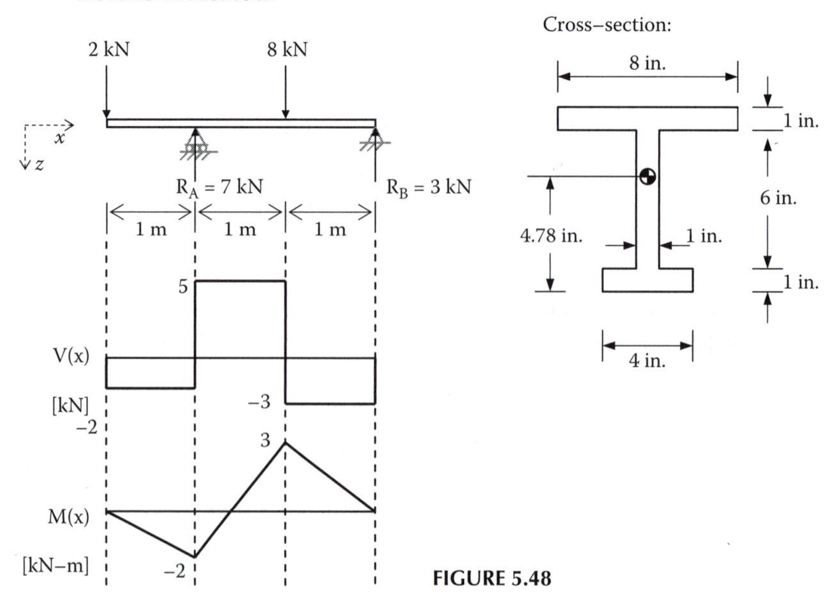

FIGURE 5.48

5.7 The shear force diagram for a beam is shown in Figure 5.49. Assuming that only forces act on the beam, determine the beam's loading and draw the bending moment diagram.

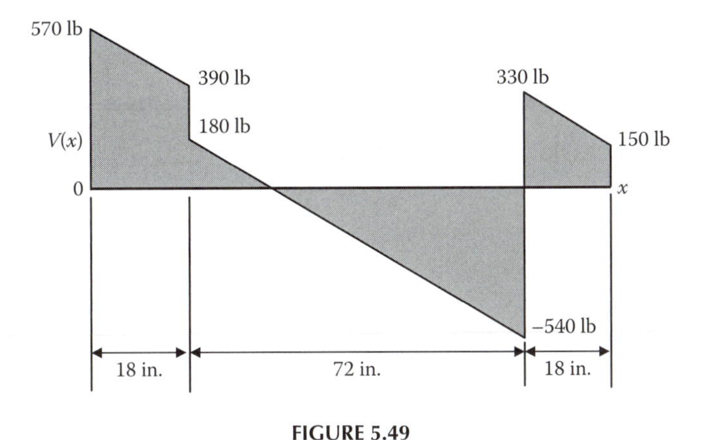

FIGURE 5.49

Case Study 3: Physiological Levers and Repairs

The human skeletal system is a natural mechanical apparatus. Our beam models can provide useful ways to explain how the musculoskeletal system works and why it sometimes breaks as well as provide a basis for repairing broken elements. In the first category, we extend Example 2.1 (from Chapter 2) to model the human forearm as a beam. Then we use a simple beam analysis to design a repair for a broken hip bone.

The Forearm Is Connected to the Elbow Joint

In Chapter 2, Example 2.1, we performed a very simple, *first-order* equilibrium analysis of the bones and bicep muscle of a human arm. Our single interest there was to find the force exerted by the biceps muscle when it supported a weight through the elbow-biceps-forearm-hand system. A more complete analysis requires a deeper consideration of that system. This is, by the way, a very old problem. Figure CS3.1 shows a diagram taken from the treatise *De Motu Animalum*, published in Italy by Giovanni Alfonso Borelli (1608–1679). Note that Borelli's work, whose English title is *On the Movement of Animals*, preceded the 1687 publication of Newton's laws of motion! Working without the benefits of Newton's laws, Borelli discovered that the forces on bones are significantly higher than the forces applied externally. Thus, the skeleton is at a mechanical disadvantage, as the following *second-order* analysis shows.

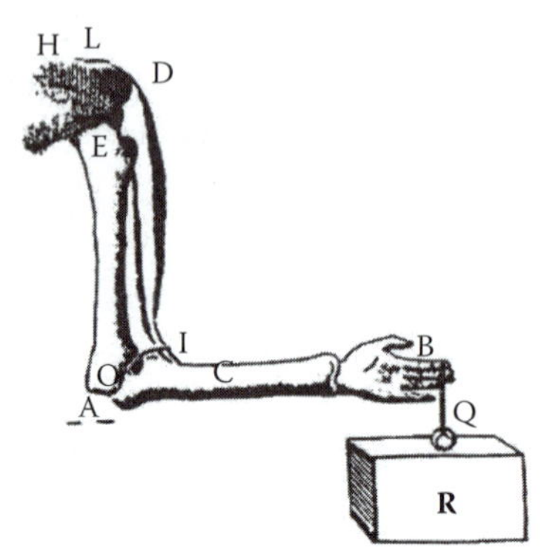

FIGURE CS3.1
The *elbow force* problem presented by the mathematician Giovanni Alfonso Borelli (1608–1679) in his treatise, *De Motu Animalum*. (From Martin, R. B., Burr, D. B., and Sharkey, N. A., *Skeletal Tissue Mechanics*, Springer-Verlag, New York, 1998. With permission.)

We show an anatomical drawing of the elbow-biceps-forearm-hand system in Figure CS3.2. The elbow joint is a complicated hinge that allows a bent arm to go straight up and down, to extend away from the body, and to rotate about an axis through the forearm. (The forearm rotations *pronation* and *supination* are much like those experienced by runners when their feet rotate with respect to their ankles and legs.) We restrict our analysis to simple lifting with no extension or rotation. Then we can assume that the three muscles shown in Figure CS3.2 act as a single "biceps-brachialis" muscle that exerts the force B shown in the FBD in Figure CS3.3. The force J is exerted by the joint on the forearm, and W is the supported weight. In this model, the elbow joint clearly acts as a simple planar hinge. The angle θ at which B acts can be determined by anatomical measurement, and the angle ϕ at which J acts is unknown or indeterminate. (The observant reader will note that the arm's own weight is left out altogether; see Problem CS3.3, Problem CS3.4, Problem CS3.5, Problem CS3.6, and Problem CS3.7.)

We now sum forces in the x and y directions and moments about an axis drawn through the elbow joint (and we ignore the small offset between the application of the muscle force B and the axis of the arm):

$$\sum F_x = B\cos\theta - J\cos\phi = 0, \tag{CS3.1a}$$

$$\sum F_y = B\sin\theta - J\sin\phi - W = 0, \tag{CS3.1b}$$

$$\sum M_z = WL - Bb\sin\theta = 0. \tag{CS3.1c}$$

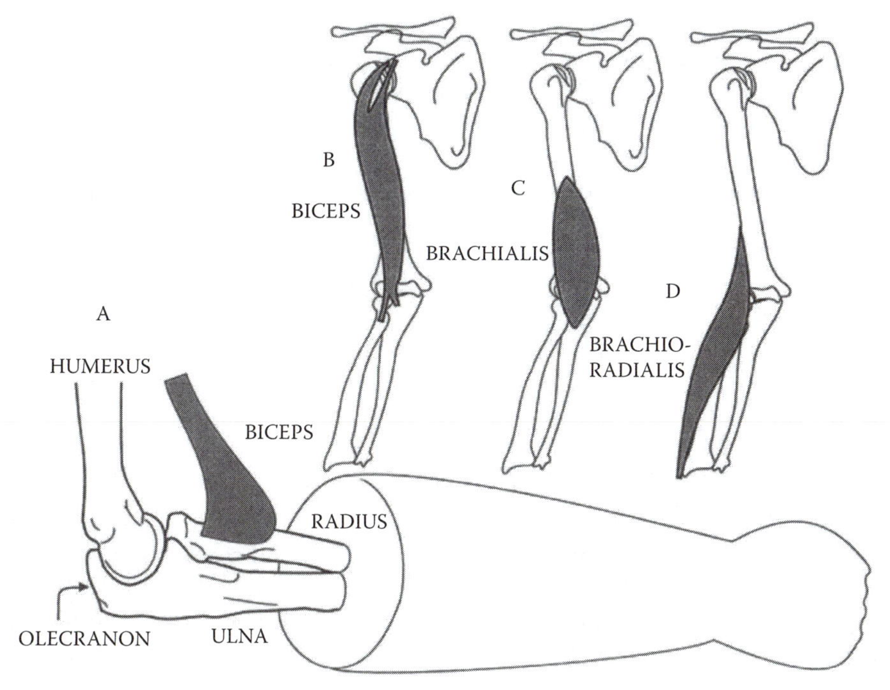

FIGURE CS3.2
Anatomical drawings of the elbow-biceps-forearm-hand system, showing some of the muscles and bones that enable the joint to flex up and down, extend in and out, and rotate about the forearm's axis. (From Martin, R. B., Burr, D. B., and Sharkey, N. A., *Skeletal Tissue Mechanics*, Springer-Verlag, New York, 1998. With permission.)

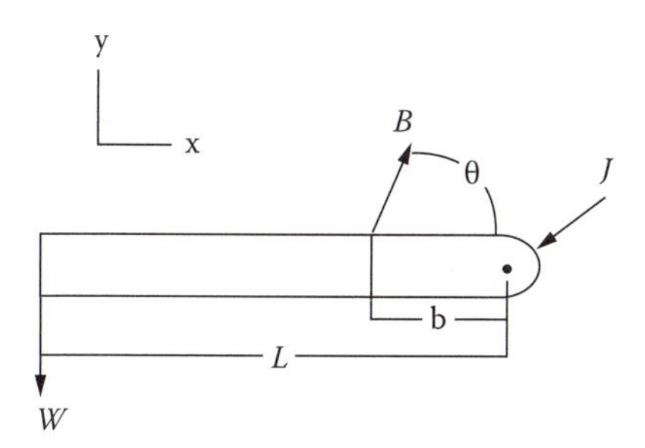

FIGURE CS3.3
An FBD of the principal forces acting on a forearm when the elbow acts as a simple (planar) hinge, raising and lowering the hand with respect to the elbow. (Adapted from Martin, R. B., Burr, D. B., and Sharkey, N. A., *Skeletal Tissue Mechanics*, Springer-Verlag, New York, 1998.)

Equation (CS3.1a), equation (CS3.1b), and equation (CS3.1c) comprise a set of three equations for three unknowns: B, J, and ϕ. They can be straightforwardly solved (Problem CS3.1) to yield

$$B = \frac{WL}{b\sin\theta}, \tag{CS3.2a}$$

$$J = \sqrt{(B\sin\theta - W)^2 + (B\cos\theta)^2}, \tag{CS3.2b}$$

$$\phi = \tan^{-1}\frac{B\sin\theta - W}{B\cos\theta}. \tag{CS3.2c}$$

The FBD in Figure CS3.3 shows that the forearm must bend like a beam and that the bending moment will be zero at the hand carrying the weight, will increase (in magnitude) linearly until it reaches its maximum value at the point where B is applied, and will then decrease to zero at the hinge. Therefore, the maximum moment carried by the ulna and radius bones is given by (see Problem CS3.2)

$$M_{max} = -W(L - b). \tag{CS3.3}$$

Now let us estimate the magnitudes of the internal forces and the moment that result from supporting the weight W. First of all, from our everyday experience, we can approximate $\sin\theta \cong 1$ because the biceps acts almost immediately adjacent to the elbow joint (or hinge). Second, in a similar estimate resulting from inspection of the geometry of the elbow-biceps-forearm-hand system, we can say that $L/b >> 1$ (see Problem CS3.8). The first consequence of these two assumptions follows from equation (CS3.2a) and is

$$B \sim W\left(\frac{L}{b}\right) >> W. \tag{CS3.4}$$

Thus, the force exerted by the biceps is an order of magnitude larger than the weight supported. The second consequence of our two assumptions follows from equation (CS3.2b) and equation (CS3.4) and states a similar result about the reaction force at the joint:

$$J \sim \sqrt{(B - W)^2 + (B)^2} \sim B >> W. \tag{CS3.5}$$

Finally, a corresponding estimate of the maximum moment in the beam resulting from the weight W at its tip follows from equation (CS3.3):

$$M_{max} \sim -WL. \tag{CS3.6}$$

This result is consistent with what we have already seen about the behavior of beams since in the limit $\sin\theta \sim 1$ the forearm is acting as a cantilever beam.

Fixing an Intertrochanteric Fracture

The *hip bone* is connected to the *thigh bone* or *femur*. The femur's *neck* and *head* comprise the familiar post-and-ball joint connecting the thigh bone to the hip bone. This ball joint allows the thigh bone to rotate and swivel (so we can sit and walk and run!). The femur's neck and head are connected to the top of the femur by the *trochanter*, an elaborate bony structure that has several parts (see Figure CS3.4). An *intertrochanteric fracture* occurs when the substantial forces transmitted from the hip to the femur cause the trochanter to break. Modeling the repair of an intertrochanteric fracture is a neat application of beam theory.

Figure CS3.5 displays a sketch of an *intertrochanteric nail plate* that has been inserted into the top of the femur. The nail plate transmits the appropriate (and substantial) forces from the hip bone, across the ball joint, to the thigh bone—when the basic trochanteric structure is no longer able to do that because it has cracked. Figure CS3.5 also shows that this substantial force of 400 N must be carried at an angle of 20° with the axis of the nail plate so

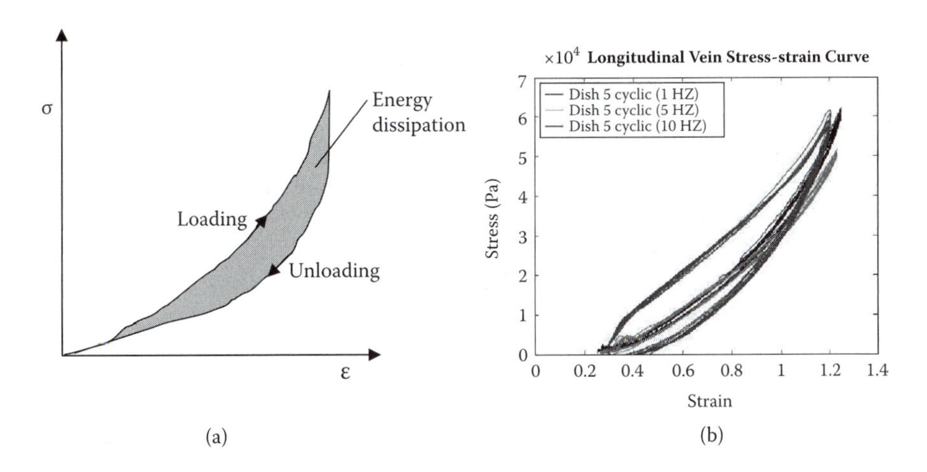

FIGURE CS3.4
The skeletal structure of the femur and its connection to the hip bone across the trochanteric structure at the top of the femur. (Adapted from *Barron's Atlas of Anatomy*, Barron's Educational Series, Hauppauge, NY, 1997.)

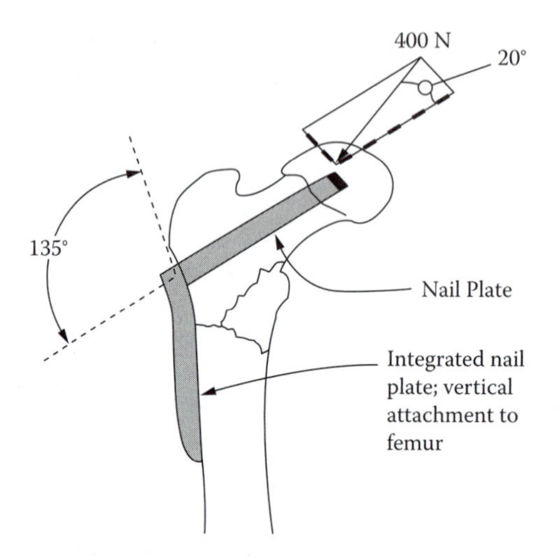

FIGURE CS3.5
The nail plate structure showing both the nail plate itself and its rigid, vertical connection to the femur. The nail plate is intended to provide the support needed after an intertrochanteric fracture. (Adapted from Enderle, J. D., Blanchard, S. M., and Bronzino, J. D., *Introduction to Biomedical Engineering*, Academic Press, San Diego, 2000.)

that the nail plate across the trochanteric structure can be modeled as a beam that also supports an axial load. We need to know the relevant stresses and strains to validate the design of this orthopedic device.

The entire nail plate structure is rigidly attached to the femur as shown in Figure CS3.5. Thus, the nail plate itself, apart from its vertical attachment to the femur, can be modeled as an axially and transversely loaded cantilever beam. The beam (or nail plate) is made of stainless steel ($E = 205$ GPa) and has the following dimensions: $b = 5$ mm, $h = 10$ mm, and $L = 60$ mm long. The FBD in Figure CS3.6 shows that the beam is subjected to an axial force of 400 (cos 20°) = 376 N and a transverse tip load of 400 (sin 20°) = 137 N. It turns out that the axial and bending behaviors can be considered as two entirely separate issues; we focus here on the beam bending (see Problem CS3.9 and Problem CS3.10).

The bending of the nail plate is modeled simply as that of a tip-loaded cantilever. Thus, with P being the load and x the distance from the tip, the shear and moment in such a beam are, respectively,

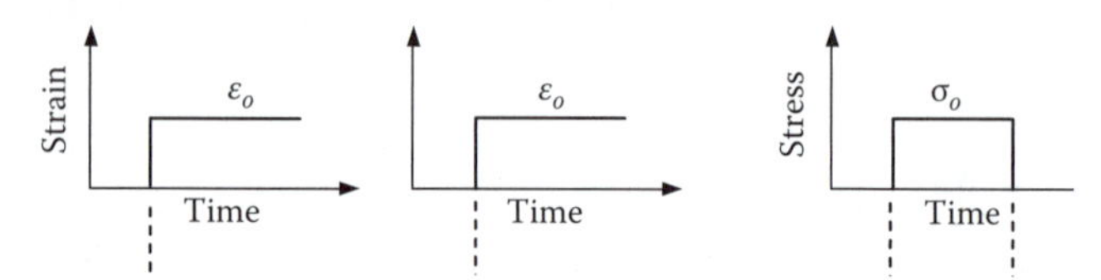

FIGURE CS3.6
FBD of the bending model of an intertrochantic nail plate structure.

$$V(x) = P. \quad \text{(CS3.7)}$$

$$M(x) = -Px. \quad \text{(CS3.8)}$$

The shear force and stress are constant over the length of the beam, and the maximum moment and bending stress will occur at the support, located at the *greater trochanter.* The maximum shear and bending stresses are, respectively,

$$\tau_{max} = 4.11 \text{ MPa} \quad \text{(CS3.9)}$$

and

$$\sigma_{max} = 197 \text{ MPa} \quad \text{(CS3.10)}$$

(see Problem CS3.11 and Problem CS3.12). Both of these stresses are much smaller than the yield stress for stainless steel,

$$\sigma_{yield} = 700 \text{ MPa},$$

so the proposed stainless steel nail plate can be considered a satisfactory design *in terms of its mechanical performance.* It is important to keep in mind that we have not considered whether, for example, the nail plate might be rejected by the body in which it was placed. There are important compatibility issues to consider when materials are selected for human implants and biomimetic devices (see Problem CS3.13).

Problems

CS3.1 Confirm that equation (CS3.2a), equation (CS3.2b), and equation (CS3.2c) are correct by solving equation (CS3.1a), equation (CS3.1b), and equation (CS3.1c).

CS3.2 Draw the moment diagram for the elbow-biceps-forearm-hand system, and determine the magnitude and location of the maximum moment.

CS3.3 How do the magnitudes of the biceps force B and joint reaction J change if the total weight w of the forearm and hand are included in the second-order analysis and are assumed to act at the midpoint of the forearm?

CS3.4 Draw the moment diagram for the elbow-biceps-forearm-hand system, and determine the magnitude and location of the maximum moment if the total weight w of the forearm and hand are included and are assumed to act at the midpoint of the forearm.

CS3.5 How large (as a fraction of the supported weight W) must the weight w of the forearm and hand be to change the analysis done under the assumption of weightlessness?

CS3.6 What sort of simple measurement or experiment could be done to determine the total weight w of the human forearm and hand?

CS3.7 How do the magnitudes of the biceps force B and joint reaction J change if the total weight w of the forearm and hand are included in the second-order analysis and are assumed to be uniformly distributed over the forearm length L?

CS3.8 Examine and measure the arms of three (or more) of your colleagues, and develop (average) estimates of the distances b and L and of the L/b ratio.

CS3.9 Determine the axial stress and strain of the axially loaded nail plate. How much does the nail plate shorten as a result of this axial response?

CS3.10 What is the maximum shear stress that is caused by the axial force on the nail plate?

CS3.11 Calculate and confirm the maximum shear stress due to the bending of the stainless steel nail plate given in equation (CS3.9).

CS3.12 Calculate and confirm the maximum bending stress of the stainless steel nail plate given in equation (CS3.10).

CS3.13 Research and identify the major materials compatibility issues that arise when devices such as the nail plate are inserted or implanted in a human being.

Notes

1. The centroid of an area is defined as its geometric center; in essence, the "center of mass" of a two-dimensional plane.
2. Key points are (1) points of application of concentrated loads and reactions; (2) points of zero shear and where the V diagram goes through zero; and (3) the endpoints of all distributed loads.
3. We do not need to consider any variation of q(x) within dx, because in the limit as dx $\rightarrow$ 0, the change in q becomes negligibly small. This is not an approximation.
4. In the immediate vicinity of the applied load, the behavior is somewhat more complex; we make use of Saint-Venant's principle to apply the assumption to the whole beam. Incidentally, this "plane sections remain plane" hypothesis for

bending was first made (with some mistakes) by influential Swiss mathematician Jacob Bernoulli (1645–1705), whose nephew Daniel was also renowned for his work in fluid mechanics.

5. That is, the only stress relative to the x, y, and z coordinates as we have defined them. The stress state on a different plane, as we well know, could look different.

6

Beam Deflections

Knowing the accurate deflection of a beam under certain loading conditions is of interest to us as designers. In some designs, for example when we design for stiffness, we are seeking to *minimize* deflection (strain). Floorboards, roof supports, and bookshelves are some examples of beams whose deflections are ideally minimized. In other cases, a functional design may *rely* on the deflections of beams; examples of this include diving boards, leaf springs, guitar strings, and thermostats. Both situations require being able to predict the deflection behavior of a beam under loading.

FIGURE 6.1
Inspiration from Bill Watterson, *Calvin and Hobbes.* Used by permission of Universal Press Syndicate.

6.1 Governing Equation

We model the deflected shape of a beam in terms of the vertical movement of the beam's neutral axis. Once again, we make use of the premise that during bending, plane sections through a beam remain plane. For simplicity, we first consider bending only about one of the principal axes of the cross section. All of this should sound familiar from the previous chapter's explanation of *pure bending*, but now we include an added generality: We permit the radius of curvature ρ of the neutral axis to vary along the span (x).

In Figure 6.2 we see a segment of a beam with a greatly exaggerated deflection w, measured from the x axis. We note that the slope of the beam's neutral axis at point A is dz/dx. This is $\tan\theta$. Since we are assuming small deformations, we say $\tan\theta \approx \theta$ in radians. The slope at point B is $-d\theta$. The change in slope between points A and B, which were originally dx apart on

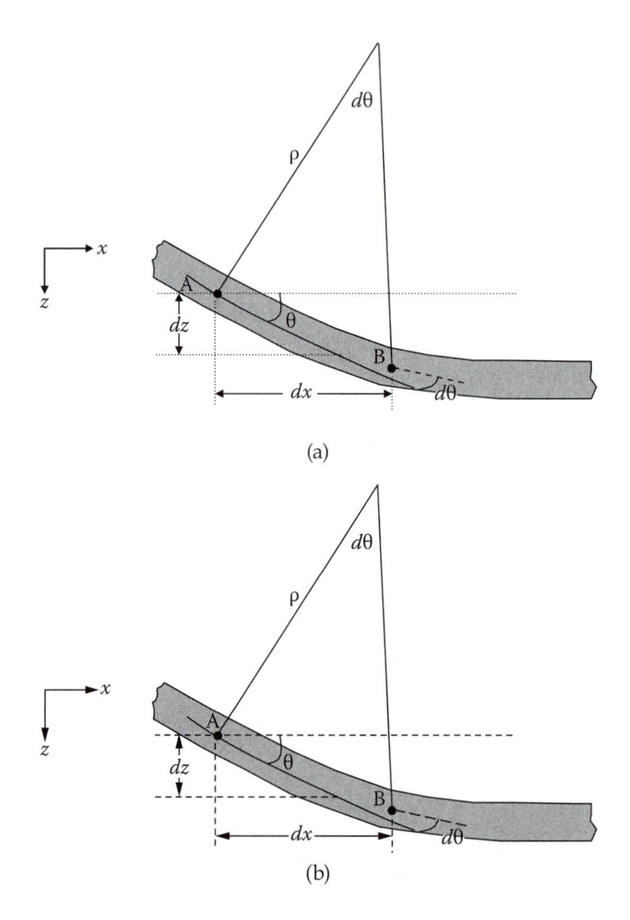

FIGURE 6.2
Beam deflected by pure bending; points A and B lie on the beam's neutral axis. A line with the slope of the neutral axis at A is extended past B to show the change in slope, $-d\theta$, between A and B.

the horizontal beam axis, is then given by $-d\theta$. The curvature of the beam, or the *rate* of change of the slope with respect to x, is

$$\frac{d}{dx}\left(\frac{dz}{dx}\right)=\frac{d}{dx}(-d\theta) \tag{6.1}$$

or

$$\frac{d^2w}{dx^2}=-\frac{d\theta}{dx}\,. \tag{6.2}$$

In Figure 6.2 we remind ourselves that the neutral axis can be thought of as a segment of a very large circle with radius ρ. The angle $d\theta$ between A (deflection w) and B ($w + dw$) is the change in θ from x to $x + dx$. In terms of ρ, this angle may be written as

$$d\theta = \frac{1}{\rho} ds, \tag{6.3}$$

where ds is the arc length given by

$$dx = ds \cos\theta = ds\,(1 - \tfrac{1}{2}\theta^2 + \ldots), \tag{6.4}$$

Again, since we are restricting ourselves to small angles θ, the higher-order terms drop out, and we have

$$dx = ds \tag{6.5}$$

and, hence,

$$\frac{d\theta}{dx} = \frac{1}{\rho}. \tag{6.6}$$

If we substitute $-\frac{d^2w}{dx^2} = \frac{d\theta}{dx}$ into this expression, we have

$$\frac{d^2w}{dx^2} = -\frac{1}{\rho}. \tag{6.7}$$

Recalling that the radius of curvature can be related to the bending moment and the beam's flexural rigidity as

$$\frac{1}{\rho} = \frac{M}{EI}, \tag{6.8}$$

we obtain a new relationship between the beam's deflection and the bending moment:

$$\frac{d^2w}{dx^2} = -\frac{M}{EI}. \tag{6.9}$$

Here, $M = M_y$ and $I = I_y$ as in the previous chapter. With this equation, we are able to calculate the deflections of beams. Our basic strategy is to determine the bending moment $M(x)$ in a beam and then to integrate this new equation twice to determine $w(x)$.

Notice that positive w is in the positive z direction: downward. This is also the direction of positive applied load $q(x)$. While somewhat surprising, this choice is informed by the fact that most beams are deflected downward by downward loads. Some textbooks choose to define positive w in the upward direction, which corresponds to our expectation that "up" is positive, but this convention results in most loads and most deflections being negative. Both sign conventions ensure agreement between "positive bending moment" and "positive curvature." Figure 6.3 illustrates this agreement.

We also observe that the governing equation for deflection can be recast in terms of the moment, shear, or load:

$$EI\frac{d^2w}{dx^2} = -M(x) \tag{6.10a}$$

$$EI\frac{d^3w}{dx^3} = -V(x) \tag{6.10b}$$

$$EI\frac{d^4w}{dx^4} = q(x) \tag{6.10c}$$

Fewer constants of integration are necessary in the lower-order equations. No matter which of these equations we choose to use, we need to use boundary conditions (BCs) to determine the constants of integration.

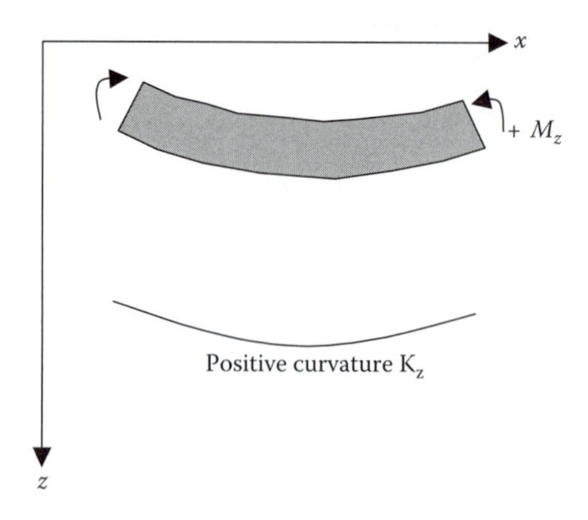

FIGURE 6.3
Positive moment, positive curvature.

6.2 Boundary Conditions

As we have seen when obtaining shear $V(x)$ and moment $M(x)$ by the integration methods of Chapter 5, Section 5.4, the conditions at the beam ends are significant. We know that the type of support at a boundary helps determine the internal forces and moments at this location, and it follows that the type of support also affects the deflection $w(x)$:

- At a *fixed* or *clamped support*, the displacement w and its slope dw/dx (negative rotation θ) must vanish. If this support is at $x = a$ as shown in Figure 6.4a, we must have

$$w(a) = 0 \qquad w'(a) = 0. \tag{6.11}$$

- At a *roller* or *pinned support* (i.e., "simple support") no deflection w or moment M can exist. So, if this support is at $x = a$ as in Figure 6.4b, we must have

$$w(a) = 0 \; M(a) = -EIw''(a) = 0. \tag{6.12}$$

- At a *free end*, the beam feels neither moment nor shear. If $x = a$ is free (Figure 6.4c), we have

$$M(a) = -EIw''(a) = 0 \qquad V(a) = -EIw'''(a) = 0. \tag{6.13}$$

- At a *guided support* like that sketched in Figure 6.4d, free vertical movement is permitted, but rotation of the end is prevented. This type of support cannot resist shear, and

$$w'(a) = 0 \qquad V(a) = -EIw'''(a) = 0. \tag{6.14}$$

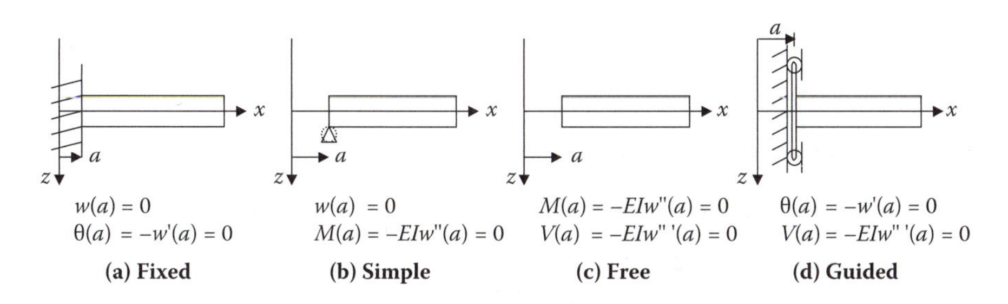

FIGURE 6.4
Homogeneous boundary conditions for beams with constant EI. In (a) both conditions are kinematic; in (c) both are static; and in (b) and (d), conditions are mixed.

The boundary conditions pertaining to force quantities (V or M) are known as *static* boundary conditions. Those that describe geometrical or deformational behavior of an end (w or θ) are known as *kinematic* boundary conditions.

The boundary conditions just listed are all *homogeneous* boundary conditions (i.e., something must equal zero). It is also possible to encounter *nonhomogeneous* boundary conditions, where a given nonzero shear, moment, rotation, or displacement is prescribed. In this case, the prescribed quantity simply replaces zero in the previous conditions.

In some calculations, we uncover discontinuities in the mathematical functions for either load or stiffness along a given beam's length. These discontinuities occur at concentrated forces or moments and at abrupt changes in cross-sectional areas. When this happens, we supplement our boundary conditions with the physical requirement of *continuity of the neutral axis*.[1] Anywhere a discontinuity occurs, we must ensure that deflection and the tangent to the neutral axis remain the same when this discontinuity's point is approached from either direction. Figure 6.5 illustrates two unacceptable geometries that would have to be corrected by imposing this requirement.

This requirement is expressed as a continuity boundary condition: At a place d where two solutions meet, we must have continuity of deflection w and its tangent or slope w':

$$w_1(d) = w_2(d) \text{ and } w_1'(d) = w_2'(d) \qquad (6.15a, b)$$

We now have sufficient information to begin solving our differential equation for deflection.

FIGURE 6.5
Discontinuous configurations of the neutral axis' deflection that would have to be corrected.

6.3 Solution of Deflection Equation by Integration

If we start with the equation $EIw^{iv} = q(x)$, we must integrate this expression four times to obtain the solution for deflection $w(x)$:

$$EI\frac{d^4w}{dx^4} = EI\frac{d}{dx}\left(\frac{d^3w}{dx^3}\right) = q(x)$$

$$EI\frac{d^3w}{dx^3} = \int q(x)dx + C_1$$

$$EI\frac{d^2w}{dx^2} = \int\int q(x)dx + C_1x + C_2$$

$$EI\frac{dw}{dx} = \int\int\int q(x)dx + \frac{1}{2}C_1x^2 + C_2x + C_3$$

$$EIw(x) = \int\int\int\int q(x)dx + \frac{1}{6}C_1x^3 + \frac{1}{2}C_2x^2 + C_3x + C_4$$

The constants C_i have physical meanings. The second of these five equations is equivalent to good old

$$V = -\int qdx + C_1,$$

since we know that $-EIw''' = V$; the third equation should also look familiar. We have worked with these equations and seen that the constants C_1 and C_2 come from the end conditions on V and M; hence, these two constants come from *static boundary conditions.* When we continue on to the fourth and fifth equations, we obtain two more constants of integrations, C_3 and C_4, which describe the slope and deflection of the neutral axis. These constants come from the *kinematic boundary conditions.*

If we begin our integration at a point further down this chain, starting with $-EIw'' = M(x)$, we obtain after two integrations:

$$-EIw = \int\int M(x)dx + C_3x + C_4. \tag{6.16}$$

We once again find C_3 and C_4 from the boundary conditions.

Any one of these five equations may be used as a starting point for finding beam deflection. The choice depends entirely on the available data.

As long as the beam behaves elastically, it is possible to *superpose* solutions to determine the deflection *w(x)* in a complex loading situation. Tables such as Table 6.1 offer deflections *w(x)* for many isolated loads. Using such a table

TABLE 6.1

Deflections and Slopes of Neutral Axes for Variously Loaded Beams

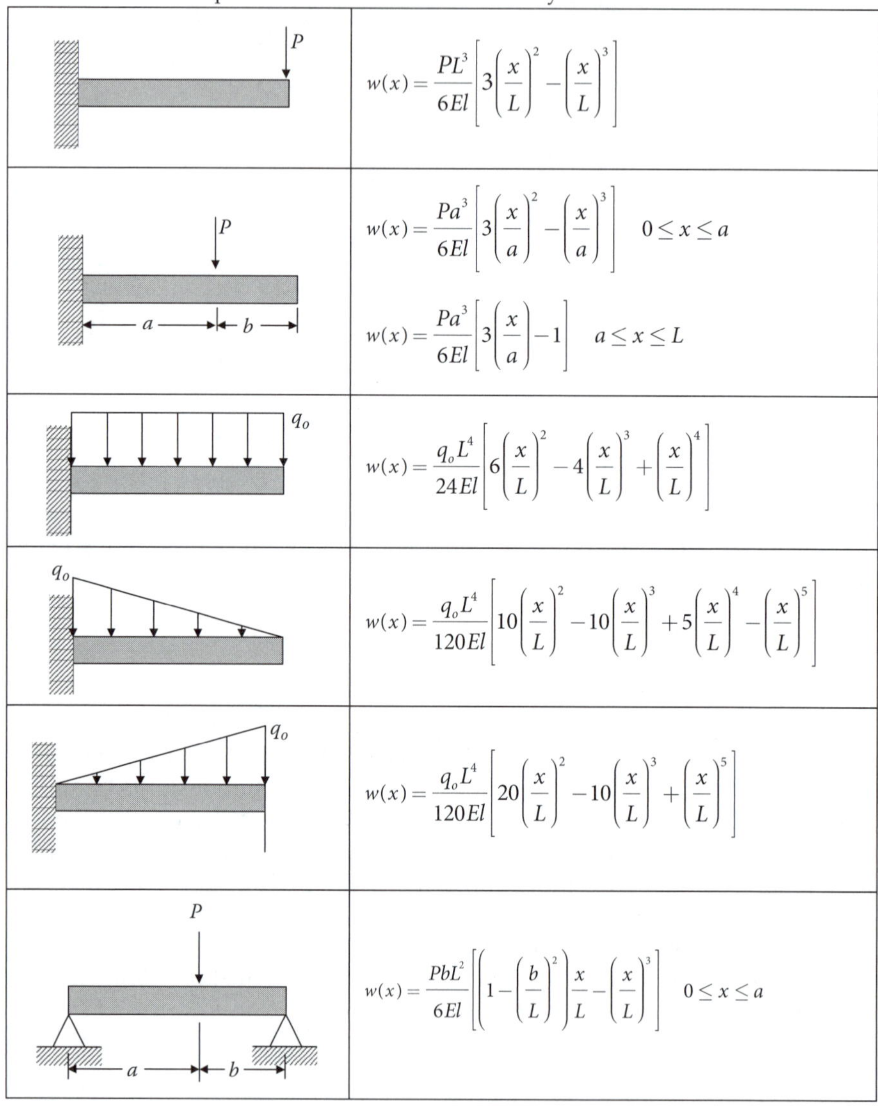

Loading	Deflection
P	$w(x) = \dfrac{PL^3}{6EI}\left[3\left(\dfrac{x}{L}\right)^2 - \left(\dfrac{x}{L}\right)^3\right]$
P, a, b	$w(x) = \dfrac{Pa^3}{6EI}\left[3\left(\dfrac{x}{a}\right)^2 - \left(\dfrac{x}{a}\right)^3\right] \quad 0 \le x \le a$ $w(x) = \dfrac{Pa^3}{6EI}\left[3\left(\dfrac{x}{a}\right) - 1\right] \quad a \le x \le L$
q_o	$w(x) = \dfrac{q_o L^4}{24EI}\left[6\left(\dfrac{x}{L}\right)^2 - 4\left(\dfrac{x}{L}\right)^3 + \left(\dfrac{x}{L}\right)^4\right]$
q_o	$w(x) = \dfrac{q_o L^4}{120EI}\left[10\left(\dfrac{x}{L}\right)^2 - 10\left(\dfrac{x}{L}\right)^3 + 5\left(\dfrac{x}{L}\right)^4 - \left(\dfrac{x}{L}\right)^5\right]$
q_o	$w(x) = \dfrac{q_o L^4}{120EI}\left[20\left(\dfrac{x}{L}\right)^2 - 10\left(\dfrac{x}{L}\right)^3 + \left(\dfrac{x}{L}\right)^5\right]$
P, a, b	$w(x) = \dfrac{PbL^2}{6EI}\left[\left(1 - \left(\dfrac{b}{L}\right)^2\right)\dfrac{x}{L} - \left(\dfrac{x}{L}\right)^3\right] \quad 0 \le x \le a$

and our previous results, we can simply add the solutions for the various loads, as in Figure 6.6. This is also an excellent way to resolve the problem of statically indeterminate beams.

In Section 6.4 and Section 6.5, we discuss two techniques that can be used to analyze more complex situations. The first method, using singularity functions, is both mathematically rigorous and very convenient. It is especially useful in resolving discontinuities along the beam's axis.

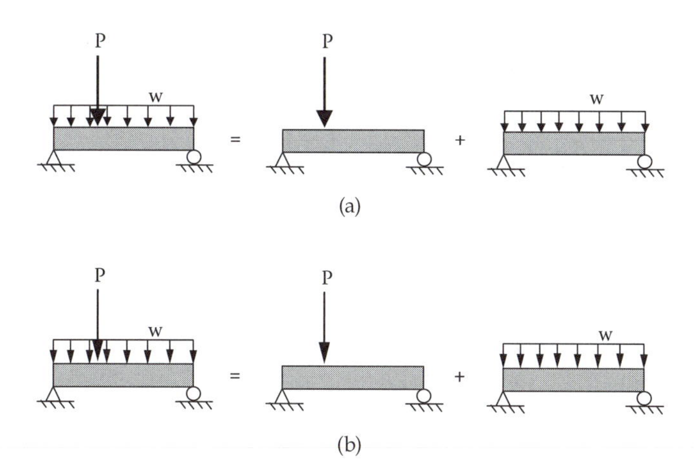

FIGURE 6.6
Finding deflection by superposition.

6.4 Singularity Functions

Determining the deflection and slope of a beam using the integration method is straightforward when we can represent the bending moment within the beam by a single analytical function *M(x)*. *Singularity functions* make it possible to characterize the shear *V* and bending moment *M* by single mathematical expressions even when the loading is discontinuous. This method is most effective for beams with a constant product *EI*. Singularity functions are particularly valuable in computational techniques.

In general, singularity functions are defined as

$$\langle x-a\rangle^n = \begin{cases} (x-a)^n \text{ when } x \geq a \\ 0 \text{ when } x < a \end{cases} \tag{6.17}$$

The singularity functions corresponding to n = 0, 1, and 2 are graphed in Figure 6.7.

Figure 6.8 shows how singularity functions can be used to express the loading on four representative beams.

Using singularity functions to describe beam deflections was first suggested in 1862 by German mathematician A. Clebsch (1833–1872), though the notation of equation (6.17) was introduced somewhat later by W. H. Macaulay (1853–1936), a British mathematician and engineer. The angular brackets < > used to write singularity functions are often called "Macaulay's brackets" (Macaulay 1919).

Like many problem-solving approaches, the use of singularity functions is best learned by applying the technique. We strongly encourage solving the worked examples in this chapter using the method of singularity functions to confirm that the solutions obtained are equivalent to those arrived at by direct integration.

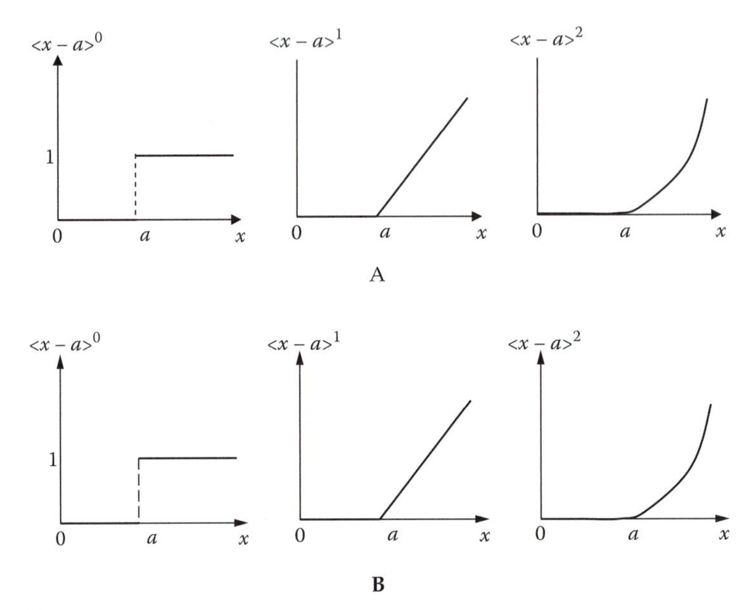

FIGURE 6.7
Singularity functions with $n = 0$, 1, and 2.

6.5 Moment Area Method

In practice, we often encounter beams whose cross-sectional areas vary and whose loading is complex. This is the typical situation with machine shafts, which have variations in shaft diameter to accommodate, for example, rotors, bearings, and collars, and is also common in aircraft and bridge construction. A graphical interpretation of the governing equations of Section 6.1 yields an alternative procedure for calculating the slope and deflection of beams that is particularly useful in the case of discontinuities in both loading and cross-sectional area—in such cases, the product EI is no longer constant.

This alternative procedure, known as the *moment area method*, was independently developed by German engineer Otto Mohr in 1868[2] and by University of Michigan professor Charles Greene in 1873. It is based on the same assumptions, and has the same limitations, as the method of direct integration.

> In addition, the deflection found by the moment area method is the deflection due only to beam flexure (bending moment); deflection due to shear is neglected.

The method makes use of two theorems based on the geometry of the neutral axis (elastic curve) and on the M/EI diagram. Constants of integration related to boundary conditions do not appear in these expressions because

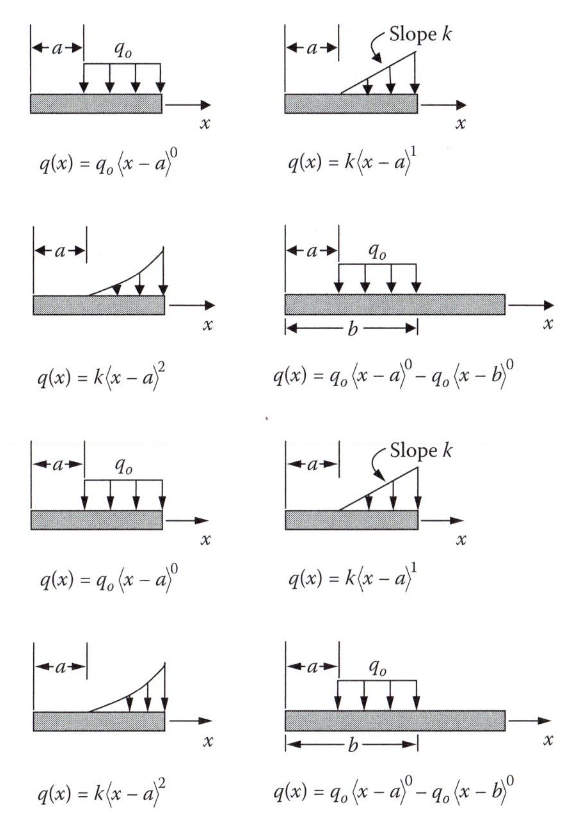

FIGURE 6.8
Basic beam loadings expressed in terms of singularity functions.

the theorems are based on definite integrals and not the indefinite "from *0* to *x*" we have used so far. We start with equation (6.9),

$$\frac{d^2w}{dx^2} = -\frac{d\theta}{dx} = -\frac{M}{EI},$$

which we can usefully rewrite as

$$d\grave{e} = \frac{M}{EI} dx \; . \tag{6.18}$$

So if we consider two arbitrary points C and D on the beam in Figure 6.9a, we can write

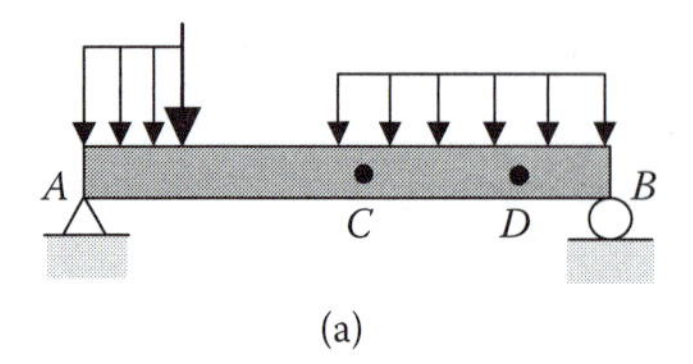

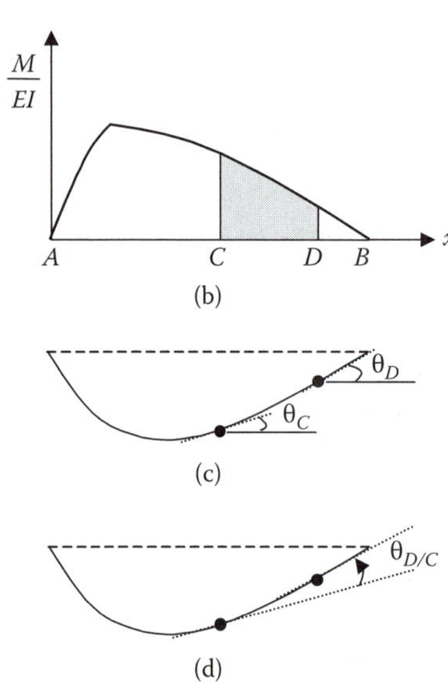

FIGURE 6.9
First moment area theorem.

$$\int_{\theta_C}^{\theta_D} d\theta = \theta_D - \theta_C = \int_{x_C}^{x_D} \frac{M}{EI} dx, \qquad (6.19)\text{v}$$

where θ_C and θ_D are the slopes of the beam at points C and D, as shown in Figure 6.9c. Their difference can be defined as $\theta_D - \theta_C = \theta_{D/C}$, the angle between the tangents to the neutral axis at C and D (Figure 6.9d). The integral on the right-hand side of this expression represents the area under the M/EI diagram between C and D, as in Figure 6.9b. This restatement equation (6.18) is the *first moment area theorem:*

$$\theta_{D/C} = \int_{x_C}^{x_D} \frac{M}{EI} dx, \qquad (6.20)$$

where $\theta_{D/C}$ is the area under the M/EI diagram between C and D. Note that the angle $\theta_{D/C}$ and the area under M/EI have the same sign. A positive area (above the x axis) corresponds to a counterclockwise rotation of the tangent to the neutral axis on the way from C to D. Similarly, a negative area corresponds to a clockwise rotation. We use this theorem to find slopes at various points: $\theta_D = \theta_C + \theta_{D/C}$.

If we now zoom in on two points between A and C, a distance dx from each other, we will be able to derive the second tenet of the moment area method. We are interested in finding a way to express the vertical distance dt in Figure 6.10. We see from the figure that $dt = x_1 d\theta$, assuming the angles involved are small. If we summed all the dt's from A to C, we

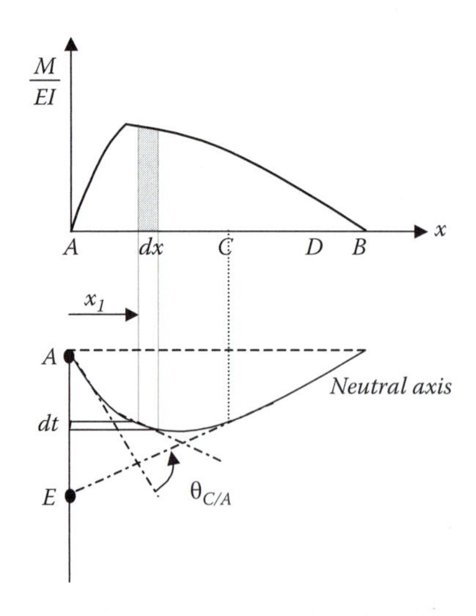

FIGURE 6.10
Second moment area theorem.

could obtain the vertical distance *AE*. This distance is the displacement of a point *A* from a tangent to the elastic curve drawn at *C*, called the tangential deviation of *A* from *C*'s tangent and denoted $t_{A/C}$.

We can express tangential deviation mathematically as

$$t_{A/C} = \int_A^C x d\theta . \tag{6.21}$$

Remembering the relationship $d\theta = (M/EI)dx$, we write the *second moment area theorem:*

$$t_{A/C} = \int_A^C \frac{M}{EI} x dx, \tag{6.22}$$

and we have a handy way to calculate this tangential deviation from the (M/EI) diagram. Although *tangential deviation* may sound like a strange geometrical quantity with questionable utility, it is actually valuable to have a way to quantify the vertical distances between points on the neutral axis, because if we know these distances along a beam's length, we can express the deflection w at any point without doing the integration to get $w(x)$.

The application of this theorem relies on our recognition that the right-hand side integral looks like the first moment of area of a shape. We remember that when we have an integral with the form $\int y dA$ it can be replaced by the quantity $y_c A$, where y_c is the distance from the reference axis to the area's centroid. So if we define $\bar{x}_1$ as the distance from the centroid of the area under the (M/EI) curve to the vertical axis through our left-hand reference point, *A*, and $\bar{x}_2$ as the distance from the centroid of the same area to the vertical axis through our right-hand reference point *C*, as shown in Figure 6.11, we have the two relations:

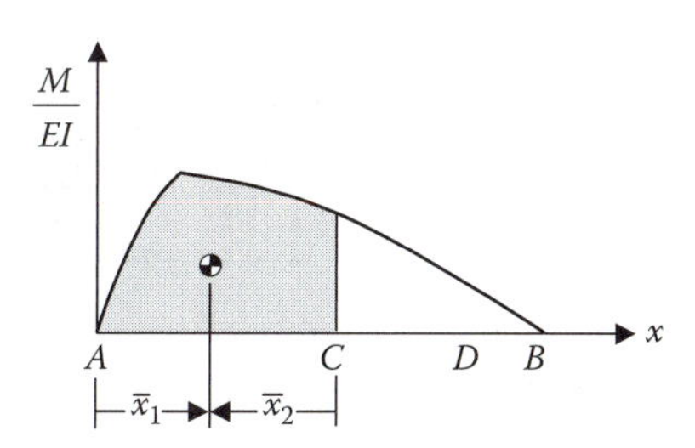

FIGURE 6.11
Useful definitions.

$$t_{A/C} = (\text{area under } (M/EI) \text{ between } A \text{ and } C) \cdot \bar{x}_1 . \tag{6.23}$$

$$t_{C/A} = (\text{area under } (M/EI) \text{ between } A \text{ and } C) \cdot \bar{x}_2 . \tag{6.24}$$

To implement the moment area method it is convenient to have the areas and centroids of several possible areas at hand, as in Table 6.2.

TABLE 6.2

Parts of *M/EI* Curve

Shape		Area	c
Rectangle		bh	$b/2$
Triangle		$bh/2$	$b/3$
Parabolic spandrel	$y = kx^2$	$bh/3$	$b/4$
General spandrel	$y = kx^n$	$bh/n{+}1$	$b/n{+}2$

6.6 Beams with Elastic Supports

What happens in the case of a support that is not rigid? How does a flexible or elastic support affect a beam's deflection? For example, consider a uniformly loaded beam for which the left end is fixed but the right end is supported on a spring of stiffness k_s, as in Figure 6.12a.

This problem is indeterminate because the magnitude of the reaction force applied through the spring is unknown, as is the beam deflection at that point. This problem can be decomposed so that the respective moments and tip deflections may be found. However, our consistency or compatibility condition for this application of the force (flexibility) method requires that we recognize the deflection at $x = 0$ due to the spring. Thus, the present compatibility condition requires that

$$\left(\delta_{tip/qo} = \frac{q_o L^4}{8EI}\right) - \left(\delta_{tip/R} = \frac{RL^3}{3EI}\right) = \frac{R}{k_s}, \quad (6.25)$$

from which it follows that an equation for the redundant R emerges as

$$R\left(\frac{1}{k_s} + \frac{L^3}{3EI}\right) = \frac{q_o L^4}{8EI} \quad (6.26)$$

Note that this equation for the redundant (spring) force requires that we add the flexibility coefficients for both the spring and the tip-loaded cantilever. It is also interesting to observe that if we define the following two stiffness coefficients,

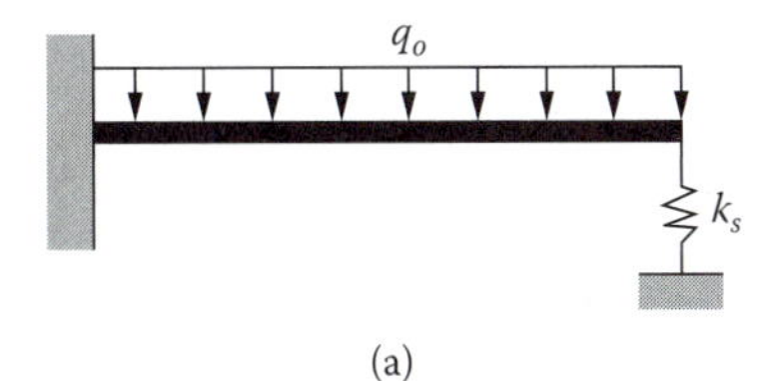

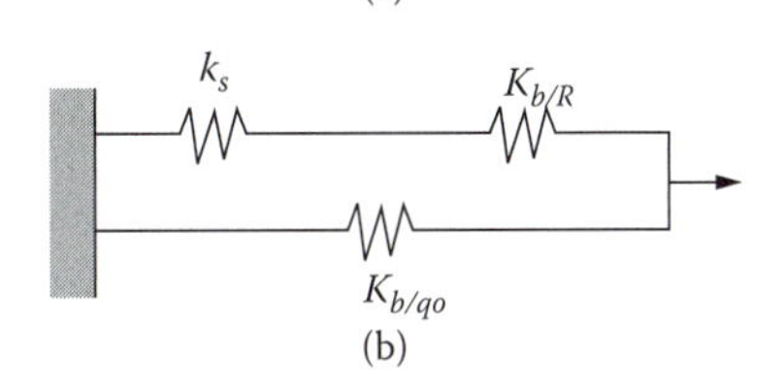

FIGURE 6.12
An indeterminate beam with elastic support (a) and its equivalent mechanical circuit (b).

$$k_{b/qo} = \frac{q_o L}{\delta_{tip/qo}} = \frac{8EI}{L^3}, \quad (6.27)$$

$$k_{b/qo} = \frac{R}{\delta_{tip/R}} = \frac{3EI}{L^3}, \quad (6.28)$$

then the equation for the redundant can be cast as

$$R\left(\frac{1}{k_s} + \frac{1}{k_{b/R}}\right) = q_o L\left(\frac{1}{k_{b/qo}}\right). \quad (6.29)$$

This result is just what we would expect from an *equivalent mechanical circuit* for this problem (shown in Figure 6.12b). The load carried by the discrete spring at the tip of the cantilever can be found to be

$$R = q_o L\left(\frac{k_{b/R}}{k_{b/qo}}\right)\left(\frac{k}{k_s + k_{b/R}}\right). \quad (6.30)$$

Clearly, if there is no discrete spring at the tip, there will be no reaction at the tip. Further, if the discrete spring is allowed to become infinitely stiff, we can take the appropriate limit in equation (6.30) and can use the prior definitions to show that there is a reaction whose magnitude is

$$\lim_{k_s \to \infty} R = q_0 L \left(\frac{k_{b/R}}{k_{b/q_0}} \right) = \frac{3q_0 L}{8}. \tag{6.31}$$

Elastic supports arise fairly often in practice, so it is useful to have the capacity to model their behavior. Two simple examples are pictured in Figure 6.13, and their corresponding stiffness coefficients for a supporting cantilever (Figure 6.13a) and for cable support (Figure 6.13b) are, respectively,

$$k_{\text{cantilever}} = \frac{3E_1 I_1}{L_1^{\ 3}}, \tag{6.32}$$

$$k_{\text{cable}} = \frac{A_2 E_2}{L_2}. \tag{6.33}$$

In practice, of course, elastic or yielding supports may be more complicated, but they can often be modeled as simple extensional or rotational springs.

6.7 Strain Energy for Bent Beams

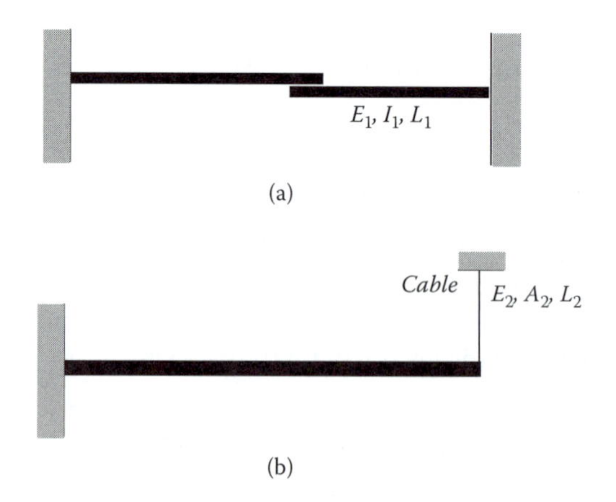

FIGURE 6.13
Examples of elastic supports.

As we remember from our discussion in Chapter 2, Section 2.11, strain energy represents the energy absorbed by a material during a loading process. This is sometimes referred to as *internal work*, or *potential energy*. Strain energy is a useful concept for determining the response of structures to static (and dynamic!) loads. We begin our discussion of strain energy in beam bending by restating results for a simply supported, uniformly loaded beam and extending them to reinforce the validity of our basic assumptions. First of all, the deflected shape of this simple beam may be found to be

$$w(x) = -\frac{q_o L^4}{24EI}\left[\left(\frac{x}{L}\right)^4 - 2\left(\frac{x}{L}\right)^3 + \left(\frac{x}{L}\right)\right]. \tag{6.34}$$

We can now go backward and use this result to calculate the moment and shear force as

$$M(x) = -EIw''(x) = \frac{q_o L^2}{2}\left[\left(\frac{x}{L}\right)^2 - \left(\frac{x}{L}\right)\right], \tag{6.35}$$

$$V(x) = -EIw'''(x) = \frac{q_o L}{2}\left[2\left(\frac{x}{L}\right) - 1\right], \tag{6.36}$$

and the bending (normal) and shear stresses for this problem as

$$\sigma_{xx} = \frac{M(x)z}{I} = \frac{q_o L^2 h}{4I}\left(\frac{z}{h/2}\right)\left[\left(\frac{x}{L}\right)^2 - \left(\frac{x}{L}\right)\right], \tag{6.37}$$

$$\hat{\sigma}_{xz} = \frac{V(x)}{2I}\left(\frac{h^2}{4} - z^2\right) = \frac{q_o L h^2}{16I}\left[1 - \left(\frac{z}{h/2}\right)^2\right]\left[2\left(\frac{x}{L}\right) - 1\right]. \tag{6.38}$$

Note first of all that we can compare the maximum values of the bending and shear stresses. Accordingly,

$$\left(\sigma_{xx}\right)_{\max} = \sigma_{xx}\left(\tfrac{L}{2}, \tfrac{h}{2}\right) = \frac{q_o L^2 h}{16I}, \tag{6.39}$$

$$\left(\tau_{xz}\right)_{\max} = \left|\tau_{xz}(0,0)\right| = \frac{q_o L h^2}{16I}, \tag{6.40}$$

from which it follows that

$$\frac{\left(\tau_{xz}\right)_{\max}}{\left(\sigma_{xx}\right)_{\max}} = \frac{h}{L}, \tag{6.41}$$

so that for slender beams, the shear stress is much smaller than the bending stress. We can then infer that the shear strain is also much smaller than the bending strain, so the fact that our kinematics assumptions produce zero shear strain should not bother us too much.

However, we can go one step further to confirm this result. In the same way that we can calculate the energy stored in a simple spring, we can calculate the energy stored in an elastic beam due to different kinds of deformation. More specifically, we can calculate the energy stored due to bending deformation and the energy stored due to shear deformation. In general, as we showed in Chapter 2, Section 2.11, the strain energy stored in an elastic solid can be calculated as

$$U = \tfrac{1}{2} \int_{\text{volume } \mathcal{V}} \sigma_{ij}\varepsilon_{ij} d\mathcal{V}. \tag{6.42}$$

The dimensions of this expression are clearly those of work, as they should be, and the details of this calculation result from a straightforward analysis of the work done on a volumetric element $dxdydz$ by a set of stresses σ_{ij} acting through the corresponding gradients of deformation or strain ε_{ij}. For our problem there are only two nonzero terms to examine. In bending,

$$U_{bending} = \frac{1}{2}\int_{\mathcal{V}} \sigma_{xx}\varepsilon_{xx} bdxdz = \frac{1}{2}\int_{\mathcal{V}} \frac{\sigma_{xx}}{E} bdxdz, \tag{6.43}$$

while for shear,

$$U_{shear} = \frac{1}{2}\int_{\mathcal{V}} 2\tau_{xz}\varepsilon_{xz} bdxdz = \frac{1}{2}\int_{\mathcal{V}} \frac{\tau_{xz}^2}{G} bdxdz. \tag{6.44}$$

If we substitute the stresses into the strain energy expressions, we can then do some algebra to find that

$$U_{bending} = \frac{q_o^2 L^5}{240EI}, \tag{6.45}$$

$$U_{shear} = \frac{q_o^2 L^5}{240EI}\left[2(1+\nu)\left(\frac{h}{L}\right)^2\right], \tag{6.46}$$

so that the ratio of these two strain energy terms is

$$\frac{U_{shear}}{U_{bending}} = 2(1+\nu)\left(\frac{h}{L}\right)^2. \tag{6.47}$$

The ratio of the energy stored in shear to that stored in bending is proportional to the square of the thickness-to-length ratio. Thus, we can neglect deformation due to shear when we are modeling the bending of long, slender beams.

6.8 Flexibility Revisited and Maxwell-Betti Reciprocal Theorem

We propose now to find the deflections at two points on a cantilever that is carrying two additional point loads (Figure 6.14). This is a second-degree indeterminate problem.

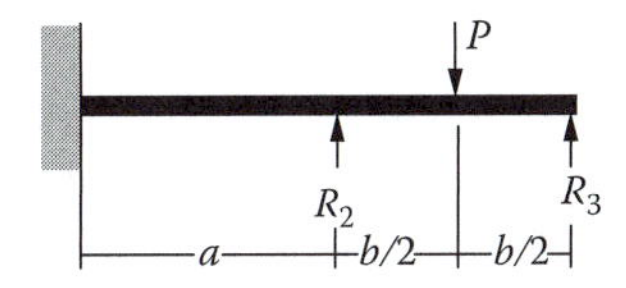

FIGURE 6.14
Cantilever with two supports: statically indeterminate.

We approach it using the method of *flexibility coefficients*[3]: We decompose the indeterminate problem into three determinate ones. We use a generic notation for the displacement δ at a point on the beam in which a subscript denotes the location and a superscript identifies the position of the load. Thus, the deflection at $x = x_1$ due to a load P applied at x_2 is denoted by $\delta_1^{P_2}$. Flexibility coefficients relate deflection to applied load: The deflection at location i due to a force P_j applied at j can be expressed in terms of a flexibility coefficient f_{ij} as

$$\delta_i^j = f_{ij} P_j \,.$$

For the loading in Figure 6.14, using decomposition, we find that

$$\delta_2^P = \left(\frac{Pa^2}{12EI}\right)(3L+a) \tag{6.48a}$$

$$\delta_3^P = \left(\frac{P}{48EI}\right)(5L^3 + 9L^2a + 3La^2 - a^3) \tag{6.48b}$$

$$\delta_2^{R_3} = \left(\frac{a^3}{3EI}\right)R_2 \tag{6.49a}$$

$$\delta_3^{R_3} = \left(\frac{a^2(3L-a)}{6EI}\right)R_2 \tag{6.49b}$$

$$\delta_2^{R_3} = \left(\frac{a^2(3L-a)}{6EI}\right)R_3 \tag{6.50a}$$

$$\delta_3^{R_3} = \left(\frac{L^3}{3EI}\right)R_3 \tag{6.50b}$$

The principle of consistent deformation (or *compatibility*) requires that when we reassemble our component problems into the original beam (i.e., apply superposition) the structure must hold together without violating any geometric or other constraints. Since we are applying compatibility at the two (redundant) supports, this means that

$$\delta_2 = \delta_2^P + \delta_2^{R_2} + \delta_2^{R_3} = 0 \tag{6.51a}$$

$$\delta_3 = \delta_3^P + \delta_3^{R_2} + \delta_3^{R_3} = 0. \tag{6.51b}$$

From this pair of equations, we can now determine the redundant reactions. In fact, let's write the equations in matrix form; in so doing we introduce the idea of a matrix of flexibility coefficients and the fact that such flexibility coefficients are symmetric for elastic structures. First up: The *flexibility coefficients*, denoted as f_{ij}, are expressed in the component results above, that is,

$$\begin{bmatrix} f_{22} & f_{23} \\ f_{32} & f_{33} \end{bmatrix} = \begin{bmatrix} \left(\dfrac{a^3}{3EI}\right) & \left(\dfrac{a^2(3L-a)}{6EI}\right) \\ \left(\dfrac{a^2(3L-a)}{6EI}\right) & \left(\dfrac{L^3}{3EI}\right) \end{bmatrix}. \tag{6.52}$$

This expression clearly shows the property of symmetry, which we could also have noted in the prior component results, namely, $\delta_2^{R_3} = \delta_3^{R_2}$. We note now that $f_{ij} = f_{ji}$.

The matrix form of our deflections' dependence on the applied loads can now be written as

$$\begin{aligned} \begin{Bmatrix} \delta_2 \\ \delta_3 \end{Bmatrix} &= \begin{Bmatrix} \delta_2^P \\ \delta_3^P \end{Bmatrix} + \begin{Bmatrix} \delta_2^{R_2} \\ \delta_3^{R_2} \end{Bmatrix} + \begin{Bmatrix} \delta_2^{R_3} \\ \delta_3^{R_3} \end{Bmatrix} \\ &= \begin{Bmatrix} \delta_2^P \\ \delta_3^P \end{Bmatrix} + \begin{bmatrix} f_{22} & f_{23} \\ f_{32} & f_{33} \end{bmatrix} \begin{Bmatrix} R_2 \\ R_3 \end{Bmatrix} = \begin{Bmatrix} 0 \\ 0 \end{Bmatrix}, \end{aligned} \tag{6.53}$$

or

$$\begin{Bmatrix} R_2 \\ R_3 \end{Bmatrix} = -\begin{bmatrix} f_{22} & f_{23} \\ f_{32} & f_{33} \end{bmatrix}^{-1} \begin{Bmatrix} \delta_2^P \\ \delta_3^P \end{Bmatrix}. \tag{6.54}$$

The details needed to find the redundants and complete the solution are evident, so we'll let that go for now. Suffice it to say that we have (a) extended the force (flexibility) method to include an arbitrary number of redundants, (b) have formulated a structural problem in matrix notation, and (c) have found that the (structural) flexibility coefficients form a symmetric matrix. This representation is very powerful for numerical work, and it is utilized in the finite element method (FEM) for structural computation.

All right, we know that the flexibility coefficients are symmetric. We can go just a little further and consider the flexibility coefficients as reflecting the deflection per unit load on the structure. If we take the applied loads as *unit loads*, the flexibility coefficients become the *influence coefficients*:

$$\begin{bmatrix} c_{22} & c_{23} \\ c_{32} & c_{33} \end{bmatrix} = \begin{bmatrix} \left(\dfrac{a^3}{3EI}\right) & \left(\dfrac{a^2(3L-a)}{6EI}\right) \\ \left(\dfrac{a^2(3L-a)}{6EI}\right) & \left(\dfrac{L^3}{3EI}\right) \end{bmatrix} \tag{6.55}$$

The symmetry of the influence coefficients can be demonstrated in general terms and is known as the Maxwell-Betti Reciprocal Theorem, which is stated as

> For linear elastic solids, the work done by a system of forces A acting through displacements caused by a second system of forces B equals the work done by the second system of forces B acting through displacements caused by the first system of forces A.

In terms of the energy formulation, the Maxwell-Betti Reciprocal Theorem can be cast in the form

$$U = \tfrac{1}{2}\int_{V} \sigma_{ij}^{A}\varepsilon_{ij}^{B}dV = \tfrac{1}{2}\int_{V} \sigma_{ij}^{B}\varepsilon_{ij}^{A}dV, \tag{6.56}$$

while in terms of the total work done on the beam (which must therefore equal the total strain energy stored in the bent beam),

$$U = \frac{1}{2}\left(c_{22}R_2^2 + 2c_{23}R_2R_3 + c_{33}R_3^2\right). \tag{6.57}$$

These relationships were obtained by manipulating the corresponding energy and work terms. Their symmetric form is appealing and is a useful reminder of how powerful these and other energy results can be for the analysis of structures. To close on this note, we point out that from equation (6.57), we can calculate (once again) the deflections of the beam at points under the loads as

$$\frac{\partial U}{\partial R_2} = c_{22}R_2 + c_{23}R_3 = \left|w(a)\right| \tag{6.58a}$$

$$\frac{\partial U}{\partial R_3} = c_{32}R_2 + c_{33}R_3 = \left|w(L)\right| \tag{6.58b}$$

This very famous result from 1879 is called Castigliano's Second Theorem.[4] This theorem is worthy of restatement:

> The deflection of a structure at any point where a load is applied can be obtained from the partial derivative of the strain energy function with respect to that load.

This has an elegant mathematical form relating the load at point i, P_i, to the deflection of the structure at that position, δ_i, through the strain energy U,

$$\frac{\partial U}{\partial P_i} = \delta_i, \tag{6.59}$$

and is quite useful to us as we endeavor to determine the deflection of beams and other structures.

6.9 Examples

Example 6.1

A bending moment M_1 is applied at the free end of a cantilever of length L and constant flexural rigidity EI (Figure 6.15). Find an expression for $w(x)$.

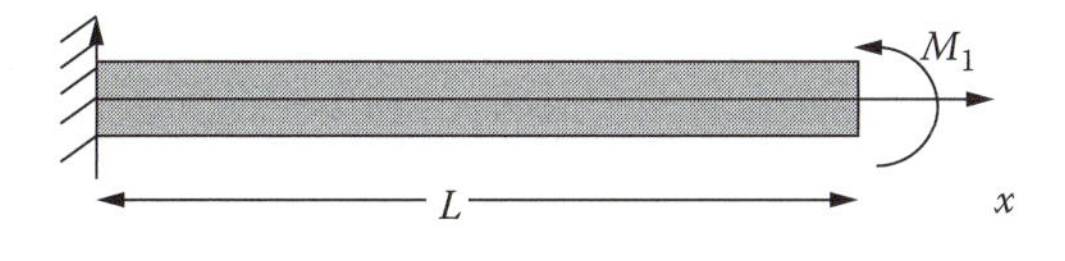

FIGURE 6.15

Given: Load applied to beam.

Find: Deflection, or "equation of elastic curve."

Assume: Hooke's law applies; beam has constant and uniform properties E, I.

Solution

We start with a free-body diagram (FBD) and the external reaction forces or moments. In this case, this procedure is immensely straightforward: The fixed support exerts a reaction moment equal and opposite to M_1 on the beam. If we made an imaginary cut at any x and used the method of sections to find the local internal bending moment, we would similarly find that at each x, the internal bending moment was M_1. $M(x) = M_1 =$ constant, as shown in the bending moment diagram in Figure 6.16.

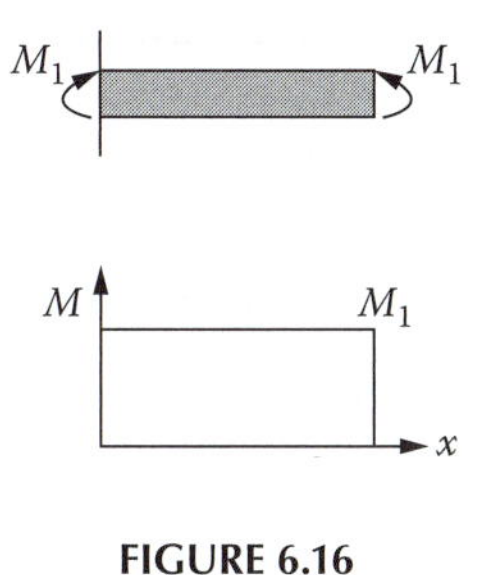

FIGURE 6.16

At the fixed end ($x = 0$), we know that deflection and slope are both zero; at the free end ($x = L$), we know the moment is M_1 and the shear is zero. We can thus begin integrating the second-order equation for deflection $w(x)$:

$$EI\frac{d^2w}{dx^2} = -M(x) = -M_1$$

$$\int \rightarrow \qquad EI\frac{dw}{dx} = -M_1x + C_3 .$$

Apply BC:

$$\theta(0) = -\frac{dw}{dx}(0) = 0 \qquad \rightarrow C_3 = 0 \text{ (fixed end)}$$

$$EI\frac{dw}{dx} = -M_1x$$

$$\int \rightarrow \qquad EI\ w(x) = -\frac{1}{2}M_1x^2 + C_4$$

Apply BC:

$$w(0) = 0 \qquad \rightarrow C_4 = 0 \text{ (fixed end),}$$

so

$$w(x) = -\frac{M_1x^2}{2EI} .$$

This deflection is negative, which means that the deflection due to M_1 is upward. The maximum deflection is at $x = L$, and the neutral axis has the general shape sketched in Figure 6.17.

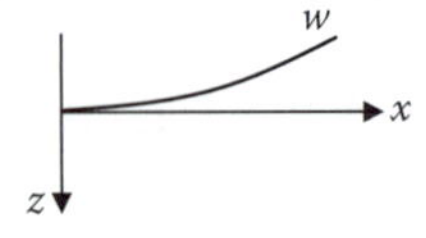

FIGURE 6.17

Example 6.2

For the beam with the given loading in Figure 6.18, with a maximum load intensity of q_o, find (a) the reaction at A; (b) the equation of the elastic curve, that is, $w(x)$; and (c) the slope at A.

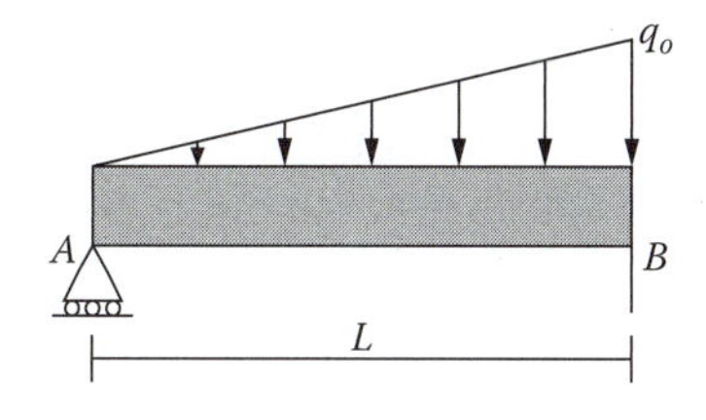

FIGURE 6.18

Given: Loading and support conditions, length of beam.

Find: Reactions, deflection $w(x)$, slope of neutral axis at A.

Assume: Hooke's law applies; beam has constant and uniform properties E, I.

Solution

We start with an FBD of the system (Figure 6.19):

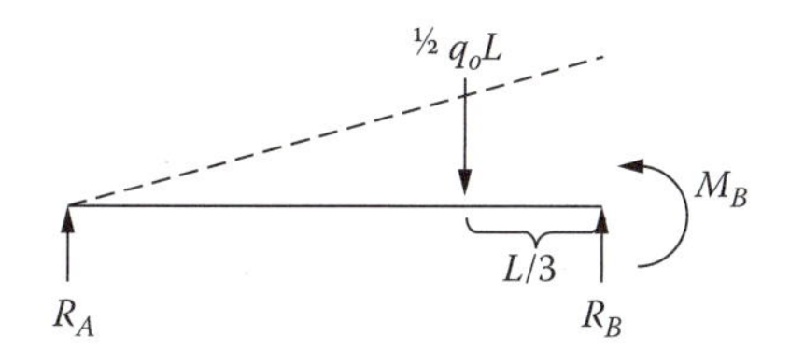

FIGURE 6.19

$$\Sigma F_y = 0 = R_A + R_B - \tfrac{1}{2}\, q_o L.$$

$$\Sigma M_A = 0 = M_B + R_B L - (\tfrac{1}{2}\, q_o L)(2L/3).$$

$$\Sigma M_B = 0 = M_B - R_A L + (\tfrac{1}{2}\, q_o L)(L/3).$$

We have three unknowns (R_A, R_B, and M_B) and only two relevant equilibrium equations. This problem is statically indeterminate! We proceed with the solution for $w(x)$, leaving the reactions as unknowns, and hope that our boundary conditions for V, M, θ, and w may help us out. First, we make an imaginary cut at some x to determine the form of $M(x)$ (Figure 6.20).

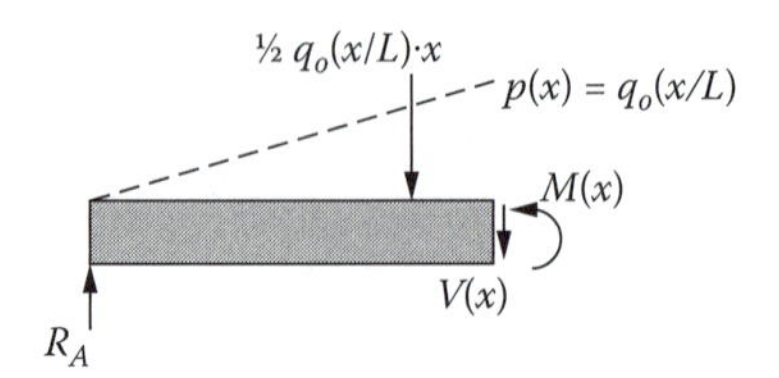

FIGURE 6.20

We require moment equilibrium about our point x, that is, $M(x) + (\frac{1}{2} q_o x^2/L)(x/3) - R_A x = 0$. Thus, $M(x) = R_A x - q_o x^3/6L$. Having this expression for internal bending moment as a function of x allows us to integrate the second-order equation for deflection $w(x)$:

$$EI\frac{d^2w}{dx^2} = -R_A x + \frac{q_o x^3}{6L}$$

$$\int \rightarrow \qquad EI\frac{dw}{dx} = -EI\theta = -\frac{1}{2}R_A x^2 + \frac{q_o x^4}{24L} + C_1$$

$$\int \rightarrow \qquad EIw(x) = -\frac{1}{6}R_A x^3 + \frac{q_o x^5}{120L} + C_1 x + C_2 .$$

Note: The numbering scheme for our constants of integration is not tied to the numbered C_i cited in Section 6.3. Although this scheme was followed in the previous example, there is no need to stick to it. In working problems, we most often integrate the second-order equation and so only have two constants to find, so they may be named in any manner the problem solver deems appropriate.

We now need some boundary conditions to find the constants C_1 and C_2 above. At A, where $x = 0$, we have a pin support, at which we are sure both moment and deflection are zero. The one of these that helps us is $w(x = 0) = 0$. At B, or $x = L$, we have a fixed support, where deflection and slope must both be zero. Applying these three BCs gets us

$$w(x=0)=0 \quad \rightarrow \quad C_2 = 0 .$$

$$w(x=L)=0 \quad \rightarrow \quad -\frac{1}{6}R_A L^3 + \frac{q_o L^4}{120} + C_1 L = 0 .$$

$$\theta(x=L)=0 \quad \rightarrow \quad -\frac{1}{2}R_A L^2+\frac{q_o L^3}{24}+C_1=0.$$

At last, we have two equations and two unknowns, a soluble system. We choose arbitrarily to solve for R_A first, and do this by multiplying the slope boundary condition by L and then subtracting the deflection condition:

$$-\left(\frac{1}{2}R_A L^3+\frac{q_o L^4}{24}+C_1 L\right)$$

$$-\left(\frac{1}{6}R_A L^3+\frac{q_o L^4}{120}+C_1 L\right)$$

$$-\frac{1}{3}R_A L^3+\frac{q_o L^4}{30}+0=0.$$

The resulting equation must equal zero, since both boundary expressions were zero. This allows us to solve for

$$R_A=\frac{1}{10}q_o L,$$

which is an upward force as assumed in the FBD, and which we note is independent of EI. By substituting this R_A into either condition at $x = L$, we are able to find that the constant

$$C_1=-\frac{1}{120}q_o L^3.$$

Putting both of these into our expression for the deflection of the neutral axis, we have

$$EIw(x)=-\frac{1}{6}\left(\frac{1}{10}q_o L\right)x^3+\frac{q_o x^5}{120L}-\left(-\frac{1}{120}q_o L^3\right)x,$$

or

$$w(x)=\frac{q_o}{120EIL}\left(x^5-2L^2x^3+L^4x\right).$$

We could then find a general expression for the slope θ of the neutral axis along the beam and find the slope at A as requested in part (c):

$$\theta(x=0)=\frac{q_o L^3}{120EI}.$$

Note: We could also have solved this complex problem by recognizing the loading on the beam as the superposition of two more straightforward conditions (Figure 6.21):

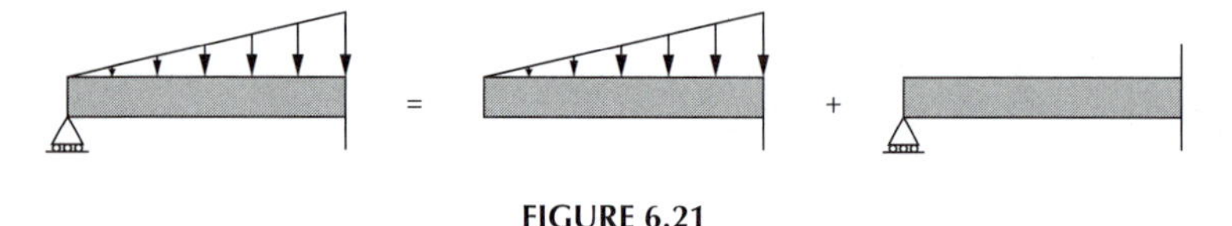

FIGURE 6.21

The superposition of $w(x)$ for both these loading conditions is exactly the result achieved above. Superposition is quite a useful technique for finding the deflections of beams.

Example 6.3

A simple beam supports a concentrated downward force P at a distance a from the left support. The flexural rigidity EI is constant. Find $w(x)$.

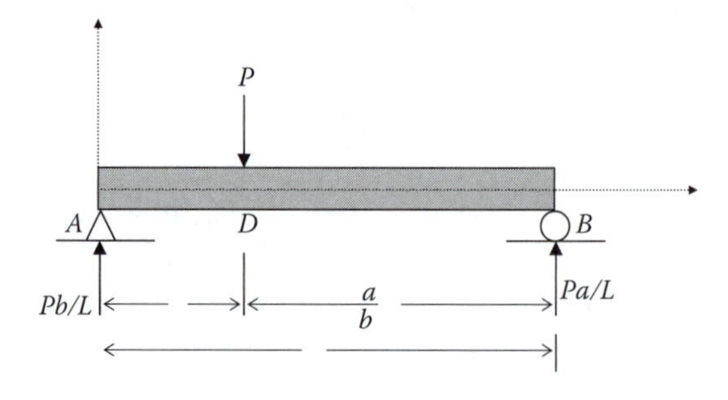

FIGURE 6.22

Given: Loading conditions, reaction forces, length of beam.

Find: Deflection $w(x)$.

Assume: Hooke's law applies; beam has constant and uniform properties E, I.

Solution

We want to integrate the internal moment to find the deflection $w(x)$. We have simple supports, so we know that at A, $w(x = 0) = 0$ and $M(x = 0) = 0$, and at B, $v(x = L) = 0$ and $M(x = L) = 0$. We use the method of sections to find $M(x)$ (Figure 6.23 and Figure 6.24).

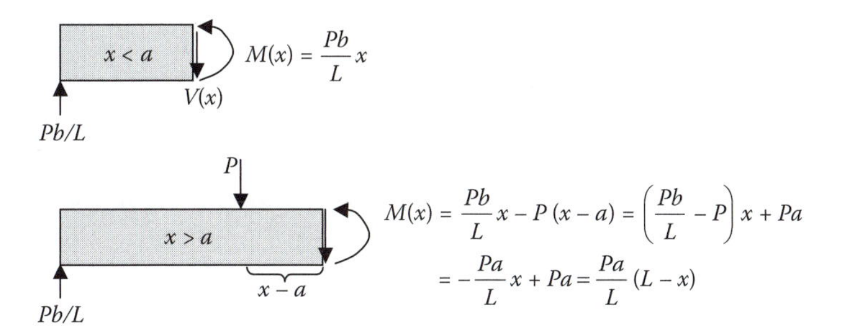

FIGURE 6.23

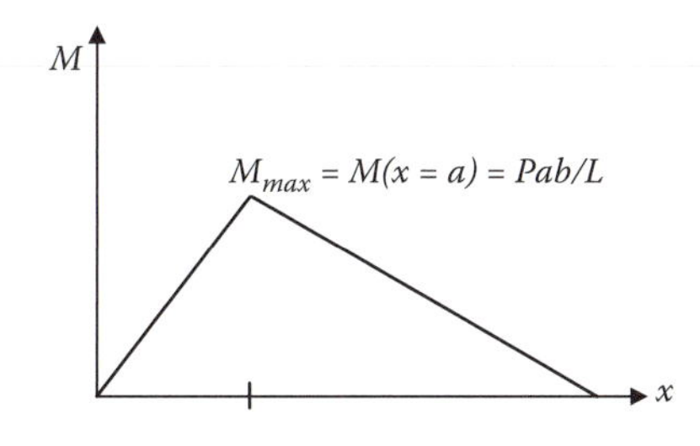

FIGURE 6.24

We note that there is a discontinuity at $x = a$; on either side of a, we have two distinct $M(x)$ expressions. Although $M(x)$ may be discontinuous in this way, neither the slope nor the deflection is allowed to be discontinuous. We can therefore integrate the two distinct $M(x)$ expressions for the deflections of the two portions of the beam and match the two solutions at $x = a$.

$$0 \le x \le a$$

$$\frac{d^2w_1}{dx^2} = -\frac{M}{EI} = -\frac{Pb}{EIL}x$$

$$\frac{dw_1}{dx} = -\frac{Pb}{2EIL}x^2 + A_1$$

$$w_1(x) = -\frac{Pb}{6EIL}x^3 + A_1x + A_2$$

$$a \le x \le L$$

$$\frac{d^2w_2}{dx^2} = -\frac{M}{EI} = -\frac{Pa}{EIL}(L-x) = -\frac{Pa}{EI} + \frac{Pax}{EIL}$$

$$\frac{dw_2}{dx} = -\frac{Pa}{EI}x + \frac{Pax^2}{2EIL} + B_1$$

$$w_2(x) = -\frac{Pax^2}{2EI} + \frac{Pax^3}{6EIL} + B_1x + B_2$$

To find the constants of integration A_i and B_i, we will apply the end BCs as well as the continuity condition: Both w and θ must be continuous at $x = a$, so that

$$w_1(a) = w_2(a)$$

and

$$\left.\frac{dw_1}{dx}\right]_{x=a} = \left.\frac{dw_2}{dx}\right]_{x=a}.$$

Beginning with the end conditions we have

$$w(x = 0) = w_1(0) = 0 = A_2$$

$$w(x = L) = w_2(L) = 0 = \frac{PaL^2}{3EI} + B_1L + B_2$$

$$w_1(a) = w_2(a) \rightarrow -\frac{Pa^3b}{6EIL} + A_1a = -\frac{Pa^3}{2EI} + \frac{Pa^4}{6EIL} + B_1a + B_2$$

$$\left.\frac{dw_1}{dx}\right]_{x=a} = \left.\frac{dw_2}{dx}\right]_{x=a} \rightarrow -\frac{Pa^2b}{2EIL} + A_1 = -\frac{Pa^2}{EI} + \frac{Pa^3}{2EIL} + B_1$$

Here we have three equations for three unknown constants, so we can solve the equations simultaneously and obtain the remaining constants:

$$A_1 = \frac{Pb}{6EIL}(L^2 - b^2) \text{ and } B_1 = \frac{Pa}{6EIL}(2L^2 + a^2),$$

so

$$A_1 = 0 \text{ and } B_2 = -\frac{Pa^3}{6EI}.$$

So the deflection of the beam (after some aesthetic rearrangements) is given by

$$0 \le x \le a \qquad w_1(x) = \frac{Pbx}{6EIL}(L^2 - b^2 - x^2)$$

$$a \le x \le L \qquad w_2(x) = \frac{Pa}{EIL}\frac{x^3}{6} - \frac{Pa}{EI}\frac{x^2}{2} + \left[\frac{Pa}{6EIL}(2L^2 + a^2)\right]x - \frac{Pa^3}{6EI}.$$

Note: The deflection at the point of application of force P may be determined by substituting $x = a$ into either of the above expressions and is $Pa^2b^2/3EIL$.

Example 6.4

The beam shown in Figure 6.25 has uniform elastic modulus E and second moment of area I. Determine (a) the reactions at the left wall; (b) the beam's deflection w as a function of x; and (c) the maximum allowable value of load intensity q_o if the beam has a square cross section with sides of 4 in. and length $L = 96$ in. and is made from a material with $E = 15 \times 10^6$ psi and maximum normal stress 110 ksi.

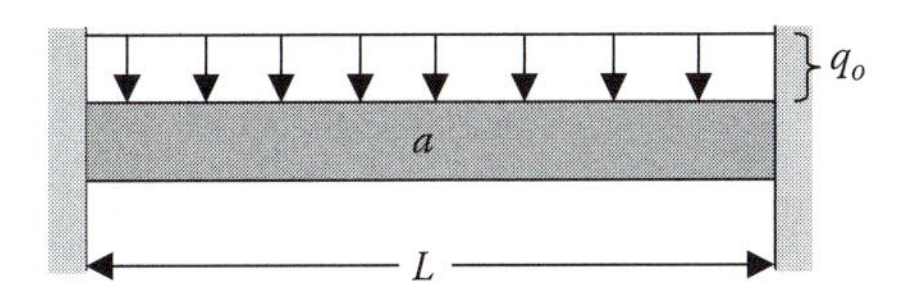

FIGURE 6.25

Given: Loading conditions; properties of beam.

Find: Reactions, deflection, maximum allowable intensity q_o.

Assume: Hooke's law applies; beam has constant and uniform properties E, I.

Solution

Since there are no applied axial loads, we assume that the supports exert no axial forces on the beam. A free-body diagram of the system can thus be constructed (Figure 6.26):

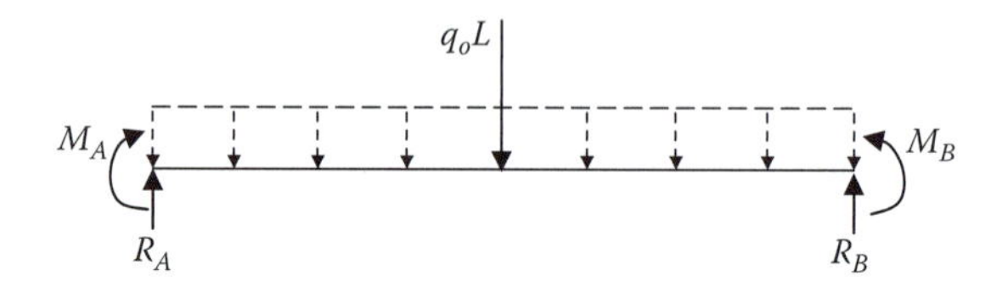

FIGURE 6.26

And, summing forces and moments, we have

$$\sum F_z = 0 = q_o L - R_A - R_B$$

$$\sum M_A = 0 = M_B - M_A + R_B L - \frac{q_o L^2}{2}.$$

By symmetry, we can reasonably assume that $R_A = R_B$ and $M_A = M_B$; however, this assumption will not help us solve the equations of statics for the reaction moments. We need more than just statics to find all four reactions. As in Example 6.2, we proceed with the solution for deflection $w(x)$ and hope that the boundary conditions helps us identify our unknowns. At the two fixed supports, we know that both deflection and slope must equal zero, that is, $w(x = 0) = w(x = L) = 0$ and $\theta(x = 0) = \theta(x = L) = 0$.

We make a "cut" at a distance x to find the internal bending moment $M(x)$ (Figure 6.27).

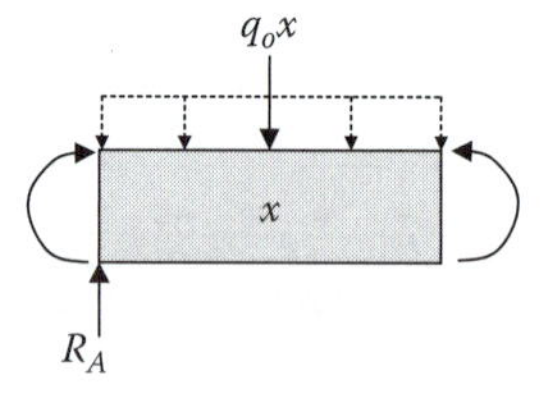

FIGURE 6.27

Balancing moments on this x-long segment, we have

$$M(x) = M_A + R_A x - q_o x^2/2\,.$$

Next, we integrate for the deflection $w(x)$:

$$EI\frac{d^2w}{dx^2} = -M(x) = -M_A - R_A x + \frac{q_o x^2}{2}$$

$$EI\frac{dw}{dx} = -M_A x - \frac{R_A x^2}{2} + \frac{q_o x^3}{6} + C_1$$

$$EIw(x) = \frac{M_A x^2}{2} + \frac{R_A x^3}{6} - \frac{q_o x^4}{24} - C_1 x + C_2.$$

Applying our BCs we have

$$w(x = 0) = 0 \rightarrow C_2 = 0$$

$$w'(x = 0) = \theta(x = 0) = 0 \rightarrow C_1 = 0$$

$$w(x = L) = 0 \rightarrow \frac{M_A L^2}{2} + \frac{R_A L^3}{6} - \frac{q_o L^4}{24} = 0$$

$$w'(x = L) = -\theta(x = L) = 0 \rightarrow M_A L + \frac{R_A L^2}{2} - \frac{q_o L^3}{6} = 0.$$

We solve these last two equations together with the two equilibrium equations for our four unknowns and find that

$$R_A = R_B = \frac{q_o L}{2}$$

$$M_A = M_B = -\frac{q_o L^2}{12}.$$

We can now substitute these values into the expression for $w(x)$ above:

$$w(x) = \frac{1}{EI}\left[\frac{q_o L^2 x^2}{24} - \frac{q_o L x^3}{12} + \frac{q_o x^4}{24}\right],$$

or

$$w(x) = \frac{q_o x^2}{24EI}\left[L^2 - 2Lx + x^2\right].$$

To find the maximum allowable load intensity q_o based on the given normal stress limitation, we must calculate the maximum normal stress induced in the beam in terms of q_o. Because normal stress is linearly proportional to bending moment, we do this by finding the maximum internal bending moment in the beam. We return to our general equation for *M(x)*:

$$M(x) = M_A + R_A x - \frac{q_o x^2}{2}$$

Now that we know both M_A and R_A, this takes a somewhat friendlier form:

$$M(x) = \frac{q_o L^2}{12} + \frac{q_o Lx}{2} - \frac{q_o x^2}{2}.$$

The maximum *M(x)* occurs where $dM/dx = 0$:

$$\frac{dM}{dx} = \frac{q_o L}{2} - q_o x = 0 \quad \text{at } x = \frac{L}{2}.$$

The maximum bending moment is then *M(L/2)*, or

$$M\left(x = \frac{L}{2}\right) = \frac{5q_o L^2}{24}.$$

In Figure 6.28, we sketch the form of *M(x)*.

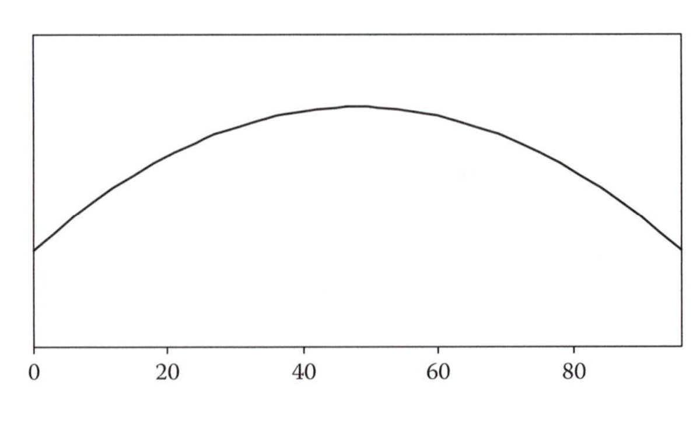

FIGURE 6.28

The second moment of area of the given cross section is $I = bh^3/12 = (4\text{ in.})^4/12 = 21.3\text{ in.}^4$ The maximum normal stress is given by

$$\sigma_{max} \geq \frac{Mc}{I},$$

so

$$M = \frac{5q_o L^2}{24} \leq \frac{\sigma_{max} I}{c}$$

$$q_o \leq \frac{(110{,}000\text{ psi})\cdot(21.3\text{ in.}^4)\cdot 24}{5\cdot(2\text{ in.})\cdot(96\text{ in.}^2)}$$

$$q_o \leq 58{,}575\text{ lb/in.} = 4.88\text{ kips/ft.}$$

Note: This result is independent of the Young's modulus of the beam, E.

6.10 Problems

6.1 A simply supported beam 5 m long is loaded with a 20 N downward force at a point 4 m from the left support (Figure 6.29). The second moment of area of the cross section of the beam is $4I_1$ for segment AB and I_1 for the remainder of the beam. Determine the deflection $w(x)$ of the neutral axis.

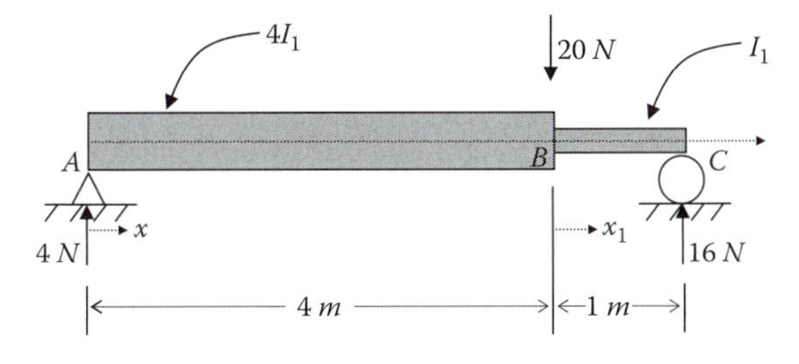

FIGURE 6.29

6.2 A beam fixed at both ends supports a uniformly distributed downward (positive!) load q_o (Figure 6.30). EI for the beam is constant. (a) Find the expression for $w(x)$ using the fourth-order governing differential equation. (b) Verify the result of (a) using the second-order differential equation.

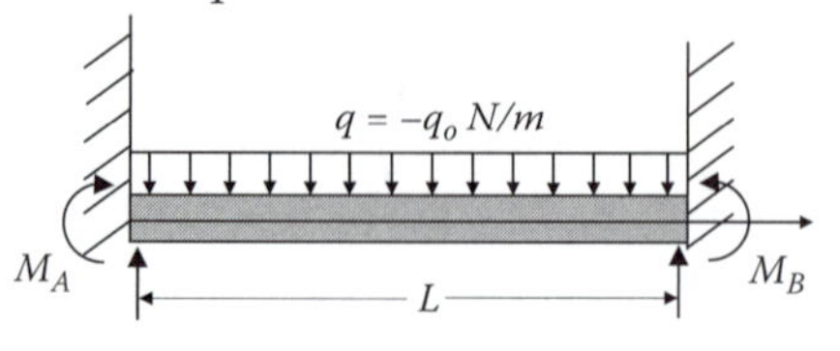

FIGURE 6.30

6.3 Consider an aluminum cantilever beam 1600 mm long, with a 10 kN force applied 400 mm from the free end (Figure 6.31). For a distance of 600 mm from the fixed end, the beam has $I_1 = 50 \times 10^6$ mm^4. For the remaining 1000 mm of its length, the beam has $I_2 = 10 \times 10^6$ mm^4. Find the deflection and the angular rotation of the free end. Neglect the weight of the beam, and use $E = 70$ GPa.

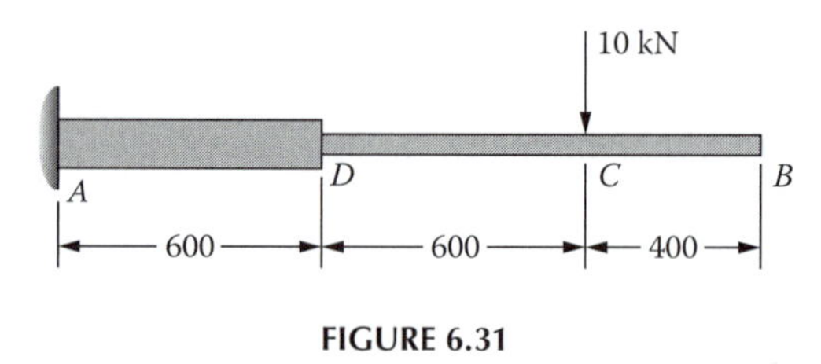

FIGURE 6.31

6.4 An aluminum ($E = 70$ GPa) cantilever must carry an end load of 1.4 kN as shown in Figure 6.32. However, in this design, when the beam is loaded the end of the cantilever *A* has to have the same elevation as point C (that is, the net deflection of *A* must be zero). A hydraulic jack may be used to raise point *B* to achieve this. Determine the amount that B should be raised and the reaction at B (when the load is applied and *B* has been raised). Do *not* consider the weight of the beam. The cross section of the beam is *half* of an I beam as shown in the figure. The properties given in the box are for the *whole* I beam. The position of the centroid of the half-section (relative to the top of the section) is shown on the figure.

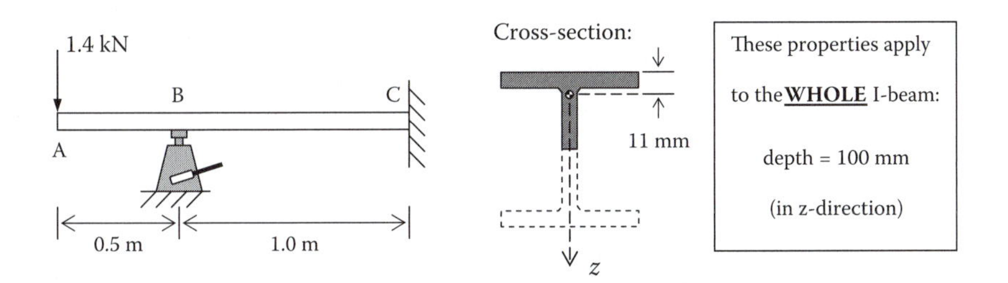

FIGURE 6.32

6.5 A cantilever beam *AB* has a rigid (i.e., its deformation is negligible relative to that of the beam) bracket *AC* attached to its free end and a vertical load *P* applied at point *C* (Figure 6.33). Find the ratio a/L required so that the deflection at point *A* will be zero. *E* and *I* are constant along the beam.

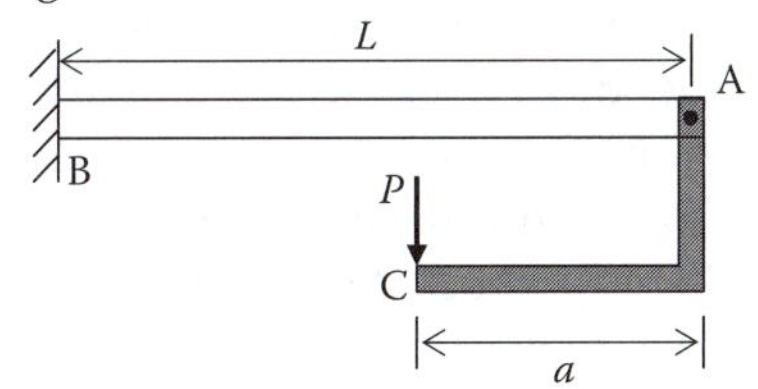

FIGURE 6.33

6.6 A cantilever beam *AB* supports a uniform load of intensity *q* acting over part of the span and a concentrated load *P* acting at the free end, as shown in Figure 6.34. Determine the deflection δ_B and slope θ_B at end *B* of the beam. The beam has length *L* and constant flexural rigidity *EI*.

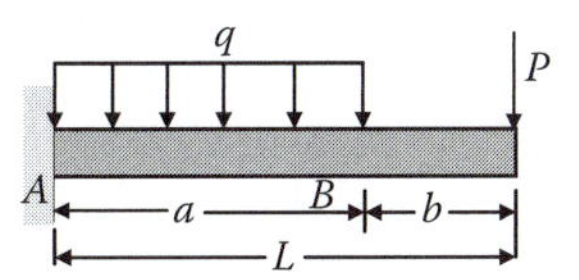

FIGURE 6.34

6.7 A simply supported beam of length *L* is subjected to loads that produce a symmetric deflection curve with maximum deflection at the midpoint of the span. How much strain energy *U* is stored in the beam if the deflection curve is (a) a parabola, or (b) a half wave of a sine curve?

6.8 A beam with a constant *EI* is loaded as shown in Figure 6.35. (a) Determine the length *a* of the overhang such that the elastic curve would be horizontal over support *B*. (b) Determine the maximum deflection between the supports.

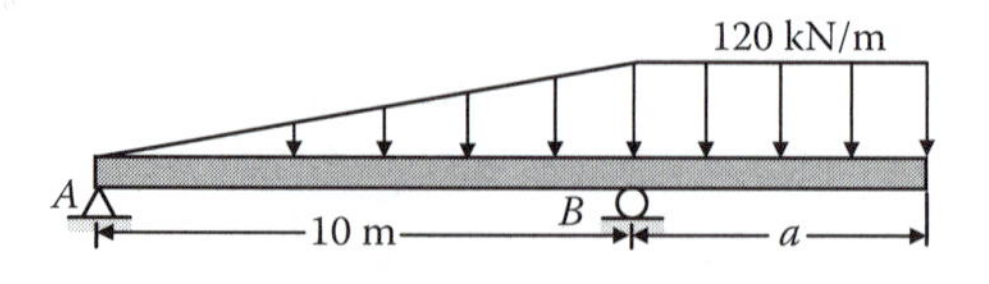

FIGURE 6.35

6.9 A gold alloy microbeam attached to a silicon wafer behaves like a cantilever beam subjected to a uniformly distributed load q. The beam has length $L = 25\ \mu m$ and a rectangular cross section of width $b = 15\ \mu m$ and thickness $t = 0.87\ \mu m$. The total load on the beam is $44\ \mu N$. (a) If the deflection at the end of the beam is measured to be $1.3\ \mu m$, what is the modulus of elasticity of the gold alloy? (b) If the load were instead applied as a point force at the end of the beam, what maximum deflection would you expect to measure?

6.10 Beam *ABCDE* has simple supports at *B* and *D* and symmetrical overhangs at each end, as shown in Figure 6.36. The center span has length L and each overhang has length b. A uniform load of intensity q acts on the beam. (a) Determine the ratio b/L such that the deflection δ_C at the midpoint of the beam is equal to the deflections δ_A and δ_E at the ends. (b) For this value of b/L, what is the deflection δ_C at the midpoint?

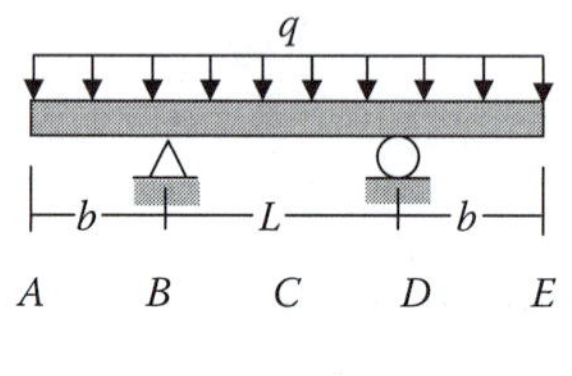

FIGURE 6.36

Notes

1. The length of the neutral axis is sometimes called the elastic curve; hence, some texts call this requirement continuity of the elastic curve.
2. This was 28 years before his eponymous circle of stress transformation.
3. This is the "force method" for resolving statically indeterminate problems, as introduced in Chapter 2, Section 2.8.
4. Castigliano's first theorem makes a similar case for the forces being calculated as the partial derivatives of strain energy with respect to the appropriate deflections!

7

Instability: Column Buckling

In our analysis of the internal response to external loading on beams, pressure vessels, and shafts in torsion, we have had two primary concerns: the *stiffness* and the *strength* of the structure. By strength, of course, we mean the ability of our structure to support the required loads without experiencing excessive stress; by stiffness, we mean its ability to support the required loads without undergoing excessive deformations. (Recall our initial discussion of strength and stiffness in Chapter 2.) In practice, we have a third concern: the *stability* of our structure, by which we mean our designs' ability to support the required loads without experiencing a sudden change in configuration.

The instability known as *buckling* typically occurs when forces much lower than those necessary to exceed material yield stresses are applied to beams. Buckling can occur whenever a slender[1] structural member is subjected to compression. These forces are applied axially, as shown in Figure 7.1. Here, by holding a hacksaw blade between his palms, a man has been able to induce instability, and the blade fails as a structural element.

The most common occurrence of this kind of loading, and of buckling instability, is in columns. Figure 7.2 shows some examples of structural columns; Figure 7.3 shows failed columns.

7.1 Euler's Formula

Consider a column of length L supported by pin supports at both ends, subjected to a compressive axial load P, as in Figure 7.4a. We would like to determine the critical value, P_{cr}, for which the initial position is no longer stable. Once P exceeds P_{cr}, any small perturbation or misalignment causes the column to buckle, taking on the sort of curvature illustrated in Figure 7.4b. Our method of finding P_{cr} is to determine the conditions under which the geometry of Figure 7.4b is possible.

We approach this column as a vertical beam subjected to an axial load and use x to denote the distance from the top, along the beam's initial elastic curve. The column's deflection w in the z direction denotes the lateral deflection of the elastic curve from its original position, just as it did for beams. We make an imaginary cut at some point C along this curve, as in Figure 7.5, and observe that at this point the internal axial force is P and the internal bending moment is $M = Pw$.

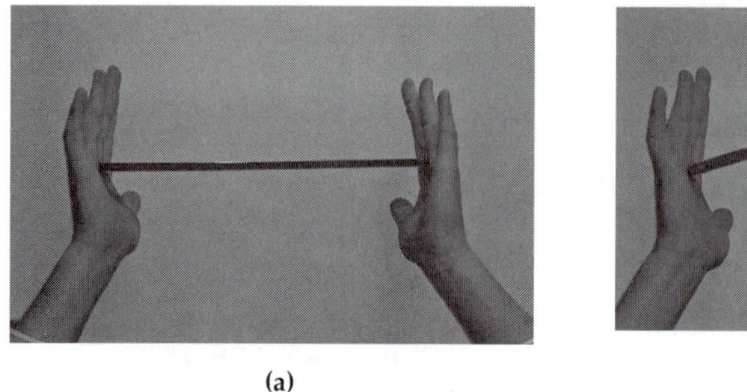

(a)

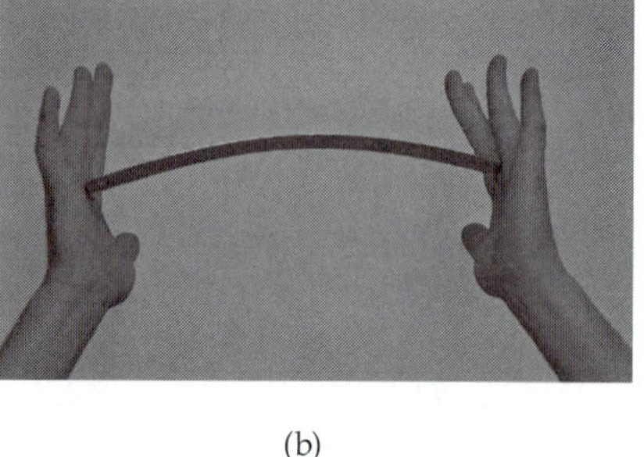

(b)

FIGURE 7.1
At left, application of compressive axial force to hacksaw blade. At right, a small compressive load causes the blade to "buckle."

FIGURE 7.2
Examples of columns: Parthenon (left), Markle Hall, Lafayette College (right).

We understand that this internal bending moment may be related to the deflection w of the column's axis:

$$\frac{d^2w}{dx^2} = -\frac{M}{EI} = -\frac{P}{EI}w. \tag{7.1}$$

We rearrange this as

$$\frac{d^2w}{dx^2} + \frac{P}{EI}w = 0 \tag{7.2}$$

and find that it is an ordinary differential equation with whose solution we are (or were once) familiar:

FIGURE 7.3
Examples of column (and pressure vessel) failure by buckling.

$$w(x) = A\sin\sqrt{\frac{P}{EI}}x + B\cos\sqrt{\frac{P}{EI}}x\,. \tag{7.3}$$

To move from this general solution to a specific expression for our buckling column, we apply the relevant boundary conditions. At the bottom of our beam, $x = 0$, and we have $w = 0$ since the pin support does not allow any deflection. We also have $w = 0$ at the top support, where $x = L$. The first condition, $w(x = 0) = 0$, requires that $B = 0$. To have $w(x = L) = 0$, we require

$$A\sin\sqrt{\frac{P}{EI}}L = 0\,. \tag{7.4}$$

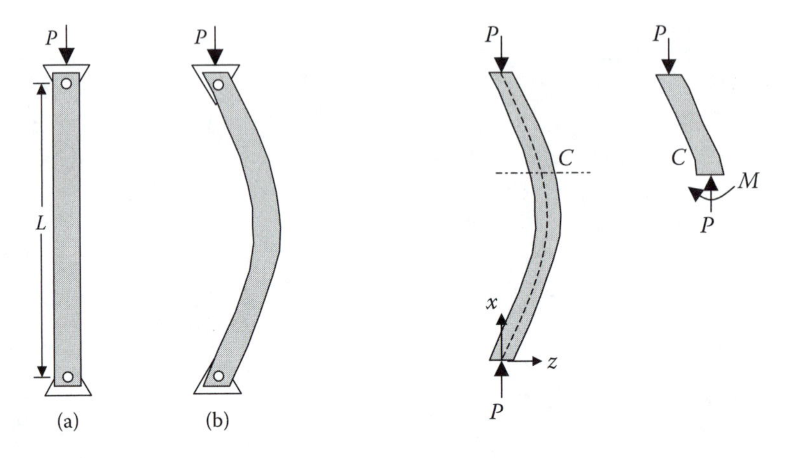

FIGURE 7.4
(a) Beam/column under compression; (b) buckled geometry.

FIGURE 7.5
Method of sections on buckling column.

This statement holds if either $A = 0$, or $\sin \sqrt{\frac{P}{EI}}L = 0$. If $A = 0$, our general solution is $w = 0$, and the column remains straight. Since we are modeling the buckling phenomenon, we must instead satisfy the second condition. Due to the nature of $\sin x$, this requires that

$$\sqrt{\frac{P}{EI}}L = n\pi.$$

Solving for the force P that will make this happen, we find

$$P = \frac{n^2\pi^2 EI}{L^2}. \tag{7.5}$$

This suggests that there are many *modes* of buckling, each with a different value of n, as shown in Figure 7.6. We are particularly interested in the first mode, the smallest load that can cause buckling, which corresponds to $n = 1$. Therefore, the critical load P_{cr} for the pinned-pinned column of Figures 7.4 and 7.5 is

$$P_{cr} = \frac{\pi^2 EI}{L^2}. \tag{7.6}$$

This result is known as *Euler's formula*, as Swiss mathematician Leonhard Euler first derived it in 1744. Applying this force makes it possible for the axis of the column to be described by $w = A \sin \frac{\pi x}{L}$. Note that we have not determined the value of the coefficient A, which is the column's maximum deflection w_{max}.

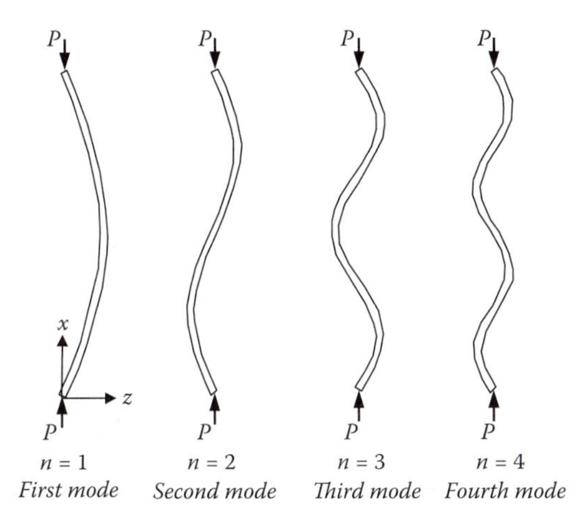

FIGURE 7.6
Deflection distributions in first four buckling modes.

The second moment of area I in Euler's formula should be taken about the axis around which the column bends. This typically is apparent from the way the column is supported; when it is not, we recognize that a buckling column bends about the principal axis of its cross section with the smaller second moment of area and make our calculations accordingly.

We note that Euler's formula (equation 7.6) as just derived applies to the particular case of a column with two pinned ends. The column ends are thus free to rotate at the ends where the loads P are applied; in other words there are no reactions at the ends other than P. This affected the boundary conditions we used in obtaining P_{cr}. For different supports, and thus different boundary conditions, the value of P_{cr} is different. We codify these differences by using an *effective length* L_e in the place of L in Euler's formula, where the relationship between L_e and L depends on the end supports:

$$P_{cr} = \frac{\pi^2 EI}{L_e^2}. \tag{7.7}$$

Values of effective length for a variety of supports are tabulated in Table 7.1.

The value of normal stress corresponding to the critical load is called the critical stress, σ_{cr}. We simply divide Euler's formula by the column's cross-sectional area:

$$\sigma_{cr} = \frac{P_{cr}}{A} = \frac{\pi^2 EI}{L_e^2 A}. \tag{7.8}$$

TABLE 7.1

Effective Length L_e as Function of Supports

End Conditions	Effective Length
Fixed-Free	$Le = 2L$
Pinned-Pinned	$Le = L$
Fixed-Pinned	$Le = 0.7L$
Fixed-Fixed	$Le = 0.5L$

Next, we set the second moment of area $I = Ar^2$, where r is the cross-sectional area's *radius of gyration*. We obtain the radius of gyration of various shapes using its definition, $r = (I/A)^{1/2}$:

$$\sigma_{cr} = \frac{\pi^2 EAr^2}{L_e^2 A} = \frac{\pi^2 Er^2}{L_e^2} = \frac{\pi^2 E}{(L_e / r)^2}. \tag{7.9}$$

The quantity L/r is known as the column's *slenderness ratio*. The critical stress is proportional to the elastic modulus of the material used and is inversely proportional to the square of this ratio. For sufficiently slender columns, σ_{cr} can be much lower than the material's yield stress, and the column almost certainly fails due to buckling. If this critical buckling stress is greater than the material's yield stress, the column in question likely yields in compression before it has the opportunity to buckle—this is often true for short, stubby columns.

In practice, loads are rarely applied as we have modeled our P—a perfectly aligned axial load. To more realistically assess the likelihood of buckling, we must develop a model that includes the effects of load eccentricity.

7.2 Effect of Eccentricity

The lines of action of applied forces are generally not through the cross section's centroid, as we had hoped in the previous section. We now analyze the potential for buckling when an eccentric load is applied, again beginning with a beam/column that is free to rotate at the ends (i.e., both ends are pinned). We see that this off-center load P applies a moment to the column, as illustrated in Figure 7.7a: The force P has a moment arm equal to its eccentricity, e. We can thus replace the off-center P by a centric load, also with magnitude P, and a moment $M = Pe$, as shown in Figure 7.7b. No matter how small either P or e is, this moment M will cause some bending of the column. In a sense, we are calculating not *how* to make the column stay straight but

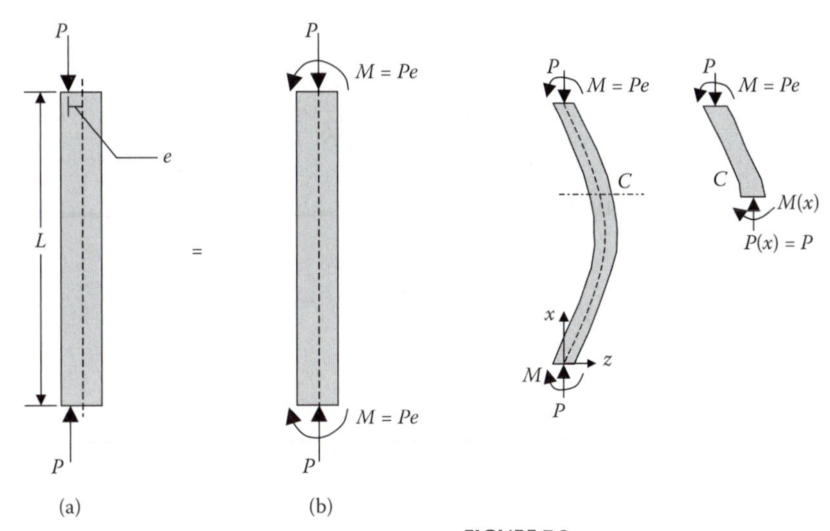

FIGURE 7.7
Modeling an eccentric load P.

FIGURE 7.8
Method of sections for eccentric P.

how much bending is permissible to maintain a normal stress $\sigma < \sigma_{cr}$ and a tolerable deflection w_{max}.

Again, we want to obtain the equation of the column's elastic curve. We begin with the method of sections in an effort to find the internal bending moment at some arbitrary position x. Figure 7.8 indicates that the internal bending moment necessary to keep this section in equilibrium is $M(x) = Pw + Pe$. We proceed with the second-order equation for the column's deflection $w(x)$:

$$\frac{d^2w}{dx^2} = -\frac{M}{EI} = -\frac{P}{EI}w - \frac{P}{EI}e, \tag{7.10}$$

or

$$\frac{d^2w}{dx^2} + \frac{P}{EI}w = -\frac{P}{EI}e\,. \tag{7.11}$$

The left-hand side of this equation is the same as the homogeneous ordinary differential equation (o.d.e.) we solved for centric loading, whose solution we already know. We add to this general solution the constant $-e$ that solves the nonhomogeneous equation, and have:

$$w(x) = A\sin\sqrt{\frac{P}{EI}}x + B\cos\sqrt{\frac{P}{EI}}x - e\,. \tag{7.12}$$

Again, we make use of our boundary conditions to identify the unknown constants. At $x = 0$, we have $w = 0$, which requires that $B = e$. At $x = L$, we also have $w = 0$, so that

$$A\sin\sqrt{\frac{P}{EI}}L = e\left(1-\cos\sqrt{\frac{P}{EI}}L\right). \tag{7.13}$$

We make use of the trigonometric identities

$$\sin\sqrt{\frac{P}{EI}}L = 2\sin\sqrt{\frac{P}{EI}}\frac{L}{2}\cos\sqrt{\frac{P}{EI}}\frac{L}{2} \quad \text{and} \quad 1-\cos\sqrt{\frac{P}{EI}}L = 2\sin^2\sqrt{\frac{P}{EI}}\frac{L}{2}$$

to write

$$A = e\tan\sqrt{\frac{P}{EI}}\frac{L}{2}, \tag{7.14}$$

which allows us to write the equation of the elastic curve:

$$w(x) = e\left(\tan\sqrt{\frac{P}{EI}}\frac{L}{2}\sin\sqrt{\frac{P}{EI}}x + \cos\sqrt{\frac{P}{EI}}x - 1\right). \tag{7.15}$$

We obtain the value of the maximum deflection, w_{max}, by evaluating this expression at $x = L/2$:

$$w_{max} = e\left[\sec\left(\sqrt{\frac{P}{EI}}\frac{L}{2}\right) - 1\right]. \tag{7.16}$$

The nature of the secant curve tells us that the value of w_{max} becomes infinite when

$$\sqrt{\frac{P}{EI}}\frac{L}{2} = \frac{\pi}{2}.$$

While the column deflection does not actually become infinite, it becomes unacceptably large at this condition. We can therefore find the critical P_{cr} that lets

$$\sqrt{\frac{P}{EI}}\frac{L}{2} = \frac{\pi}{2}.$$

It is

$$P_{cr} = \frac{\pi^2 EI}{L^2}, \tag{7.17}$$

which is Euler's formula for the buckling of a column under centric loading. Knowing this allows us to recast the maximum deflection in terms of this critical load:

$$w_{max} = e\left[\sec\left(\frac{\pi}{2}\sqrt{\frac{P}{P_{cr}}}\right) - 1\right]. \tag{7.18}$$

The maximum normal stress in the column occurs where the bending moment is maximized, that is, at $x = L/2$. We obtain this stress by superposing the stress due to P with the bending stress,

$$\sigma_{max} = -\frac{P}{A} + \frac{M_{max}c}{I}, \tag{7.19}$$

where M_{max} is simply $Pw_{max} + Pe = P(w_{max} + e)$. We plug in our expression for w_{max} and have

$$\sigma_{max} = -\frac{P}{A}\left[1 + \frac{ec}{r^2}\sec\left(\sqrt{\frac{P}{EI}}\frac{L}{2}\right)\right], \tag{7.20}$$

or

$$\sigma_{max} = -\frac{P}{A}\left[1 + \frac{ec}{r^2}\sec\left(\frac{\pi}{2}\sqrt{\frac{P}{P_{cr}}}\right)\right]. \tag{7.21}$$

Since the applied force P is compressive, the maximum normal stress is compressive, as reflected by the negative sign in the previous expressions.

If the end conditions for a particular column differ from the pinned-pinned supports assumed in this model, L should be replaced by the appropriate effective length L_e.

7. 3 Examples

Example 7.1

An aluminum column of length L and rectangular cross section has a fixed end B and supports a centric axial load at A (Figure 7.9). Two smooth and rounded fixed plates restrain end A from moving in one of the vertical planes of symmetry but allow it to move in the other plane. (a) Determine the ratio a/b of the two sides of the cross section corresponding to the most efficient design against buckling. (b) Design the most efficient cross section for the column, knowing that $L = 50$ cm, $E = 70$ GPa, $P = 22$ kN and that a safety factor of 2.5 is required.

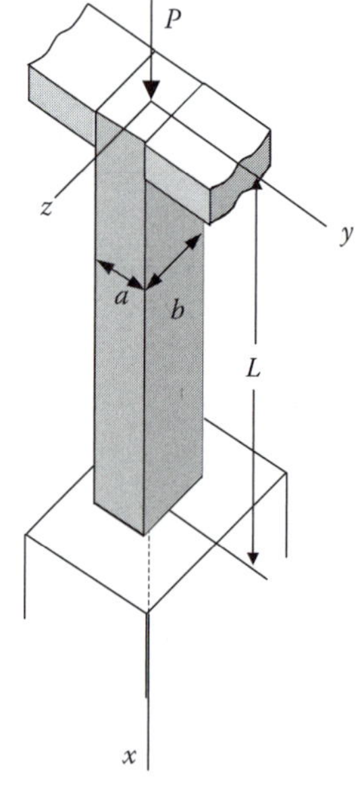

FIGURE 7.9

Given: Loading and support conditions for column; safety factor; length and elastic modulus.

Find: Optimal cross section of column.

Assume: Hooke's law applies (have assumed constant E to integrate y-M o.d.e. in derivations).

Solution

Figure 7.9 indicates that we must consider buckling in both the xy and xz planes, and that due to the nature of the support at A, the critical load and the prospect of buckling will be different in the two planes. As the supports allow end A to move freely in the z direction, for buckling in the xz plane we have a fixed-free support combination; the supports constrain motion in the

y direction but do not provide a reaction moment so that in the xy plane we have a fixed-pinned support.

Buckling in the xy Plane

Due to the fixed-pinned support combination, we find from Table 7.1 that the effective length of the column with respect to buckling in this plane is $L_e = 0.7L$. We obtain the radius of gyration r_z of the cross section by writing

$$I_z = \tfrac{1}{12}ba^3 \quad A = ab,$$

and, since $I_z = Ar_z^2$,

$$r_z^2 = \frac{I_z}{A} = \frac{\frac{1}{12}ba^3}{ab} = \frac{a^2}{12}, \text{ so } r_z = \frac{a}{\sqrt{12}}.$$

The slenderness ratio of the column with respect to buckling in the xy plane is then

$$\frac{L_e}{r_z} = \frac{0.7L}{a/\sqrt{12}}.$$

Buckling in the xz Plane

Again, in this plane the column sees a fixed-free support situation, so the effective length is $Le = 2L$. We find the radius of gyration r_y much as we found r_z:

$$r_y^2 = \frac{I_y}{A} = \frac{\frac{1}{12}ab^3}{ab} = \frac{b^2}{12}, \text{ so } r_y = \frac{b}{\sqrt{12}}.$$

The slenderness ratio of the column with respect to buckling in the xz plane is then

$$\frac{L_e}{r_y} = \frac{2L}{b/\sqrt{12}}.$$

Most Efficient Design

The most efficient design is that for which the critical stresses corresponding to the two possible modes of buckling are equal; neither mode is preferred. This is the case if the two slenderness ratios are equal. So,

$$\frac{0.7L}{a/\sqrt{12}} = \frac{2L}{b/\sqrt{12}}.$$

Solving for the ratio a/b, we have

$$\frac{a}{b} = 0.35.$$

Design for Given Parameters

Since a safety factor of 2.5 is required, we must have

$$P_{cr} = 2.5P = (2.5)(22\text{ kN}) = 55\text{ kN}.$$

Using the ratio a/b found above, $a = 0.35b$, we have $A = ab = 0.35b^2$, and

$$\sigma_{cr} = \frac{P_{cr}}{A} = \frac{55{,}000\text{ N}}{0.35b^2}.$$

We also know that the critical stress must satisfy

$$\sigma_{cr} = \frac{\pi^2 E}{(L_e/r)^2}.$$

We are free to use either slenderness ratio, since we have forced their equivalence. Plugging in $L = 50$ cm, we can find $L_e/r_y = (2\cdot 50\text{ cm})/(b/\sqrt{12}) = 138.6/b$ and write

$$\frac{P_{cr}}{A} = \frac{\pi^2 E}{(L_e/r)^2},$$

$$\frac{55{,}000\text{ N}}{0.35b^2} = \frac{\pi^2(7.0\times 10^6\text{ N/cm}^2)}{(346.4/b)^2}.$$

Solving for b we find that $b = 4.06$ cm and $a = 0.35b = 1.42$ cm.

Example 7.2

An 8 ft length of structural tubing has the illustrated cross section (Figure 7.10). Using Euler's formula and a safety factor of 2, determine the allowable centric load for the column and the corresponding normal stress. Assuming that this allowable load is applied as shown at a point 0.75 in. from the geometric axis of the column, determine the horizontal deflection of the top of the column and the maximum normal stress in the column. Use $E = 29 \times 10^6$ psi.

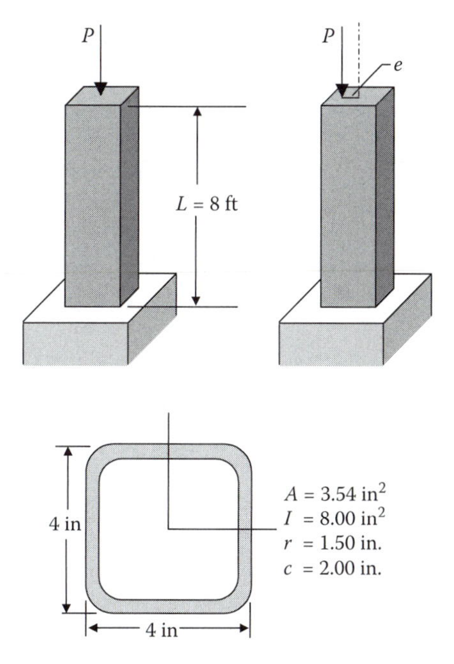

FIGURE 7.10

Given: Geometry of column, safety factor.

Find: P_{cr} and σ_{cr}; w_{max} and σ_{max} if $P_{cr}/2$ applied eccentrically.

Assume: Hooke's law applies.

Solution

Since the column has one fixed and one free end, the effective length is $L_e = 2L = 16$ ft $= 192$ in. Using Euler's formula, we find the critical load to be

$$P_{cr} = \frac{\pi^2 EI}{L_e^2} = \frac{\pi^2 (29 \times 10^6 \text{ psi})(8.00 \text{ in.}^4)}{(192 \text{ in.})^2} = 62.1 \text{ kips}.$$

Since we are asked to use a safety factor of 2, our allowable centric load is then

$$P_{all} = \frac{P_{cr}}{2} = 31.1 \text{ kips.}$$

The corresponding normal stress is

$$\sigma = \frac{P_{all}}{A} = \frac{31.1 \text{ kips}}{3.54 \text{ in.}^2} = 8.79 \text{ ksi.}$$

We have been asked for the horizontal deflection at the top of the column, which, given the supports seen here, is the maximum deflection w_{max}. As long as we have used the correct L_e to obtain P_{cr}, we are able to use the secant formulas for eccentric loading on columns with any type of supports:

$$w_{max} = e\left[\sec\left(\frac{\pi}{2}\sqrt{\frac{P}{P_{cr}}}\right) - 1\right] = (0.75 \text{ in.})\left[\sec\left(\frac{\pi}{2}\sqrt{\frac{1}{2}}\right) - 1\right]$$

$$= (0.75 \text{ in.})(2.252 - 1) = 0.939 \text{ in.}$$

The maximum normal stress is calculated as

$$\sigma_{max} = -\frac{P}{A}\left[1 + \frac{ec}{r^2}\sec\left(\frac{\pi}{2}\sqrt{\frac{P}{P_{cr}}}\right)\right]$$

$$= -\frac{31.1 \text{ kips}}{3.54 \text{ in.}^2}\left[1 + \frac{(0.75 \text{ in.})(2 \text{ in.})}{(1.50 \text{ in.})^2}\sec\left(\frac{\pi}{2\sqrt{2}}\right)\right]$$

$$= -(8.79 \text{ ksi})[1 + 0.667(2.252)]$$

$$= -22.0 \text{ ksi.}$$

Again, this is a compressive stress.

7.4 Problems

7.1 An I beam with the proportions shown is to be used as a long column. There is a concern about buckling, so two reinforcing plates are to be welded along the length of the column. Two options for the resulting cross section are shown in Figure 7.11. Which will increase the critical buckling load more effectively? Explain why.

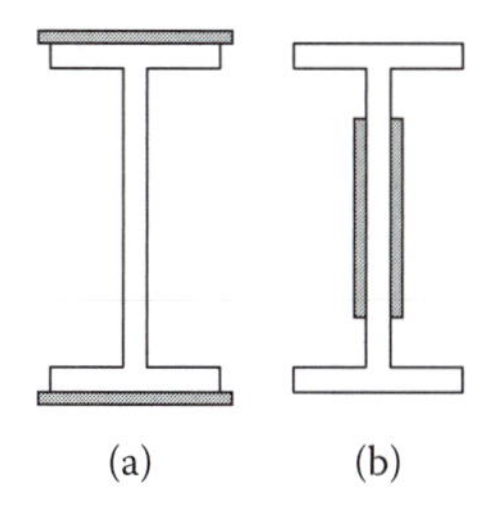

FIGURE 7.11

7.2 A slender vertical bar AB with pinned ends and length L is held between immovable supports. What increase ΔT in the temperature of the bar will produce buckling?

7.3 In more than one paragraph but less than a page, discuss some of the failure modes experienced in the collapse of the World Trade Center and how they might have been prevented.

7.4 A truss ABC supports a load W at joint B, as shown in Figure 7.12. The length L_1 of member AB is fixed, but the length of strut BC varies as the angle θ is changed. Strut BC has a solid circular cross section. Assuming that collapse occurs by Euler buckling of the strut, determine the angle θ for minimum allowable weight of the strut.

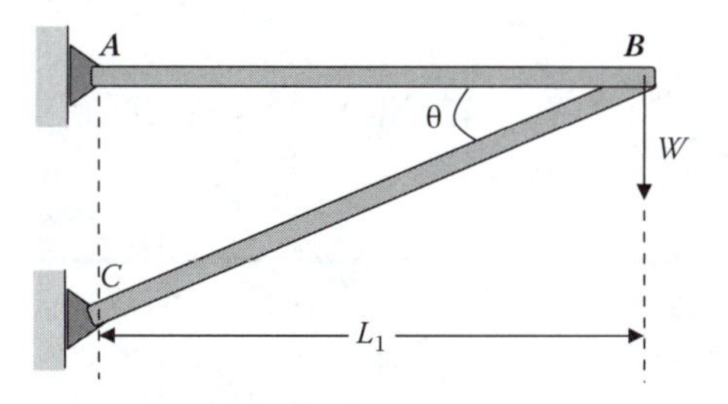

FIGURE 7.12

7.5 For a deck, supports are proposed to be built from aluminum with a Young's modulus of 72 GPa and a yield stress of 480 MPa. A cylindrical design is proposed with outer diameter d and wall thickness t. If the design specs require that each column support a load of 100 kN with a safety factor of 3, find the necessary column thickness t.

7.6 A rectangular brass column is loaded as shown in Figure 7.13, with a load of $P = 1500$ lb applied 0.45 in. off its centroidal axis. Find the longest permissible length L of the column if the deflection of its free end cannot exceed 0.12 in.

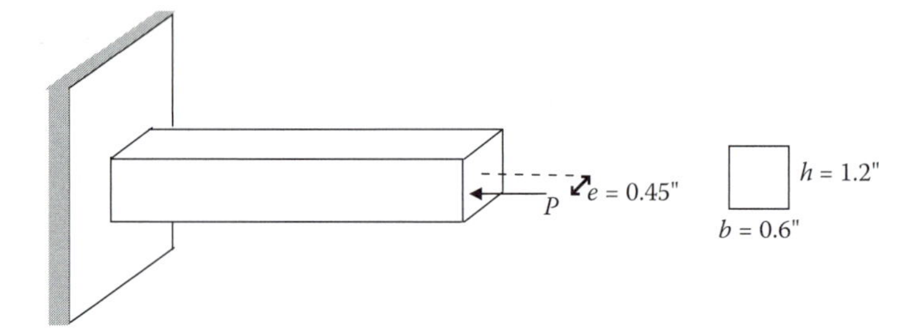

FIGURE 7.13

Case Study 4: Hartford Civic Arena

A new arena in Hartford, Connecticut, was approved in 1970 and built in 1973. The facility suffered a catastrophic failure in January 1978, when its roof collapsed only hours after a large crowd had attended a University of Connecticut hockey game. The resulting damage is seen in Figure CS4.1. The center of the roof appears sunken in, while the corners have been thrust upward.

FIGURE CS4.1
Damage at the Hartford Civic Arena, 1978. (From Feld, J. and Carper, K., *Construction Failure*, Wiley & Sons, New York, 1997. With permission.)

Tasked with saving money for the city of Hartford, the architect and engineering firm created an innovative design for the arena's roof. The proposed roof consisted of two main layers arranged in 30 × 30 ft grids composed of horizontal steel bars 21 ft apart. Diagonal bars 30 ft in length connected the nodes of the upper and lower layers, and, in turn, were braced by a middle layer of horizontal bars. The 30-ft bars in the top layer were also braced at their midpoint by intermediate diagonal bars. The space frame, shown in Figure CS4.2, looks like a set of linked pyramid-shaped trusses.

This was not a conventional space frame roof design. Many of its unique features contributed to the vulnerability of the structure. In particular, the configuration of the four steel angles did not provide good resistance to buckling. The cross-shaped section has a much smaller radius of gyration than either an I section or a tube section (Figure CS4.3). Also, the top horizontal bars intersected at a different point than the diagonal bars rather than at the same point, making the roof especially susceptible to buckling as this load eccentricity induced bending stresses. And the space frame was not cambered. Computer analysis predicted a downward deflection of 13 in. at the midpoint of the roof and an upward deflection of 6 in. at the corners.

To save time and money, the roof frame was assembled on the ground. While it was on the ground the inspection agency notified the engineers that it had measured excessive deflections. No changes or repairs were made. Hydraulic jacks were used to lift the completed roof into position. Once the frame was in its final position but before the roof deck (which would support the final roofing material) was installed, the roof frame's deflection was measured to be twice that predicted by computer analysis, and the engineers were notified. However, they expressed little concern

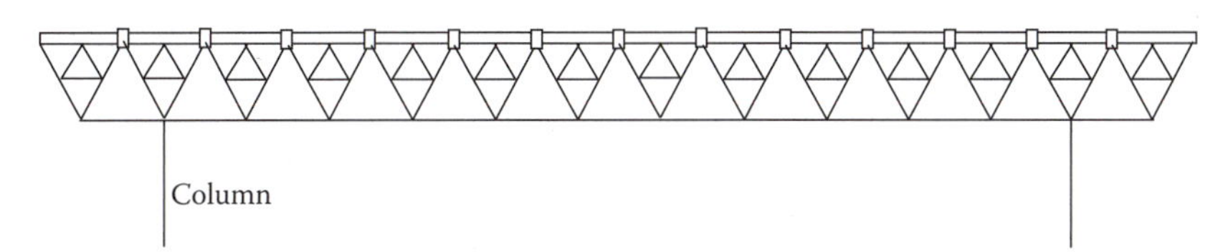

FIGURE CS4.2
Sketch of roof design. (Adapted from Levy, M. and Salvadori, M., *Why Buildings Fall Down,* Norton, New York, 1992.)

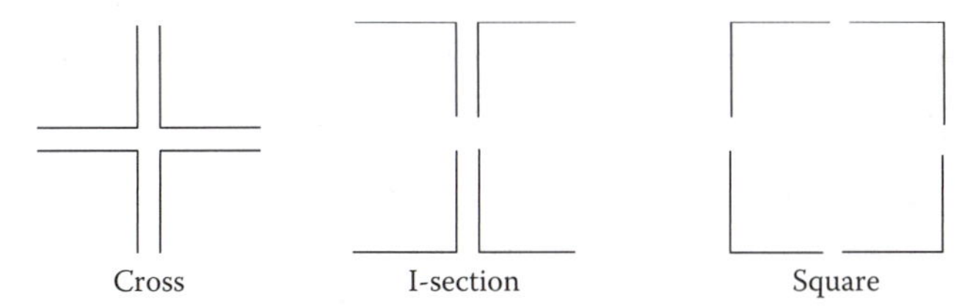

FIGURE CS4.3
Cross-shaped member cross section, as used in the Hartford Arena roof frame; more conventional I section and tube cross-section shapes.

and responded that such discrepancies between the actual and the theoretical should be expected. The subcontractor fitting the steel frame supports for fascia panels onto the outside of the truss ran into difficulties due to the excessive deflections of the frame, but as directed by the contractor, he backcut some panels and remade others so that they would all fit together more closely.

The engineers, contractor, and members of the Hartford City Council made public statements attesting to the safety of the structure. And the roof survived for five years before the heavy snow of January 1978 triggered its catastrophic failure. At 4:15 a.m. on January 18, witnesses reported hearing a loud crack and seeing the center of the roof begin to sink in before the explosive chaos of the rapid collapse. Because the hockey crowd had left hours earlier, no one was hurt in the collapse.

In the subsequent investigation (performed by an appointed panel and an outside failure analysis agency), it was determined that the roof of the Hartford Arena had begun failing as soon as it was completed due to an underestimation of the "dead load"[2] the roof would need to support and also because of three design errors that resulted in a significant overloading of structural components:

- The top layer's exterior compression members on the east and the west faces were overloaded by 852%.
- The top layer's exterior compression members on the north and the south faces were overloaded by 213%.
- The top layer's interior compression members in the east–west direction were overloaded by 72%.

In addition, the support braces in the middle layer had been installed at 30-ft-intervals rather than the designed 15 ft, reducing the structure's ability to withstand loading—particularly such dramatic overloading. The most overstressed members in the top layer buckled under the added weight of the snow, causing the other members to buckle. This changed the forces acting on the lower layer from tension to compression, causing them to buckle as well.

The investigators also determined that several departures from the engineers' design contributed to, but did not cause, the collapse: (1) the slenderness ratio of the built-up members violated the American Institute of Steel Construction (AISC) code provisions; (2) the members with bolt holes exceeding 85% of the total area violated the AISC code; (3) spacer plates were placed too far apart in some members, allowing individual angles to buckle; (4) some of the steel did not meet specifications; and (5) there were misplaced diagonal members.

A second investigation blamed the failure not on lateral buckling but on torsional buckling of diagonal members that could not support the live load of the heavy snowfall.[3] A third investigation pinned the blame on a faulty weld securing the scoreboard to the roof.

It was noted that the roof, despite its many flaws, had apparently survived for five years before its dramatic failure. One study analyzed the *progressive failure* of the roof, which was a 5-year-long process. When a member of a frame structure buckles, it transfers its load to adjacent bars. These bars eventually buckle under the increased load and continue the load-transferring domino effect until the entire roof structure cannot withstand any greater load and begins to give way. This sort of progressive failure can be triggered by even a minor structural flaw unless the design includes *redundancy*—as Levy and Salvadori (1992) put it, "structural insurance." Analyses have shown that relatively few additional braces in the Hartford roof would have prevented bar buckling.

The assessment of responsibility for the collapse was as complicated as determining the reasons. The fact that five independent subcontractors constructed the arena made assessing responsibility especially tricky. The lack of any one body with ownership and oversight of the entire project created a fragmented system in which no one examined the "big picture." Six years after the collapse, all of the parties involved reached an out-of-court settlement.

It's also worth noting that potential problems with the Hartford arena design were brought before the engineers several times during the construction of the arena. The engineers, confident in their designs (and, perhaps willfully unaware that what was built might not be precisely what they'd designed) and in their computations (from which they'd omitted buckling as a possible failure mode), did not heed warnings or reexamine their work. In fact, unanticipated deformations can indicate a flawed design and are generally worth investigating.

Notes

1. We have just developed a theory of beam bending and deflection that applies to slender beams, for which the cross-section dimensions are much less than the axial length; for columns, we continue to work under this assumption, and we quantify a measure of slenderness.
2. Structures we design must withstand both "dead" and "live" loads. The dead load is simply the weight of the structure itself; live load is the anticipated weight it must also be able to support. For example, bridges must be able to support a predicted traffic load of cars and trucks; buildings must support the weight of the people and furniture in them; and all structures must also withstand loading due to wind, rain, and snow.
3. Several other roof collapses in the Northeast were attributed to heavy snows in 1978, including the roof of the auditorium at C.W. Post College on Long Island.

8

Connecting Solid and Fluid Mechanics

We are now familiar with the response of solids to external loading. We have learned about the stress tensor, the strain tensor, and the individual components of these measures of *internal response to external loads.* Time and time again, we have returned to the essence of continuum mechanics:

- Kinematics (i.e., compatibility)
- Definition of stress
- Constitutive law (stress–strain relationship)
- Equilibrium

Solids, we remember, are *continua*—their densities may be mathematically defined. Fluids (i.e., gases and liquids) may also satisfy this definition, so these concepts of stress and strain also apply to them. As we have done for solids, we would now like to contemplate the response of fluids to loading, and to consider how stress may be related to the material's deformation.

Remember that a fluid may be called a *continuum* if the Knudsen number, Kn, is less than about 0.1. The Knudsen number is defined as

$$Kn = \frac{\lambda}{L}, \tag{8.1}$$

where L is a problem-specific characteristic length, such as a diameter or width, and λ is the material's *mean free path.* We have already considered what is and what is not a continuum at some length.

When this assumption of a material's continuity is made, the properties of a material—solid or fluid—may be assumed to apply uniformly in space and time. That is, the density ρ may vary in space and time, but it is always definable and is a continuous function of x, y, z, and t.

Fluids are usually defined and distinguished from solids as materials that deform continuously under shear stress. This is true no matter how small the applied shear stress is. Also, when normal stress is applied (i.e., when a fluid is squeezed in one direction), the fluid flows in the other two directions. This can be observed when you squeeze a hose in the middle and see water flow from its ends. Fluids cannot offer permanent resistance to these kinds of loads.

If we now consider fluid mechanics with the ideas of solid mechanics fresh in our minds, we can see many connections and analogies between the two fields. Fluids have their own measures of *elasticity, resistance* to loads, and *deformation*. In the following sections, we discuss some of the important properties of fluids. If density variation or heat transfer is significant, these fluid properties must be supplemented with additional information.

Once again we rely on the fundamentals of (1) kinematics, (2) the definition of stress; (3) constitutive law; and (4) equilibrium, or, more generally, Newton's second law. Put another way, also by now familiar, we ensure equilibrium (or Newton's second law), compatibility, and a constitutive law are satisfied at all times. In this chapter, we first consider the kinds of stress that may be experienced by a volume of fluid; next, we discuss a fluid's constitutive law. Finally, we develop a way to talk about the kinematics of deformation of a fluid, this time using strain *rate* rather than strain as we did for solids. In Chapter 9, we use these three definitions to enforce equilibrium.

8.1 Pressure

In fluids, pressure results from a normal compressive force acting on an area, as shown in Figure 8.1. It is written as

$$p = \lim_{\Delta A \to 0} \frac{\Delta F_n}{\Delta A} \tag{8.2}$$

and has units of N/m^2 or psi. We recognize that this is also the definition of a normal stress. In fact, if this compression were the only force acting on an element of the fluid, the element's stress tensor could be written as

$$\underline{\underline{\sigma}} = \sigma_{ij} = \begin{pmatrix} -p & 0 & 0 \\ 0 & -p & 0 \\ 0 & 0 & -p \end{pmatrix}, \tag{8.3}$$

where the negative sign is present because positive pressure is compressive, and compression is represented by negative normal stress. We will see that in reality a variety of forces may act on a fluid element but that pressure is always an important part of its stress state.

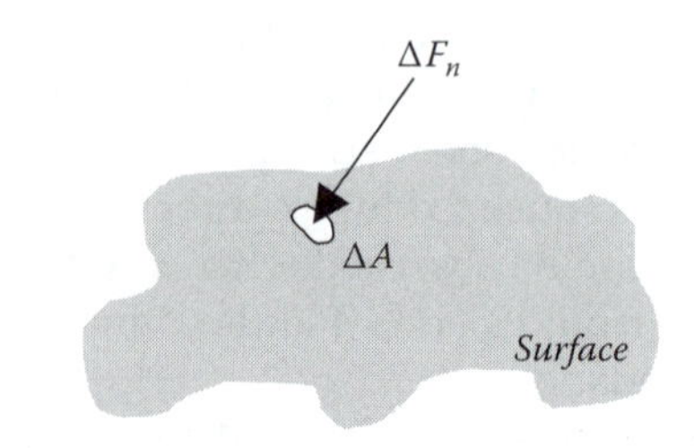

FIGURE 8.1
Definition of pressure.

As in our discussion of pressure vessels, we generally speak of a *gauge pressure* that is measured relative to local atmospheric pressure:

$$p_{\text{gage}} = p_{\text{absolute}} - p_{\text{atm}}.$$

Pressure, as shown in Figure 8.1, is a *surface force,* acting on boundaries of a fluid through direct contact. Shear forces and stresses also fit this description. Fluids may also be acted on by *body forces,* which are applied without physical contact and distributed over the entire fluid volume. The total body force is in fact proportional to the fluid volume. Gravitational and electromagnetic fields impart body forces to fluids.

8.2 Viscosity

A fluid's viscosity can be thought of as a measure of how well the fluid *flows.* Water and maple syrup, for example, flow differently, at different rates; the difference is reflected in their viscosities. The rate of deformation of a fluid is directly linked to the fluid's viscosity. If we consider a fluid element of area $dx \times dy$ under application of a shearing stress τ, as in Figure 8.2, we see that the shear strain angle $d\theta$ grows continuously as long as τ is maintained. Remember that this is what differentiates fluids from solids: that they deform continuously under shear. There is therefore a time dependence in their constitutive law. The rate at which this deformation occurs depends on many factors, and particularly on the fluid's properties.

In Figure 8.2, we see that a plate sliding with speed du over our initially rectangular fluid element induces some angular deformation $d\theta$ in a time dt. When this experiment is performed on common fluids like water, oil, and air, the experimenters observe that the shear stress τ is proportional to the rate of angular deformation $d\theta/dt$. The constant of proportionality is the fluid's viscosity μ.

We use the geometry of Figure 8.2 to manipulate this experimentally observed relationship into its more useful form:

$$\propto \frac{d\theta}{dt} \tag{8.4a}$$

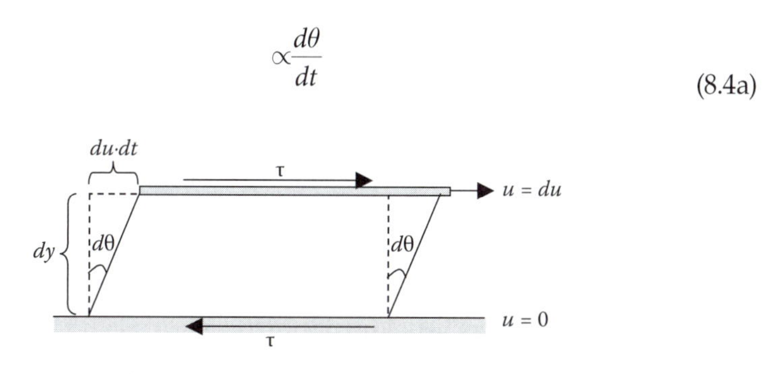

FIGURE 8.2
Sliding plate inducing shear stress τ. Note that this is τ_{yx}.

$$\tan d\theta = \frac{dudt}{dy} \tag{8.4b}$$

For small angles, tan $d\theta \approx d\theta$, and we can rearrange this to have

$$\frac{d\theta}{dt} = \frac{du}{dy}.$$

Finally, we have

$$\tau = \mu \frac{du}{dy}. \tag{8.5}$$

Fluids for which this linear proportionality exists, for which viscosity μ does not itself depend on the strain rate, are called *Newtonian*, and we see that this is analogous to the behavior of a *Hookean* solid. Isaac Newton first referred to the *slipperiness* of fluids and wrote down the essence of equation (8.5) in his *Principia* in 1687. In both cases, we have *Stress* = (*constant*) · (*Strain* or *Strain rate*), whether this constant is E, G, or μ. The dimensions of viscosity are time·force/area, or Pa·s (N·s/m^2) in the International System of Units (SI). The viscosity of a fluid, we see, measures its ability to resist deformation due to shear stress or to resist flow. In a sense, it measures the fluid's stiffness, just as E and G did for solids.

Although we are struck by the analogy between the constitutive laws for solids and fluids, we also note the key difference: the dependence on strain for solids and on strain rate for fluids. Remembering Hooke's initial source of inspiration—the extension of a spring—we arrive at another comparison:

Component	Constitutive Law	Material	Constitutive Law
(spring)	$F = kx$	Solid	$\tau = G\gamma$
(dashpot)	$F = c\dfrac{dx}{dt}$	Fluid	$\tau = \mu\dfrac{du}{dy}$

If we think of solids as behaving more like springs and fluids as behaving more like dashpots, we can relate these constitutive laws to ones with which we are familiar. We can also foresee the introduction of other materials whose behavior is neither purely solid nor purely fluid (e.g., non-Newtonian

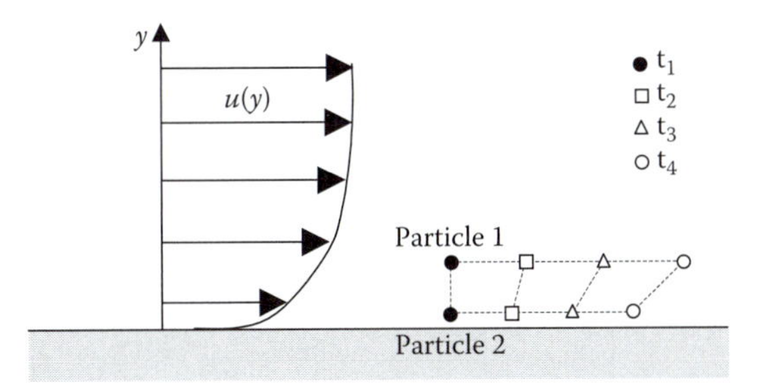

FIGURE 8.3
Relative motion of two fluid particles in the presence of shear stress.

fluids), which may be modeled by the series or parallel combination of these spring and dashpot elements. We can even visualize the gamut of constitutive behavior as a spectrum with springs (Hookean elastic solids) at one end and dashpots (Newtonian fluids) at the other, with myriad variations between.[1] Please see Case Study 5 at the end of this chapter for a discussion of the many types of material behavior possible in between these two idealized extremes.

The du/dy term that appears in the definition of viscosity (equation 8.5) was derived in terms of the angular deformation, or shear strain, of the fluid element per time—that is, the strain rate. It also represents a *gradient of velocity*, as shown in Figure 8.3. Note that if a fluid is not flowing, shear stresses cannot exist, and only normal stress (pressure) is considered.

Viscosity varies with temperature, as shown in Figure 8.4. For a liquid, the temperature dependence can be approximated by an exponential equation, $\mu(T) = c_1 \cdot \exp\,[c_2/T]$, where the constants c_1 and c_2 are determined from measured data. Figure 8.4 demonstrates that the viscosity of a gas is much less dependent on temperature; this is because in liquids the shear stress is due in greater part to intermolecular cohesive forces than to thermal motion of molecules, and these cohesive forces decrease with *T*.

For non-Newtonian fluids (Figure 8.5) the viscosity may also depend on the type or rate of loading applied to the fluid. *Dilatants* such as quicksand or slurries become more resistant to motion as the strain rate increases. A mixture of cornstarch and water is a dilatant and, as you can experimentally verify, "feels" harder the harder (faster!) you pound it. *Pseudoplastics* become less resistant to motion with increased strain rates. Examples of this include ketchup and latex paint. *Bingham plastics*, or viscoplastics, require a minimum shear stress to cause motion but after this threshold behave like Newtonian fluids. Toothpaste is a Bingham plastic.[2]

Viscosity causes fluid to adhere to surfaces; this is called the *no-slip condition*, and it means that the fluid adjacent to any surface moves with the same velocity as the surface itself. Incidentally, μ is more formally called the *dynamic viscosity* of a given fluid. We may also wish to think in terms of a fluid's *kinematic viscosity*, denoted by ν:

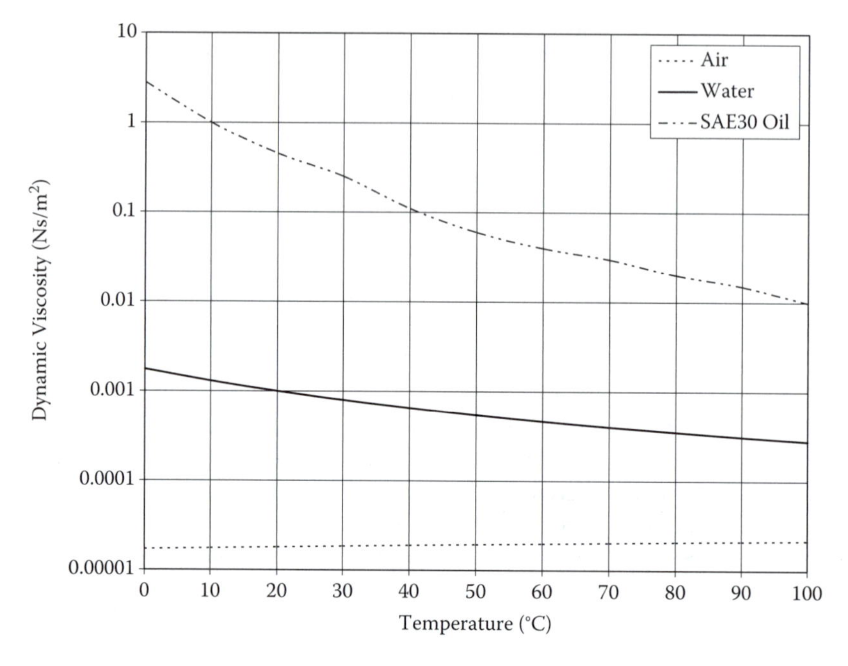

FIGURE 8.4
Viscosity versus temperature for (a) liquids and (b) gases. (Adapted from Fox, R. W. and McDonald, A. T., *Introduction to Fluid Mechanics*, Wiley, 1978.)

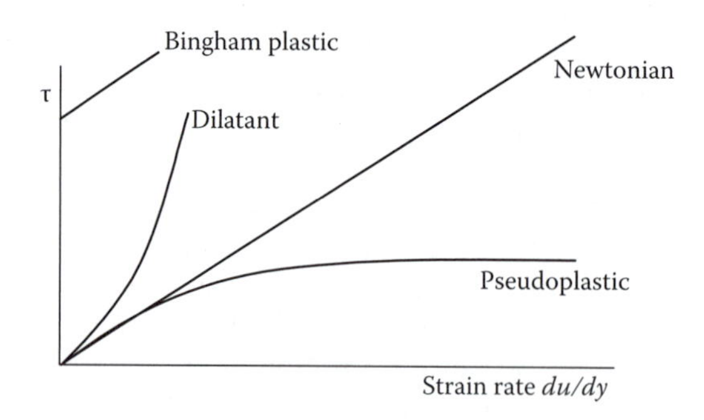

FIGURE 8.5
Newtonian and non-Newtonian fluids. The slope of the stress–strain rate diagram, representing the viscosity or resistance to deformation, is constant for a Newtonian fluid.

$$\nu \equiv \frac{\mu}{\rho}, \tag{8.6}$$

which is of special interest as it reflects a fluid's tendency to *diffuse* velocity gradients. The SI units of kinematic viscosity are m^2/s.

8.3 Surface Tension

The attractive forces between fluid molecules result in *surface tension*. Molecules deep within the fluid are closely packed and are bound by cohesive forces. But the molecules at surfaces are less densely packed, and—because half their neighbors are missing—have nothing to balance their cohesive forces. The result is an inward force, or contraction, at the surface.

Surface tension is generally represented with a lower-case sigma, but to avoid confusion with normal stress components we denote surface tension with s. It is measured in (N/m) or in (lb/ft) and depends on the two fluids in contact and on their temperature.

The scale of a given problem determines which forces (inertia, pressure, viscosity, or surface tension) are involved in its physics. Though in traditional fluid mechanics textbooks the importance of this last force, surface tension, is often minimized, it has enormous relevance in emerging microscale and nanoscale fields.

Because inertia (the ma term in $F = ma$) scales as the volume of an object, when objects get smaller, inertia decreases by a power of 3. But the force due to surface tension only goes as the length of a given surface so that the same reduction in size causes it to decrease by only a power of 1. This scaling means that surface tension dominates the microscale physics, and inertia hardly enters the picture. However, the importance of surface tension has not always been well understood. Surface tension was seen as a major problem when researchers first began designing microelectromechanical (MEMS) devices. The slightest amount of moisture beneath a miniature cantilever beam would pull the beam down to the substrate, welding it in place. The first micromotors could be rendered inoperable by the moisture in a single drop of water. Now that it is better understood, surface tension can be harnessed to create motion if it is increased locally and decreased somewhere else. Researchers do this by adding a surfactant (e.g., soap, which lowers surface tension), by raising the temperature at one point (which decreases surface tension), or by applying an electrical potential.

8.4 Governing Laws

Newton's laws of motion apply to fluids just as they do to solids. Newton's second law, $\underline{F} = m\underline{a}$, will be especially useful to us as we consider the combined effects of all forces (due to, e.g., pressure, viscosity, surface tension) on a fluid and require their resultant to equal ma.

8.5 Motion and Deformation of Fluids

The motion and deformation of a fluid element depend on the velocity field. The relationship between this motion and the forces causing the motion depends on the acceleration field (via $F = ma$). We use an *Eulerian* description, in which we concentrate on a spatial point x and consider the flow through and around this point, rather than the *Lagrangian* method of description sometimes used to track individual fluid particles.

8.5.1 Linear Motion and Deformation

If all points in a given fluid element have the same velocity, the element simply translates from one point to the next. However, we typically have *velocity gradients* present so that the element is deformed and rotated as it moves. We write the velocity $V = (u, v, w)$ in Cartesian (x, y, z) coordinates. A sample fluid element, a cube with infinitesimal volume $dV = dxdydz$, is shown in Figure 8.6. This element is part of a flow with velocity gradient

$$\frac{\partial u}{\partial x},$$

that is, the x velocity is varying with x. In a time interval dt, the change in the element's volume is given by

$$\left(\frac{\partial u}{\partial x}dxdt\right)\left(dydz\right)=\frac{\partial u}{\partial x}dtd\mathcal{V}. \tag{8.7}$$

The rate at which the volume dV is changing, per unit volume, due to $\frac{\partial u}{\partial x}$ may be written

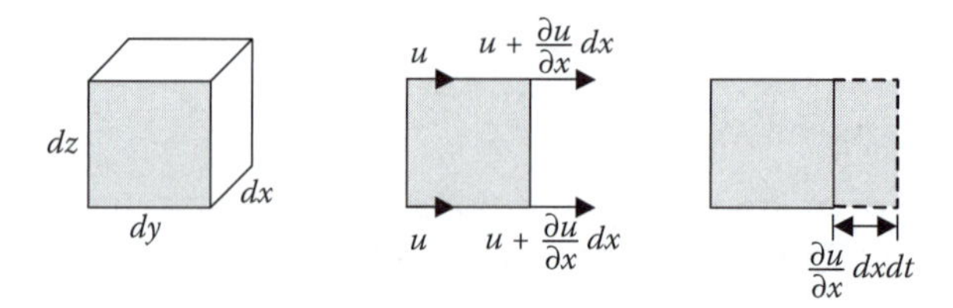

FIGURE 8.6
Linear deformation of fluid element by $\frac{\partial u}{\partial x}$.

$$\frac{1}{dV}\frac{d(dV)}{dt} = \lim_{dt \to 0}\left[\frac{\frac{\partial u}{\partial x}dt}{dt}\right] = \frac{\partial u}{dx}. \tag{8.8}$$

And if velocity gradients $\frac{\partial v}{\partial y}$ and $\frac{\partial w}{\partial z}$ are also present, we have the rate of change in volume (per unit volume):

$$\frac{1}{dV}\frac{d(dV)}{dt} = \frac{\partial u}{\partial x} + \frac{\partial v}{\partial y} + \frac{\partial w}{\partial z} = \nabla \cdot \underline{V}, \tag{8.9}$$

where we have recognized the sum of the partial derivatives of V's components as the *divergence* of V.

Notice that the isolated effect of each of these velocity gradients causes a one-dimensional, normalized change in length, or *normal strain rate*, which can be written in the same way:

$$\varepsilon_{xx} = \varepsilon_x = \frac{\partial_y}{\partial_x}. \tag{8.10a}$$

$$\varepsilon_{yy} = \varepsilon_y = \frac{\partial v}{\partial y}, \tag{8.10b}$$

$$\varepsilon_{zz} = \varepsilon_z = \frac{\partial w}{\partial z}. \tag{8.10c}$$

The quantity $\nabla \cdot \underline{V}$ derived in equation (8.9) for the entire volume is known as the *volumetric strain rate*. For an incompressible fluid, the volume of a fluid element cannot change, and we must have $\nabla \cdot \underline{V} = 0$.

8.5.2 Angular Motion and Deformation

In addition to undergoing normal strain rates, a fluid element may experience angular motion and deformation. We measure this with a *shear strain rate*, derived from the change in shape of the fluid element in Figure 8.7. The figure shows the position of an element with initial area $dxdy$ at time t and its subsequent position at time $t + dt$. We see that the initially horizontal side (initial length dx) has undergone a rotation $d\alpha$, and the initially vertical side (initial length dy) has been rotated $d\beta$. We can calculate these angles and then find an expression for the shear strain rate, defined as $(d\alpha + d\beta)/dt$. We

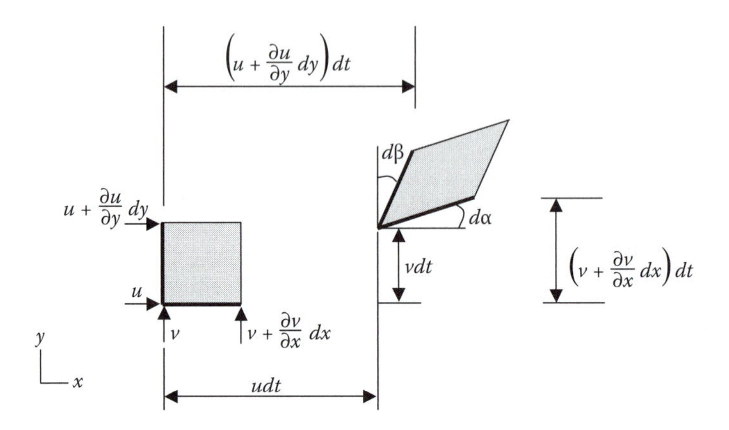

FIGURE 8.7
Translation and angular deformation of a fluid element.

also assume that the angles are small, so tan $d\alpha \approx d\alpha$. From the figure we see that

$$\tan d\alpha \approx d\alpha = \frac{\left(\frac{\partial v}{\partial x}\right) dxdt}{dx} = \frac{\partial v}{\partial x} dt. \tag{8.11}$$

We note that if

$$\frac{\partial v}{\partial x} > 0,$$

the rotation of this side is counterclockwise. Similarly, we find that

$$d\beta = \frac{\partial u}{\partial y} dt, \tag{8.12}$$

seeing that if

$$\frac{\partial u}{\partial y} > 0,$$

the rotation of this side is clockwise, and, thus, we find the shear strain rate:

$$\frac{d\alpha + d\beta}{dt} = \frac{\partial v}{\partial x} + \frac{\partial u}{\partial y}. \tag{8.13}$$

And extending this to the other two dimensions, we see that in general the *ij* component of shear strain rate may be written as

$$\gamma_{ij} = \frac{\partial u_i}{\partial x_j} + \frac{\partial u_j}{\partial x_i}. \tag{8.14}$$

However, as before, when we compose the *strain rate tensor,* these shear components must be divided by 2 to make the tensor behave like a tensor. Now, a general form for the *ij* component of the strain rate tensor, including both normal and shear components, may be written as

$$\varepsilon_{ij} = \frac{1}{2}\left(\frac{\partial u_i}{\partial x_j} + \frac{\partial u_j}{\partial x_i}\right) \tag{8.15}$$

so that the matrix form of the tensor itself looks like

$$\underline{\underline{\varepsilon}} = \begin{pmatrix} \varepsilon_x & \varepsilon_{xy} = \frac{\gamma_{xy}}{2} & \varepsilon_{xz} = \frac{\gamma_{xz}}{2} \\ \varepsilon_{yx} = \frac{\gamma_{xy}}{2} & \varepsilon_y & \varepsilon_{yz} = \frac{\gamma_{yz}}{2} \\ \varepsilon_{zx} = \frac{\gamma_{xz}}{2} & \varepsilon_{zy} = \frac{\gamma_{yz}}{2} & \varepsilon_z \end{pmatrix} = \begin{pmatrix} \frac{\partial u}{\partial x} & \frac{1}{2}\left(\frac{\partial u}{\partial y} + \frac{\partial v}{\partial x}\right) & \frac{1}{2}\left(\frac{\partial u}{\partial z} + \frac{\partial w}{\partial x}\right) \\ \frac{1}{2}\left(\frac{\partial u}{\partial y} + \frac{\partial v}{\partial x}\right) & \frac{\partial v}{\partial y} & \frac{1}{2}\left(\frac{\partial w}{\partial y} + \frac{\partial v}{\partial z}\right) \\ \frac{1}{2}\left(\frac{\partial u}{\partial z} + \frac{\partial w}{\partial x}\right) & \frac{1}{2}\left(\frac{\partial w}{\partial y} + \frac{\partial v}{\partial z}\right) & \frac{\partial w}{\partial z} \end{pmatrix}. \tag{8.16}$$

All of these components of the strain rate tensor should look strikingly similar to the components of the strain tensor derived for a solid in Chapter 3, Section 3.2. This similarity, while undeniably wondrous, should not be surprising: Both fluids and solids are *continua,* and their deformations can be written mathematically in the form of a nine-component tensor, which we have seen can be related to the nine components of stress. The difference between this strain rate tensor and the strain tensor in Section 8.4 is simply that for solids, strain is a dimensionless quantity measuring percent length change, while for fluids, we measure rate of strain so that *(u, v, w)* here are velocities rather than lengths.

8.5.3 Vorticity

To quantify the rotation of fluid elements due to a given flow, we again consider the angles $d\alpha$ and $d\beta$ as shown in Figure 8.7. We want to find an expression for the average rotation rate of this element. Again we consider both $d\alpha$ and $d\beta$, the rotations of two mutually perpendicular lines. (This is because the average of these two rotation rates is independent of the initial orientation of the pair.) To combine these two we must remember that $d\alpha$

is a counterclockwise rotation, while $d\beta$ was clockwise; thus, we find the combined effect to be

$$\text{Angular velocity of element about } z \text{ axis} = \frac{1}{dt}\left[\frac{1}{dy}\left(-\frac{\partial u}{\partial y}dydt\right)+\frac{1}{dx}\left(\frac{\partial v}{\partial x}dxdt\right)\right],$$

or

$$\omega_z = \frac{\partial v}{\partial x} - \frac{\partial u}{\partial y}. \tag{8.17}$$

This angular velocity could also be computed for rotation about the x and y axes, with similar results, giving us three components,

$$\omega_x = \frac{\partial w}{\partial y} - \frac{\partial v}{\partial z}, \tag{8.18a}$$

$$\omega_y = \frac{\partial u}{\partial z} - \frac{\partial w}{\partial x}, \tag{8.18b}$$

$$\omega_z = \frac{\partial v}{\partial x} - \frac{\partial u}{\partial y}. \tag{8.18c}$$

of what is known as the *vorticity* vector. We recognize that the vorticity may be written as the *curl* of the velocity field, or

$$\underline{\omega} = \nabla \times \underline{V}. \tag{8.19}$$

If a flow has

$$\nabla \times \underline{V} = 0,$$

the flow is called *irrotational*. For such flows, the velocity vector V can be written as the gradient of a scalar potential function

$$[\mathrm{V} = \nabla\phi],$$

since the curl of a gradient must be zero.

8.5.4 Constitutive Equation (Generalized Hooke's Law) for Newtonian Fluids

We recall that the relationship between stress and deformation in a continuum is known as a *constitutive equation*. We now seek an equation linearly relating the stress to the rate of strain in a fluid, a counterpart to the generalized form of Hooke's law for solids.

We have already seen that pressure is a normal stress on the surface of a fluid element. This contribution to the stress tensor may be written as a diagonal matrix with eigenvalues $-p$. We make use of the tensor equivalent of the identity matrix, known as the *Kronecker delta*. The Kronecker delta is a second-order, isotropic tensor whose matrix representation is

$$\underline{\underline{\delta}} = \begin{pmatrix} 1 & 0 & 0 \\ 0 & 1 & 0 \\ 0 & 0 & 1 \end{pmatrix}. \tag{8.20}$$

And so we can write the pressure's contribution to the fluid's stress state as

$$\sigma_{ij} = -p\delta_{ij}, \tag{8.21}$$

where δ_{ij}, and, hence, σ_{ij} is only nonzero when $i = j$.

Knowing that we can superpose these components of normal stress with any normal components that arise due to fluid motion, as described in Section 8.5.1, we simply add on the stress tensor that is developed by fluid motion so that the complete stress picture is given by

$$\tau_{ij} = -p\delta_{ij} + \sigma_{ij}^{d}, \tag{8.22}$$

where σ_{ij}^{d}, the part of the stress tensor due to fluid motion, is known as the *deviatoric stress tensor*. It is related to the velocity gradients, as we have seen through the construction of the strain rate tensor. We now know

$$\varepsilon_{ij} = \frac{1}{2}\left(\frac{\partial u_i}{\partial x_j} + \frac{\partial u_j}{\partial x_i}\right), \tag{8.23}$$

and we assume a linear relationship between stress and strain rate,

$$\sigma_{ij}^{d} = K_{ijmn}\varepsilon_{mn}, \tag{8.24}$$

where K_{ijmn} is a fourth-order tensor with eighty-one components, very much like the large tensor invoked in our discussion of the generalized form of

Hooke's law (Chapter 3, Section 3.5). We recall that for solids, this large tensor depended on E and G and Poisson's ratio ν. For fluids, K turns out to depend on viscosity μ and to have a very simple form for most fluids. We need only assume that the fluid is isotropic and that the stress tensor is symmetric to reduce K to a matter of only two (not eighty-one) elements.[3] In fact, the whole mess can be reduced quite nicely to

$$\tau_{ij} = -(p + \tfrac{2}{3}\mu\nabla\cdot\underline{v})\delta_{ij} + 2\mu\varepsilon_{ij}, \tag{8.25}$$

which for an incompressible fluid

$$(\nabla\cdot\underline{v} = 0)$$

reduces still further to

$$\tau_{ij} = -p\delta_{ij} + 2\mu\varepsilon_{ij}\,. \tag{8.26}$$

This is the constitutive law for an incompressible, Newtonian fluid. As we did for solids, we are able to consider a few components of this relationship at a time. But again, it is useful to see the big picture.

8.6 Examples

Example 8.1

In the center of a hurricane, the pressure can be very low. Find the force acting on the wall of a house, measuring 10 ft × 20 ft, when the pressure inside the house is 30 in. Hg and the pressure outside is 26.3 in. Hg (Figure 8.8 and Figure 8.9). Express the answer in both pounds and Newtons.

FIGURE 8.8

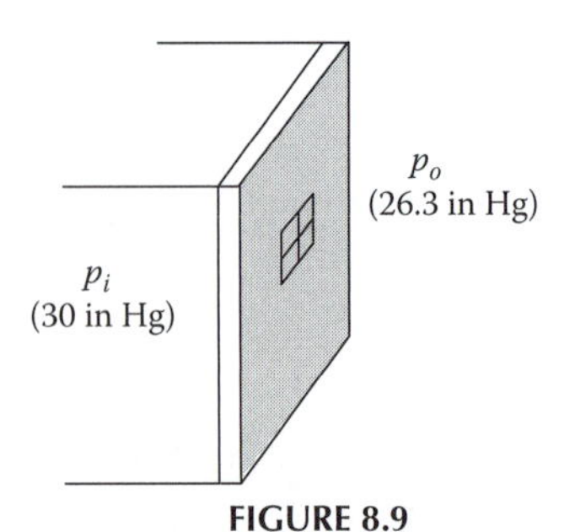

FIGURE 8.9

Given: Pressure on both sides of wall; wall dimensions.

Find: Resultant force on wall.

Assume: Uniform pressure distributions on both sides of wall.
Negligible pressure contributions from inside wall.

Solution

A mercury barometer measures the local atmospheric pressure. A standard atmosphere has a pressure of 14.7 psi, or 101.325 kPa. A mercury barometer reads this standard atmospheric pressure as 760 mm Hg, or 29.92 in. Hg. Since we are asked for a result in two different units, we must be mindful of these conversion factors.

The resultant force on the wall is simply the net pressure applied to it times its area. A quick free-body diagram (FBD) of the wall will be of use (Figure 8.10).

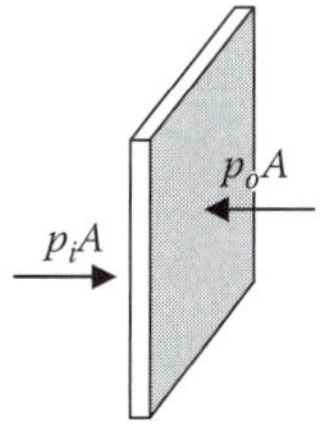

FIGURE 8.10

We see that the net force on the wall will be directed outward and that it is

$$F = (p_i - p_o)A$$

$$= (30.0 - 26.3)\ (\text{in. Hg})\frac{14.7\ \text{psi}}{29.92\ \text{in. Hg}} \cdot (120\ \text{in.})(240\ \text{in.})$$

$$= 52{,}354\ \text{lb}$$

$$= 52{,}354\ \text{lb}\ \frac{4.448\ \text{N}}{1\ \text{lb}}$$

$$= 232{,}871\ \text{N}.$$

This outward force is very large, and if the wall has not been adequately strengthened the force can explode the wall outward. If you know a hurricane is coming, it's therefore a good idea to open as many windows as possible to equalize the pressure inside and out.

Example 8.2

The flow between two parallel plates, one of which is moving with constant speed U, is known as *Couette flow* (Figure 8.11). If the fluid between the two plates is Newtonian, develop an expression for the velocity distribution in the fluid layer. If the fluid is SAE oil at 20°C, which has a viscosity of 0.26 Pa·s, and if the top plate moves with speed $U = 3$ m/s and the gap thickness is $h =$ 2 cm, what shear stress is applied to the fluid?

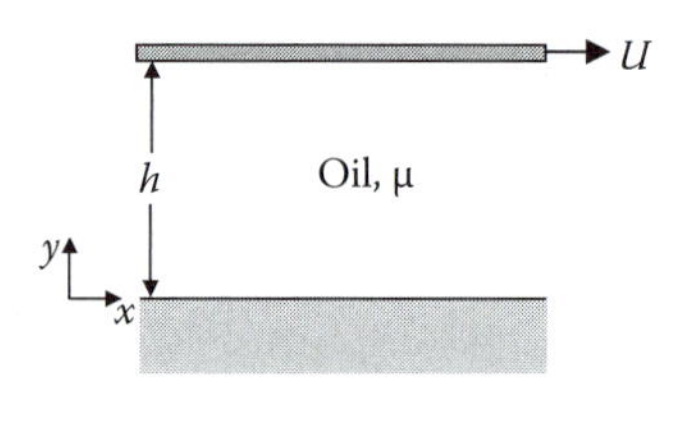

FIGURE 8.11

Given: Couette flow.

Find: Fluid velocity distribution, shear stress.

Assume: Newtonian fluid; any transient effects due to initiation of plate motion have died out and flow is steady; negligible gravity; one-dimensional flow; $u = u(y)$ only.

Solution

We know that, for a Newtonian fluid, the definition

$$\tau_{yx} = \mu \frac{du}{dy}$$

is a linear relationship with constant viscosity μ. This is a differential equation we can solve for velocity $u(y)$.

Due to equilibrium, the shear stress will be constant throughout the layer of fluid (Figure 8.12):

FIGURE 8.12

This is because there are no other forces on the fluid, so to keep a fluid element in equilibrium we must have τ_{yx} (= τ_{xy}) = constant = τ. So,

$$\tau = \mu \frac{du}{dy},$$

$$\frac{du}{dy} = \frac{\tau}{\mu} = \text{constant.}$$

If we call this constant C and then integrate, we find that the velocity must have the form $u(y) = Cy + D$. To complete the solution, we use boundary conditions. We have discussed an important property of viscous fluids: that the fluid adjacent to a solid surface moves with the same speed at that surface. This is known as the *no-slip* condition. At the lower plate, which is at rest, we have $u(y = 0) = 0$; at the upper plate which slides with speed U we have $u(y = h) = U$. Applying these boundary conditions:

$$u(y = 0) = 0 \rightarrow D = 0$$

$$u(y = h) = U \rightarrow C = U/h$$

We therefore must have (Figure 8.13)

$$\varepsilon_{xx} = \varepsilon_x = \frac{\partial_y}{\partial_x}.$$

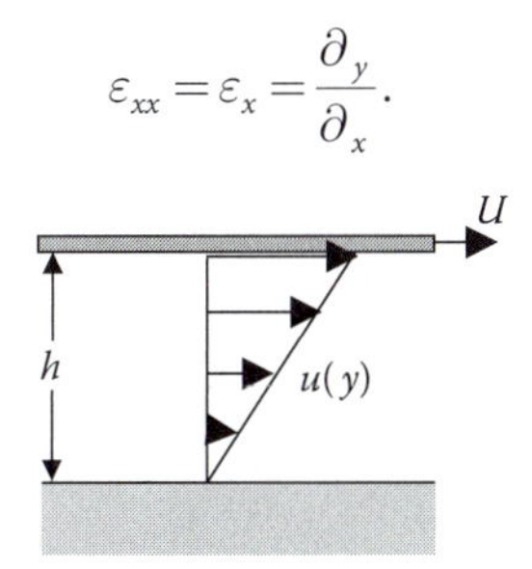

FIGURE 8.13

For the given numerical values, we find the shear stress to be

$$\tau = \mu \frac{du}{dy} = \mu \frac{U}{h} = (0.26 \text{ Pa·s}) \frac{(3 \text{ m/s})}{(0.02 \text{ m})} = 39 \text{ Pa.}$$

Example 8.3

A 5-kg cube with sides 12 cm long slides down an oil-coated incline (Figure 8.14). If the incline makes a 10° angle with the horizontal and the oil layer is 0.2-mm thick, estimate the speed with which the block slides down the incline. The viscosity of the oil is 0.1 Pa·s.

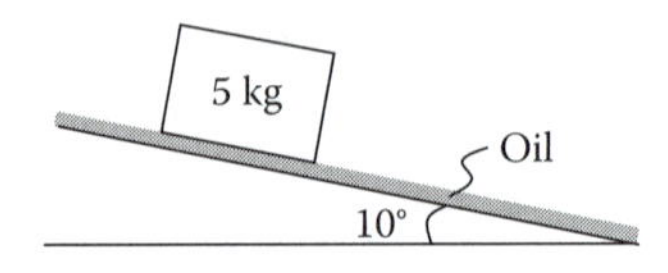

FIGURE 8.14

Given: Dimensions of cube and fluid layer.

Find: Cube's terminal velocity.

Assume: Newtonian fluid; flow is steady; negligible end effects; one-dimensional fluid flow in thin layer can be modeled as Couette flow.

Solution

We begin with an FBD of the cube (Figure 8.15):

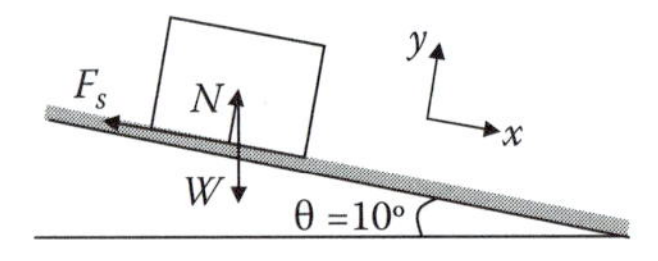

FIGURE 8.15

This contains the weight of the cube itself, a normal force upward, and a frictional resistance from the oil. This shear force F_s is simply the fluid shear stress acting over the area of contact between cube and fluid. (An equal and opposite shear force acts on the layer of oil.)

The cube is not accelerating—we are seeking its terminal velocity. So the cube is in static equilibrium. We must have the sum of forces in both x and y directions equal zero. The x direction is more useful to us:

$$\Sigma F_x = 0$$

$$0 = W \sin\theta - F_s.$$

In the layer of oil, we have a top surface (the bottom of the cube) that is moving with constant velocity, say V, in the x direction, and a bottom surface (the inclined plane) that is at rest. The oil is therefore in Couette flow! We make use of our result from Example 8.2 to express the shear force as

$$F_s = \tau A = \mu \frac{V}{h} A,$$

so

$$0 = W\sin\theta - \mu\frac{V}{h}A\,.$$

Rearranging, we have

$$V = \frac{Wh\sin\theta}{\mu\, A}\,.$$

Plugging in the given values,

$$V = \frac{(5\ \text{kg})(9.81\ \text{m/s}^2)(0.2\times 10^{-3}\ \text{m})\sin(10^\circ)}{(0.1\ \text{Pa}\cdot\text{s})(0.12\ \text{m})^2} = 1.18 \quad \text{m/s}.$$

Example 8.4

The steady flow of an incompressible fluid has the x and y velocity components $u = x^2 + y^2 + z^2$ and $v = xy + yz + z$. What form does the z component of velocity have?

Given: u, v for steady, incompressible flow.

Find: w.

Assume: Steady flow.

Solution

We know that the volumetric strain rate can be written as the divergence of the velocity field and that for an incompressible fluid or flow, this must equal zero:

$$\nabla\cdot\underline{V} = \frac{\partial u}{\partial x} + \frac{\partial v}{\partial y} + \frac{\partial w}{\partial z} = 0.$$

For the given velocity components,

$$\frac{\partial u}{\partial x} = 2x,$$

$$\frac{\partial v}{\partial y} = x + z.$$

Hence,

$$2x + x + z + \frac{\partial w}{\partial z} = 0,$$

or

$$\frac{\partial w}{\partial z} = -3x - z\,.$$

Integrating both sides in z, we have

$$w = -3xz - \frac{1}{2}z^2 + k(x, y),$$

where $k(x, y)$ may be a constant, or any function of x or y or both. The precise nature of $k(x, y)$ cannot be determined from what is known.

8.7 Problems

8.1 A beaker has the shape of a circular cone of diameter 7 in. and height 9 in. When empty, it weighs 14 oz; full of liquid, it weighs 70 oz. Find the density of the liquid in both SI and US units.

8.2 Some experimental data for the viscosity of argon gas at 1 atm are provided:

Fit these data to a power law.

T (K)	300	400	500	600	700	800
μ (N·s/m^2)	2.27×10^{-5}	2.85×10^{-5}	3.37×10^{-5}	3.83×10^{-5}	4.25×10^{-5}	4.64×10^{-5}

8.3 The space between two very long parallel plates separated by a distance h is filled with a fluid with viscosity

$$\mu = \mu_o \left(\frac{du}{dy}\right)^n,$$

where μ_o is a constant and n is a constant exponent. The top plate slides to the right with constant speed V_o, as shown in Figure 8.16.

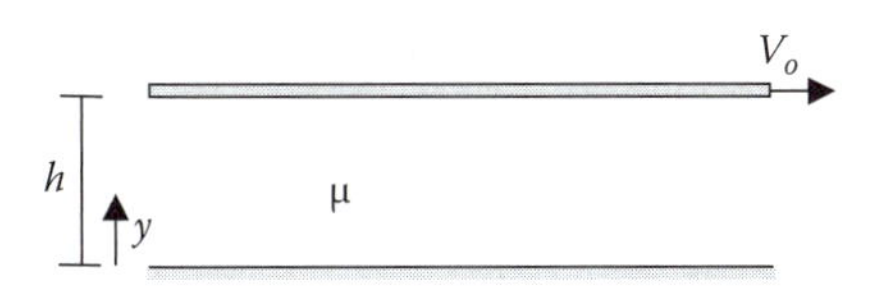

FIGURE 8.16

(a) Find the velocity distribution between the plates and an expression for the shear stress τ.

(b) Graph the shear stress versus the shear strain rate V_o/h for several values of $n > 0$ (dilatant), $n = 0$ (Newtonian), and $n < 0$ (pseudoplastic).

8.4 Many devices have been developed to measure the viscosity of fluids. One such device, known as a *rotational viscometer*, involves a pair of concentric cylinders with radii r_i and r_o, and total length L. The inner cylinder rotates at a rate of Ω rad/s when a torque T is applied. Derive an expression for the viscosity of the fluid between the cylinders, μ, as a function of these parameters.

8.5 A thin plate is separated from two fixed plates by viscous liquids with viscosity values μ_1 and μ_2. The plate spacings are h_1 and h_2 as shown in Figure 8.17. The contact area between the center plate and each fluid is A. Assuming a linear velocity distribution in each fluid, find the force F required to pull the thin plate at velocity V.

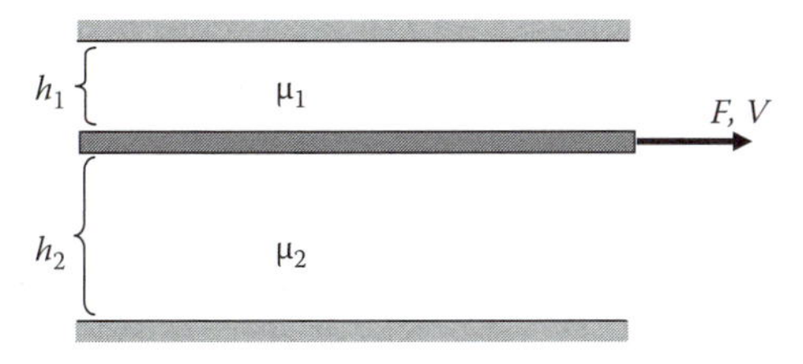

FIGURE 8.17

8.6 Magnet wire is single-strand wire with a thin insulation layer (of, e.g., enamel, varnish, glass) to prevent short circuits. In a production facility, copper (E = 120 GPa, σ_{ys} = 70 MPa) magnet wire is to be coated with varnish by pulling it through a circular die (i.e., a cylindrical tube) of 0.35-mm diameter. The wire diameter is 0.30 mm, and it is centered in the die. The varnish (μ = 0.020 Pa·s) completely fills the space between the wire and the die for a length of 30 mm. Determine the maximum speed with which the wire can be pulled through the die while ensuring a factor of safety of 3.0 with respect to yielding.

8.7 A solid cylindrical needle of diameter d, length L, and density ρ_n is able to float in liquid of surface tension s. Assuming a contact angle of 0°, derive an expression for the maximum diameter d_{max} that will be able to float in the liquid. If the needle is steel and the liquid is water, what is the value of d_{max}?

8.8 A flow is described (in Cartesian coordinates) by the velocity vector $\underline{V} = 2xy\hat{\underline{i}} - 3y2\hat{\underline{j}}$. Is the flow incompressible?

8.9 A flow is described (in Cartesian coordinates) by the velocity vector $\underline{V} = (2x^2 + 6z^2x)\hat{\underline{i}} + (y^2 - 4xy)\hat{\underline{j}} - (2z^3 + 2yz)\hat{\underline{k}}$. Is the flow incompressible?

Case Study 5: Mechanics of Biomaterials

We have discussed the properties and behavior of Hookean solids and Newtonian fluids: materials that are special cases, on either extreme of the spectrum of material behavior. These materials and most applications fit our favorite simplifying assumptions (homogeneity, isotropy, linearity, small deformations) to a tee. While a great number of engineering materials are well served by these assumptions and models, the growing category of biomaterials demands that we consider the more complex behaviors between these two idealized extremes.

Biomaterials may be natural (blood vessels, bone, cartilage, or the cornea) or artificial (joint replacements, blood vessel shunts and stents, or the results of tissue engineering). Scientific interest in biomaterials is not an exclusively modern phenomenon: Ancient technology relied on horn, tendon, and various woods and fibers. However, we are now able to analyze the biological role of biomaterials and how these complex behaviors contribute to the species that rely on them. This helps us understand the relationship between properties and applications or between structure and function. Since engineers often seek to replace or mimic biological materials, we must understand both the material behavior and the biological reasons for it.

It is critical for engineers to understand how such materials respond to loading, to mechanical stresses, and to biochemical and electrical stimuli as well. In his pioneering text on biomechanics, Y. C. Fung (1993) outlines a systematic approach to problems in biomechanics: The first step is studying organism morphology, organ anatomy, tissue histology, and structure of materials. The second is determining the mechanical properties of the materials involved before later steps (i.e., deriving the governing equations, developing boundary conditions, solving the problems, and performing experiments) follow. As Fung notes, determining the mechanical properties of biomaterials can be complicated by the difficulties of isolating the tissue for testing, extracting specimens of sufficient size, or maintaining tissue's in vivo conditions. Compounding these difficulties, biological tissues are often subjected to large deformations and exhibit nonlinear and time-dependent stress-strain behavior.

Tensile testing of the sort described in Chapter 2 and Chapter 3 has yielded an extensive array of properties for biomaterials. Values of Young's modulus are tabulated in Table CS5.1; values of the shear modulus and Poisson's ratio are shown in Table CS5.2. These values come with strong disclaimers, though, as they are only as valid as the assumptions behind them. While these parameters have familiar meanings, and while biomaterials obey many of the equations we have already derived, we must be cautious. Remember well the assumptions implicit in many results of engineering mechanics, and consider how well such assumptions describe the material of interest. Remember from Chapter 3 the handy equation (3.2) that could be used to relate E, G, and v? It does not hold for biomaterials. The measured G and E for bone would suggest a Poisson's ratio of 0.8, which is twice the measured value, and tree trunks and bamboo stalks would have Poisson's ratios of 6 or 7. The trouble is those assumptions that we've begun to make almost implicitly about materials—linearity, homogeneity, and small deformations—often don't apply to biomaterials.

TABLE CS5.1

Modulus of Elasticity for Various Biomaterials

Material	*Young's Modulus of Elasticity* **E** *(MPa)*
Aorta, cow	0.2
Aorta, pig	0.5
Nuchal ligament (mainly elastin)	1.0
Dragonfly tendon (mainly resilin)	1.8
Cartilage	20
Tendon (mainly collagen)	2,000
Tree trunks	6,400
Wood, dry, with grain	10,000
Teeth (dentine)	15,000
Bone (large mammal)	18,000
Teeth (enamel)	60,000
Kevlar (synthetic fiber)	130,000
Steel	200,000

Note: These values should be regarded as rough approximations, with wide variations depending on the rate of stretching, on the amount of deformation, and on the natural biological diversity of each material.

Source: Vogel, S., *Comparative Biomechanics: Life's Physical World*, Princeton University Press, Princeton, NJ, 2003.

TABLE CS5.2
Modulus of Rigidity and Poisson's Ratio for Various Biomaterials

Material	*Shear Modulus* **G** *(MPa)*	*Poisson's Ratio* ν
Aorta (at 100 mm Hg)	0.15	0.24
Cartilage (rabbit)	0.35	0.30
Tendon (mainly collagen)	1 (huge variation)	0.40
Tree trunks	450	0.33
Bone (large mammal)	3300–5000	0.40
Teeth (enamel)	65,000	0.30
Kevlar (synthetic fiber)	30,000	—
Steel	77,000	0.33

Note: These values should be regarded as rough approximations, with wide variations depending on the rate of stretching, on the amount of deformation, and on the natural biological diversity of each material.

Source: Vogel, S., *Comparative Biomechanics: Life's Physical World*, Princeton University Press, Princeton, NJ, 2003.

Material testing of biomaterials to obtain the values shown in Tables CS5.1 and CS5.2 is also a challenge. The properties, microstructure, and behavior of natural biomaterials change in response to the physiologic environment. This makes determining decisive experimental results or developing detailed constitutive models very difficult. The bulk of elastomechanical testing of natural biomaterials has been conducted in vitro in an experimental simulacrum of in vivo conditions, including thermal conditions and ionic concentrations, which affect smooth muscle activity. Proper specimen preparation and conditioning are vital to maintaining the integrity of the material. Debes and Fung (1995) first proposed a preconditioning of very low frequency cyclic loading for a few cycles, suggesting that the internal structure of the tissue would respond to this loading until it reached a steady state that would allow consistent mechanical response to loading.

Nonlinearity

We may have begun to take for granted the linear elasticity of most engineering materials. When working problems, we may even have been tempted to construct a rubber stamp saying, "Hooke's law applies," for the Assumptions section of our solutions. It is time, however, to reexamine that assumption. Many biomaterials have stress–strain curves that are not linear but are J-shaped—that is, curves that get increasingly steep. An example is shown in Figure CS5.1. This sort of curve signifies that the Young's modulus or stiffness of the material increases with extension. For materials with nonlinear behavior, the Young's modulus cited in tables such as Table CS5.1 and used

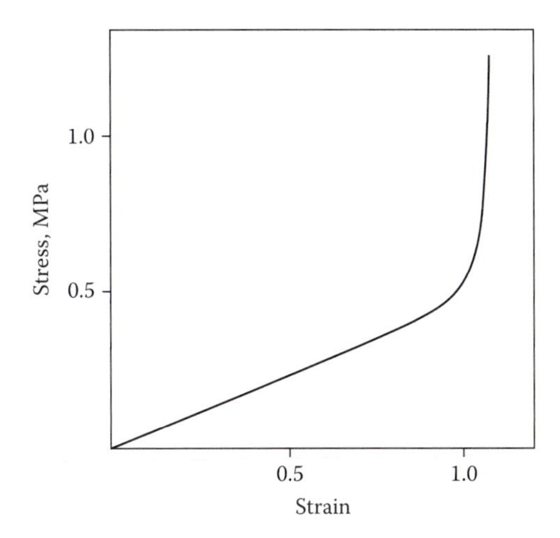

FIGURE CS5.1
Exemplary J-shaped stress–strain curve for nuchal ligament of deer. (Adapted from Vogel, S., *Comparative Biomechanics: Life's Physical World*, Princeton University Press, Princeton, NJ, 2003.)

in calculations is that of the low-strain, quasilinear "toe" region of the stress–strain curve.

In our discussion of pressure vessels we considered an abdominal aortic aneurysm, suggesting that the aneurysm exposed to high fluid pressures might remodel itself into a more spherical shape to reduce the induced stresses. The aortic wall, however, is not a Hookean metal. And it's a good thing, too: Artery walls must expand and contract with each heartbeat to accommodate the heart's pressure pulse. A Hookean material would be a poor choice for a cylinder meant to have a compliant wall, because already dilated portions of the cylinder would tend to expand further just as much as areas that had not expanded, thus creating regions of dangerously high stress.

Although Hookean linearity is the best-case scenario to simplify many of our analyses, there are many reasons that *nonlinearity* is an advantage for biomaterials in practice. Getting stiffer as it gets closer to the failure point can make a structure safer, because it then requires a disproportionate force to break. Additionally, the stress–strain curve's upward concavity reduces the area under the curve (compared with a Hookean material with the same limits), meaning less energy is released on failure of the biomaterial. Energy release drives crack propagation, and for a biomaterial (e.g., skin), we would prefer cracks not to propagate.

Composite Materials

One way to keep cracks from propagating dramatically, and disastrously, through materials is to construct composites from materials with different properties. A well-known engineering trick is to use carbon fibers, which are very stiff, to reinforce other materials (e.g., concrete, which is stronger in compression than in tension) or plastic (carbon fiber-reinforced plastic, widely known as *carbon fiber,* is widely used in modern bicycles and race cars. This addition of a hard component allows the matrix materials (concrete or plastic) to be used in a wider array of applications, and to be more durable.

Nature has made good use of composite materials, in wood and leaves of grass and in tendons (which are collagen fiber-reinforced) and bone (in which cells form osteons that reinforce the longitudinal direction against compressive loads). Table CS5.3 shows some natural composites and their components.

Blood vessels are soft tissue composed of elastin and collagen fibers, smooth muscle, and a single layer of endothelial cells lining the vessel lumen. The proportions of the fibrous proteins and vascular smooth muscle depend on the type of blood vessel and the loading it must withstand. The two types of protein fibers have important consequences for the material behavior of vessels. Elastin is a very elastic fiber with a large Hookean region in its stress–strain behavior. It provides blood vessels with the ability to expand (or *distend*) to accommodate the pressure pulse of the heartbeat. Vessels that are too stiff will not expand, and this will result in high blood pressure or hypertension. Collagen fibers form a network outside the elastin. Collagen has a very high elastic modulus and a very high ultimate strength: Its stiffness provides a *limit* to the vessel walls' distensibility. The collagen fibers are typically arranged with some slackness or "give"; in tendons this is known as *crimp.* This crimp means that each collagen fiber must be stretched taut before it begins to resist additional deformation so that the material gradually stiffens as it is stretched. In this way, each collagen fiber in turn is "recruited" to contribute to the overall behavior of the vessel. This behavior creates a J-shaped stress–strain curve, as shown in the experimental data in Figure CS5.2.

TABLE CS5.3

Natural Composite Biomaterials

Material	*Hard Component*	*Matrix*
Wood	Cellulose (polysaccharide)	Lignin, hemicelluloses
Sponge body wall	Calcareous, siliceous spicules, collagen	Miscellaneous organic
Stony corals	Aragonite ($CaCO_3$) crystals	Chitin fibril network
Mollusk shell	Calcite ($CaCO_3$), aragonite	Protein, sometimes chitin
Bird eggshell	Calcite crystals	Protein, some polysaccharide
Cartilage	Collagen fibrils	Mucopolysaccharide
Bovid horn	Keratin fibers	Wet, amorphous keratin
Bone	Hydroxyapatite ($Ca_5(PO_4)_3(OH)$)	Collagen, other organic
Tooth enamel	Hydroxyapatite	Organic

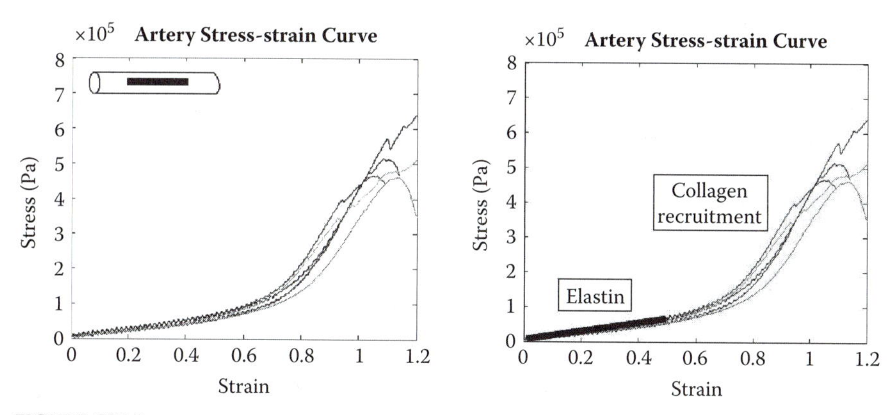

FIGURE CS5.2
Experimental stress–strain diagrams for a bovine artery, showing two distinct regimes of material stiffness and the J-shaped overall curve. (From J. S. Rossmann and B. Utela, unpublished data. With permission.)

Figure CS5.2 shows a stress–strain curve for a bovine artery with two identifiable regions: a long "toe" region of linear behavior dominated by elastin, in which large deformations result in only small stresses, and an increasingly steep region illustrating the recruitment of collagen fibers to stiffen the composite material.

Blood vessels also exhibit behavior called *cylindrical orthotropy*—different mechanical properties in the circumferential and longitudinal direction. Since we know from our study of pressure vessels that arteries and veins experience different stresses in these two directions, it is not surprising that they have responded to directionally dependent pressure loading by having directionally dependent properties. The different stiffness values in circumferential and longitudinal directions are evident from the slopes of the experimentally obtained stress–strain diagrams shown in Figure CS5.3.

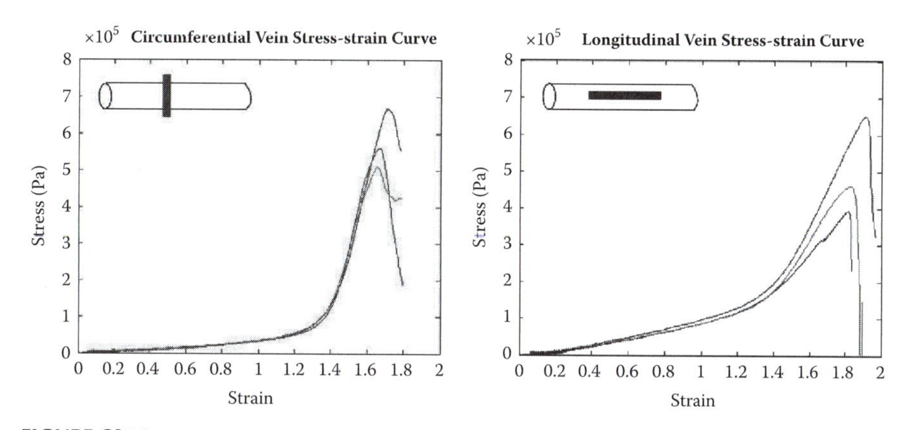

FIGURE CS5.3
Experimental data for bovine veins in the (left) circumferential and (right) longitudinal directions. The slopes of the toe (elastin-dominated) regions are E_{circ} = 30 kPa and E_{long} = 100 kPa. (From J. S. Rossmann and B. Utela, unpublished data. With permission.)

Viscoelasticity

Recall the material behavior spectrum and the special cases at its extrema: Hookean solids, for whom shear stress equals the product of a shear modulus and shear strain; and Newtonian fluids, for whom shear stress is viscosity times shear strain rate. At these extremes, these constitutive relationships are linear. We have already recognized that Hookean solids may be modeled as springs (i.e., as is the extension, so is the force) and that Newtonian fluids behave like dampers or dashpots. A spring is an elastic element, a dashpot a viscous one—so we are able to fill in the middle of the material behavior spectrum with combinations of these two elements. Materials whose behavior is best described by such a combination are known as viscoelastic materials.

For viscoelastic materials, both how much they deform and how fast they are deformed are important. Many, many biomaterials exhibit some degree of viscoelasticity. The two primary characteristics of viscoelastic behavior are *creep* and *stress relaxation*. *Creep* occurs when a material is exposed to a constant load for a long time and the material deforms increasingly: It's why a rubber band used to suspend a weight gradually lengthens and why you find that you are measurably shorter at the end of an active day during which your intervertebral cartilage has been subjected to constant compressive loading. *Stress relaxation* means that when a constant deformation is applied to a material, over time it will resist that deformation less so that the experienced loading decreases with time.

Another key feature of viscoelastic materials is *hysteresis*. This is the term used to describe the tendency of viscoelastic materials to dissipate energy rather than to store all of the energy of deformation as linearly elastic solids do. A schematic of this behavior is shown on a stress–strain diagram in Figure CS5.4a; the area between the loading and unloading curves represents dissipated or lost energy. (For a Hookean solid, the loading and unloading

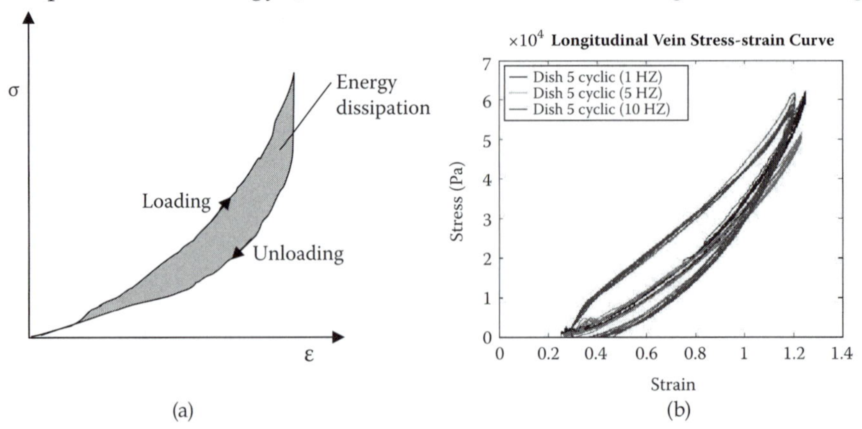

FIGURE CS5.4
Hysteresis of viscoelastic materials: (a) representative stress–strain diagram; and (b) experimentally obtained stress–strain diagram for bovine veins. (From J. S. Rossmann and B. Utela, unpublished data. With permission.)

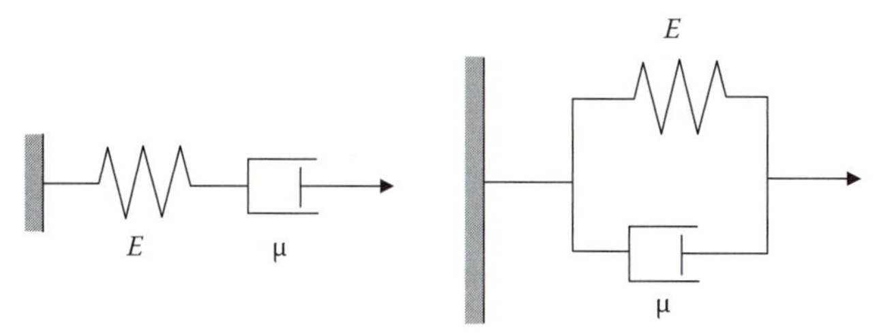

FIGURE CS5.5
Schematics of the two most common mechanical models for viscoelastic material behavior: (a) Maxwell and (b) Kelvin-Voigt materials.

curves are the same for small deformations.) Figure CS5.4b shows an experimentally obtained hysteresis curve for bovine veins in which it becomes clear that for these vessels, the amount of energy dissipated increases with increasing strain rate. This energy dissipation is what makes viscoelastic materials well suited to absorbing or cushioning shock.

The classical models for viscoelasticity represent different combinations of the spring (elastic) and dashpot (viscous) elements, as shown in the schematics of Figure CS5.5. Other combinations are possible, of course, but these two models prove remarkably effective for many materials.

For a Maxwell material, because the elastic and viscous elements are in series, they experience the same load ($\sigma_s = \sigma_d = \sigma$), and the net deformation of the material is the sum of the deformation of each element ($\varepsilon_s + \varepsilon_d = \varepsilon$).[4] The resulting constitutive law relating stress and strain is thus written,

$$E\mu\dot{\varepsilon} = \mu\dot{\sigma} + E\sigma, \tag{CS5.1}$$

where E is the elastic modulus or spring stiffness of the elastic element, and μ is the dynamic viscosity of the viscous element of the material.

In a Kelvin-Voigt material, the elements in parallel share the same deformation ($\varepsilon_s = \varepsilon_d = \varepsilon$), and the total stress is the sum of that experienced by each element ($\sigma_s + \sigma_d = \sigma$). The constitutive law for the viscoelastic material is therefore

$$E\varepsilon + \mu\dot{\varepsilon} = \sigma. \tag{CS5.2}$$

The ability of these models to capture viscoelastic behavior such as stress relaxation and creep varies, as shown in Problem CS5.2, Problem CS5.3, and Problem CS5.4. The response of each model to a step input in load or deformation, which can be deduced from the constitutive equations, is shown in Figure CS5.6.

A brief historical note: These models are attributed to James Clerk Maxwell (1831–1879), a prolific Scottish theoretical physicist who developed the Maxwell model of viscoelasticity to mathematically describe the viscous behavior

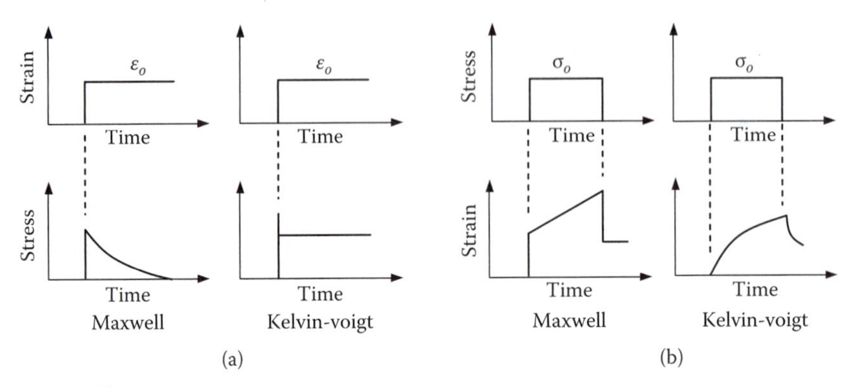

FIGURE CS5.6
Characteristic (a) stress relaxation and (b) creep responses of the Maxwell and Kelvin-Voigt models. (Adapted from Humphrey, J. D. and Delange, S. L., *Biomechanics*, Springer, New York, 2003.)

of air; Woldemar Voigt (1850–1919), a German physicist notable for his work in crystallography; and William Thomson (Lord Kelvin, 1824–1907), an Irish thermodynamicist whose interest in modeling mechanical behavior was rooted in his interest in irreversibility (i.e., the second law of thermodynamics).

Problems

CS5.1 Show that the constitutive law for a Maxwell body must have the form given in equation (CS5.1).

CS5.2 For the Kelvin-Voigt model of viscoelastic behavior, (a) find the solution of equation (CS5.2), $\epsilon(t)$, due to a step input in stress (a constant stress of magnitude unity applied beginning at $t = 0$), and (b) the solution $\sigma(t)$ due to a step input in strain. Sketch both solutions.

CS5.3 For the Maxwell model of viscoelastic behavior, (a) find the solution of equation (CS5.1), $\epsilon(t)$, due to a step input in stress (a constant stress of magnitude unity applied beginning at $t = 0$), and (b) the solution $\sigma(t)$ due to a step input in strain. Sketch both solutions.

CS5.4 Compare your results for Problem CS5.2 and Problem CS5.3 with the sketches in Figure CS5.6. Which of the two models represents creep behavior well, and which better represents stress relaxation?

Notes

1. We should pause again to empathize with Robert Hooke, whose work was suppressed by the bitterly competitive Isaac Newton but who now finds himself facing his rival on the opposite end of the material behavior spectrum.
2. The materials scientist Eugene Bingham (1878–1945) was a professor of chemistry at Lafayette College. He coined the term *rheology* for the study of fluid deformation and flow—that is, for the continuum mechanics of fluids. Bingham chose a quote from Heraclitus, "panta rei"—"everything flows"—as a suitable motto for the Society of Rheology he helped found in 1929. He and chemical engineer Markus Reiner proposed the Deborah number as a fundamental quantity of rheology, with larger Deborah numbers resulting in material behavior further toward the "solid" end of the spectrum. It was named for the prophetess Deborah, who sang, "The mountains flowed" to the defeated Philistines. We refer readers to Reiner (1960).
3. For the details of this, please see Kundu (1990, 89–93). For the mathematical justification, see Aris (1962).
4. The elements experiencing the same load translates to experiencing the same stress because the spring and dashpot are modeling two behavioral aspects of the same tissue, so they can be considered to have the same area.

9

Fluid Statics

When there is no relative motion between fluid particles, no shearing stresses exist, and the only stress present is a normal stress, the pressure. Hence $F = ma$ is a balance between the forces due to pressure and the inertia of the fluid. Our fluid, like the solids we studied in Chapters 2 through 6, is in *equilibrium*.

9.1 Local Pressure

We have defined fluid pressure as an infinitesimal normal force divided by the infinitesimal area it acts on. From our study of solid mechanics, we may suspect that the value of p will change if the orientation of this planar area changes—that we will have a different p if the xy plane is rotated to $x'y'$. However, this is not the case, as we can show by a simple analysis of a now familiar inclined plane.

If we write the equations of motion ($F = ma$) in the y and z for the element shown in Figure 9.1, we have

$$\sum F_y = p_y dxdz - p_s dxds \sin \grave{e} = \rho \frac{dxdydz}{2} a_y \tag{9.1a}$$

$$\sum F_z = p_z dxdy - p_s dxds \cos \grave{e} - \rho g \frac{dxdydz}{2} = \rho \frac{dxdydz}{2} a_z \tag{9.1b}$$

noting the geometry of the problem, $dy = ds \cos \theta$ and $dz = ds \sin \theta$, we can rewrite these equations as

$$p_y - p_s = \rho a_y \frac{dy}{2}, \tag{9.2a}$$

$$p_z - p_s = (\rho a_z + \rho g)\frac{dz}{2}. \tag{9.2b}$$

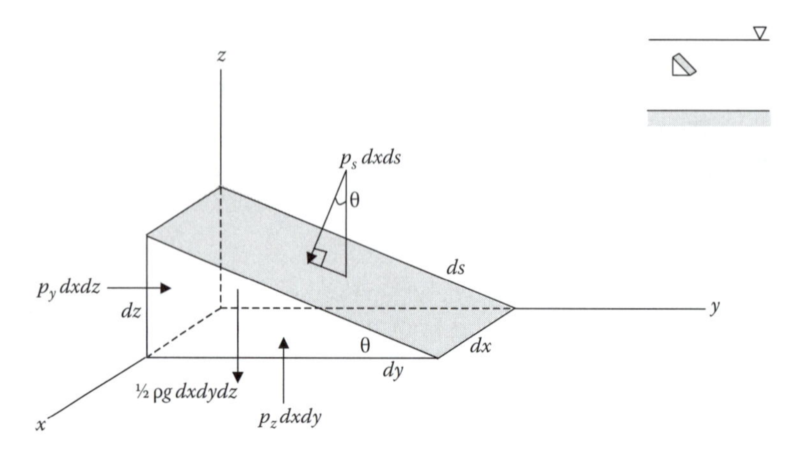

FIGURE 9.1
Forces on an arbitrary wedge-shaped element of fluid.

And since our real interest is in what's happening at a *point*, we shrink this element, taking the limit as dx, dy, and dz go to zero (while maintaining θ); hence, we must have $p_y = p_s$ and $p_z = p_s$, or

$$p_s = p_y = p_z \,. \tag{9.3}$$

Since θ was an arbitrary angle, this must be true for any θ so that we may say:

> The pressure at a point at a fluid is independent of direction as long as there are no shearing stresses present.

This result is due to the French mathematician Blaise Pascal (1623–1662) and is known as *Pascal's law.*

We could also consider the Mohr's circle of stress for a fluid with no relative motion between fluid particles. A general Mohr's circle is sketched in Figure 9.2. When shear stress is absent, Mohr's circle degenerates to a point and $p_i = p_j$. Hence, using Mohr's circle, we could have beaten Pascal to the punch.

9.2 Force Due to Pressure

We would like to be able to determine the pressure variation within a fluid. Certainly Pascal's law helps us do this. We consider another small fluid element, this time in the shape of a cube (the shape is ours to choose since Pascal's law tells us that p at the center of an element is independent of the ori-

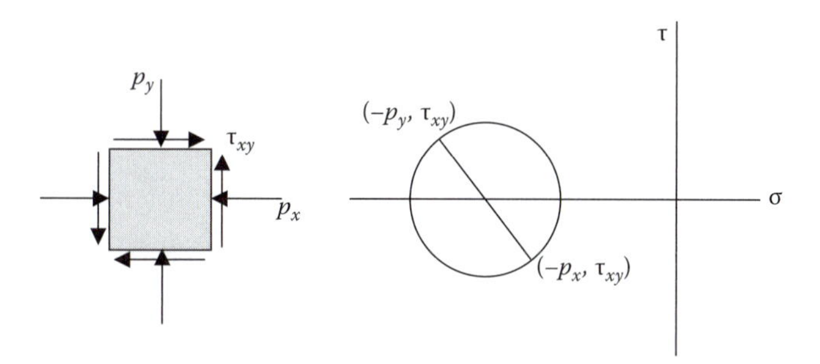

FIGURE 9.2
Mohr's circle.

entations of the element's faces, and choosing a cube simplifies the geometry). This element is shown in Figure 9.3. We write Newton's second law for this element, first finding an expression for the force due to pressure $p(x, y, z)$.

The pressure is p at the center of our element, a point with coordinates (x, y, z). Since p varies in x, y, and z, we can write the pressures at each of the element's faces using the chain rule,

$$dp = \frac{\partial p}{\partial x}dx + \frac{\partial p}{\partial y}dy + \frac{\partial p}{\partial z}dz, \tag{9.4}$$

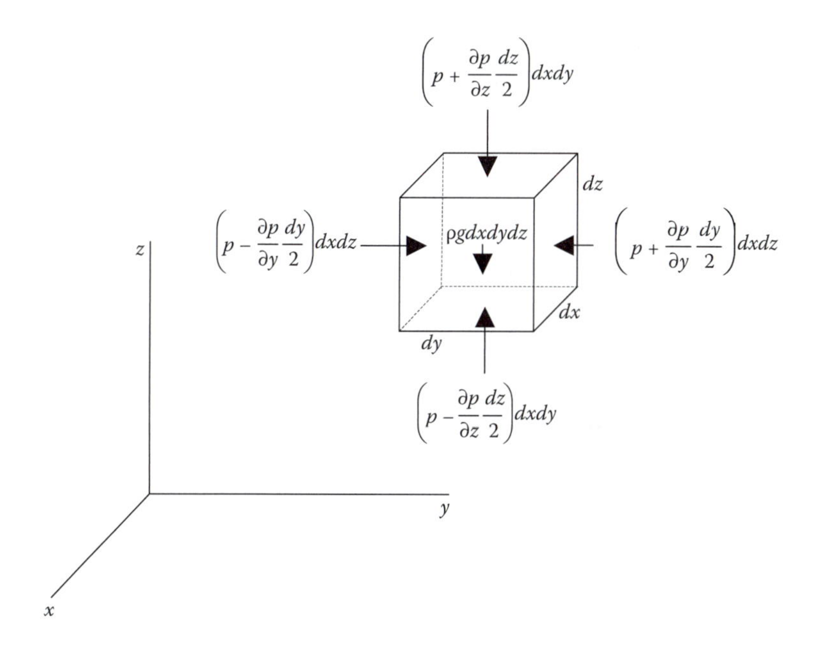

FIGURE 9.3
Forces acting on a small fluid element (x forces, not shown, have similar form).

so that the pressure at a distance ($dx/2$) from the element's center is written

$$p(x+\frac{dx}{2},y,z)=p(x,y,z)+\frac{\partial p}{\partial x}\frac{dx}{2}\,. \tag{9.5}$$

The same reasoning gives us expressions for the pressure on all six faces of the element in Figure 9.3.

The resultant forces on the element are the differences between those on top and bottom, right and left, or front and back faces. For example, the resultant force due to pressure in the y direction is

$$dF_y=-\frac{\partial p}{\partial y}dxdydz\,. \tag{9.6}$$

And the resultant surface force on the element can be written in vector form as $dF = dF_x\hat{i} + dF_y\hat{j} + dF_z\hat{k}$, or

$$d\underline{F}=-\left(\frac{\partial p}{\partial x}\hat{i}+\frac{\partial p}{\partial y}\hat{j}+\frac{\partial p}{\partial z}\hat{k}\right)dxdydz\,. \tag{9.7}$$

The group of terms in parentheses is the vector form of the *pressure gradient*, or grad p. We can thus write the resultant surface force on the element using the notation

$$\frac{d\underline{F}}{dxdydz}=-\nabla p\,. \tag{9.8}$$

Something very interesting has happened: The force on the element due to pressure has been shown to depend only on the *gradient* of pressure, or on how pressure varies in x, y, and z.

To complete $F = ma$, we combine this resultant surface force with the body force (gravity) acting on the element and set these forces equal to the inertia of the element. The body force is written as $-\rho g\, dxdydz\, \hat{k}$, and ma is written as $\rho\, dxdydz\, a$. The element volume, $dxdydz$, appears in all terms and may be divided out of the equation, leaving in vector form

$$-\nabla p-\rho g\,\hat{k}=\rho\,\underline{a}\,. \tag{9.9}$$

This is the general equation of motion for a fluid in which there are no shear stresses. It is an equation per unit volume of the fluid, since we have divided through by $dxdydz$, and each term in equation (9.9) hence has units of force per volume.

9.3 Fluids at Rest

In the special case of a fluid at rest, acceleration $a = 0$ and the governing equation reduces to

$$-\nabla p - \rho g \ \hat{k} = 0, \tag{9.10}$$

with three component equations:

$$\frac{\partial p}{\partial x} = 0, \tag{9.11a}$$

$$\frac{\partial p}{\partial y} = 0, \tag{9.11b}$$

$$\frac{\partial p}{\partial z} = -\rho \, g. \tag{9.11c}$$

The x and y components show that the pressure in this special case does not depend on x or y. The pressure $p = p(z)$ only, and its dependence is given by

$$\frac{dp}{dz} = -\rho \, g \ . \tag{9.12}$$

To use this equation to calculate pressure throughout a fluid, it is necessary to specify how the product (ρg) varies with z. In most engineering applications, variation in g is negligible, so we concern ourselves primarily with the variation of density ρ.

An *incompressible* fluid is defined as one that requires a *very large* pressure change to effect a small change in volume. This threshold is so high that in most cases, the fluid's volume and therefore its density are constant. Most liquids satisfy this requirement. When ρg can be taken to be constant, the equation for p is easily integrated:

$$\int_{p_1}^{p_2} dp = -\rho g \int_{z_1}^{z_2} dz, \tag{9.13}$$

so

$$p_1 - p_2 = \rho\, g(z_2 - z_1), \tag{9.14}$$

or, if $h = z_2 - z_1$, then $p_1 - p_2 = \rho g h$, or

$$p_1 = p_2 + \rho g h. \tag{9.15}$$

Equation (9.15) describes what is called a *hydrostatic* pressure distribution. The distance $h = z_2 - z_1$ is measured downward from the location of p_2. Hydrostatic pressure increases linearly with depth, as the pressure increases to "hold up" the fluid above it. Many devices such as hydraulic lifts exploit the hydrostatic pressure distribution.

For a *compressible* fluid—typically a gas—the fluid density can change significantly due to relatively small changes in pressure and temperature. For these fluids, the product ρg is typically quite small—for air at sea level at 60°F, ρg is 0.0763 lb/ft^3, compared with 62.4 lb/ft^3 for water at the same conditions. It therefore requires very large elevation changes h to make much difference in the pressure of compressible fluids. To account for the variation in ρg, we make use of the ideal gas law, $p = \rho RT$, to write

$$\frac{dp}{dz} = -\frac{gp}{RT}. \tag{9.16}$$

Separating variables and integrating, we get

$$\int_{p_1}^{p_2} \frac{dp}{p} = \ln\left(\frac{p_2}{p_1}\right) = -\frac{g}{R}\int_{z_1}^{z_2} \frac{dz}{T}, \tag{9.17}$$

where g and R are assumed constant over the elevation change from z_1 to z_2.

Pressure is often measured using liquid columns in vertical or inclined tubes, or *manometers*. These devices make use of the information we have just obtained: that pressure increases with depth; and that (therefore) two points at the same elevation in a continuous length of the same fluid must have the same pressure. The three most common types of manometers are U-tube and inclined-tube manometers and piezometers. Examples of manometers are shown in Figure 9.4.

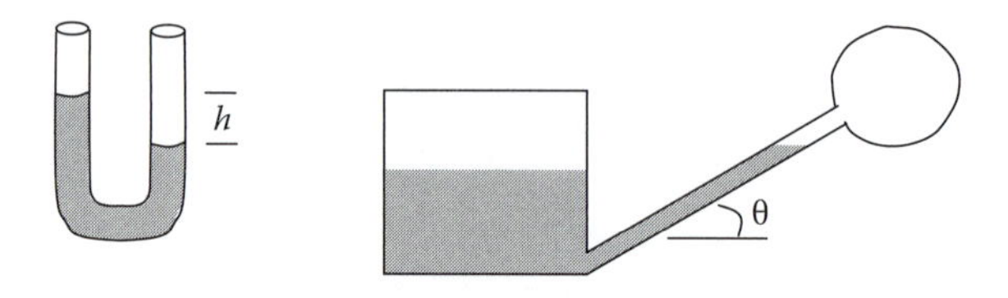

FIGURE 9.4
Types of manometers: left, U-tube; right, inclined-tube.

As you might expect, manometers are not particularly well suited for the measurement of very high pressures (since they must then include a very, very long tube) or of pressures that vary rapidly in time. Some other devices have thus been developed—and they are of special interest to us as students of continuum mechanics. This other class of measurement devices makes use of the idea that when a pressure acts on an elastic structure the structure will deform, and this deformation can be related to the magnitude of the pressure. A Bourdon tube is one example of this; it consists of a calibrated hollow, elastic curved tube that tends to straighten when the pressure inside it increases. A *pressure transducer* as in Figure 9.5 converts the reading from a Bourdon tube or other measurement device into an electrical output.

Another example of this type of device is shown in Figure 9.6. In this case the sensing element is a thin, elastic diaphragm that's in contact with the fluid. Fluid pressure causes the diaphragm to deflect, and its deflection is measured and converted into an electrical voltage. Strain gauges are attached to the reverse side of the diaphragm or to an element attached to the diaphragm. Figure 9.6a shows two differently sized strain-gauge pressure transducers, both made by Viggo-SpectraMed (now Ohmeda), commonly used to measure physiological pressures within the human body. Pressure-induced deflection of the diaphragm is measured using a silicon beam on which strain gauges and a bridge circuit have been deposited (as shown in Figure 9.6b).

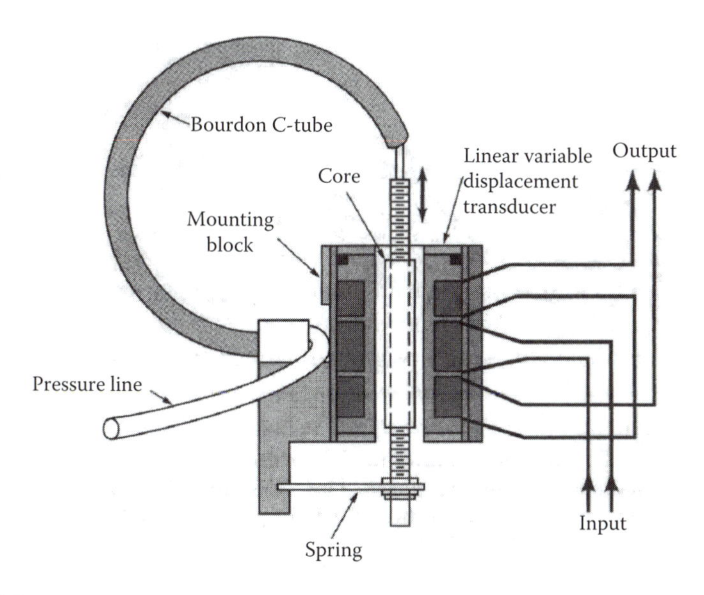

FIGURE 9.5
Bourdon tube pressure transducer. (From Munson, B. R., Young, D. F., and Okiishi, T. H., *Fundamentals of Fluid Mechanics*, Wiley, 1998. With permission.)

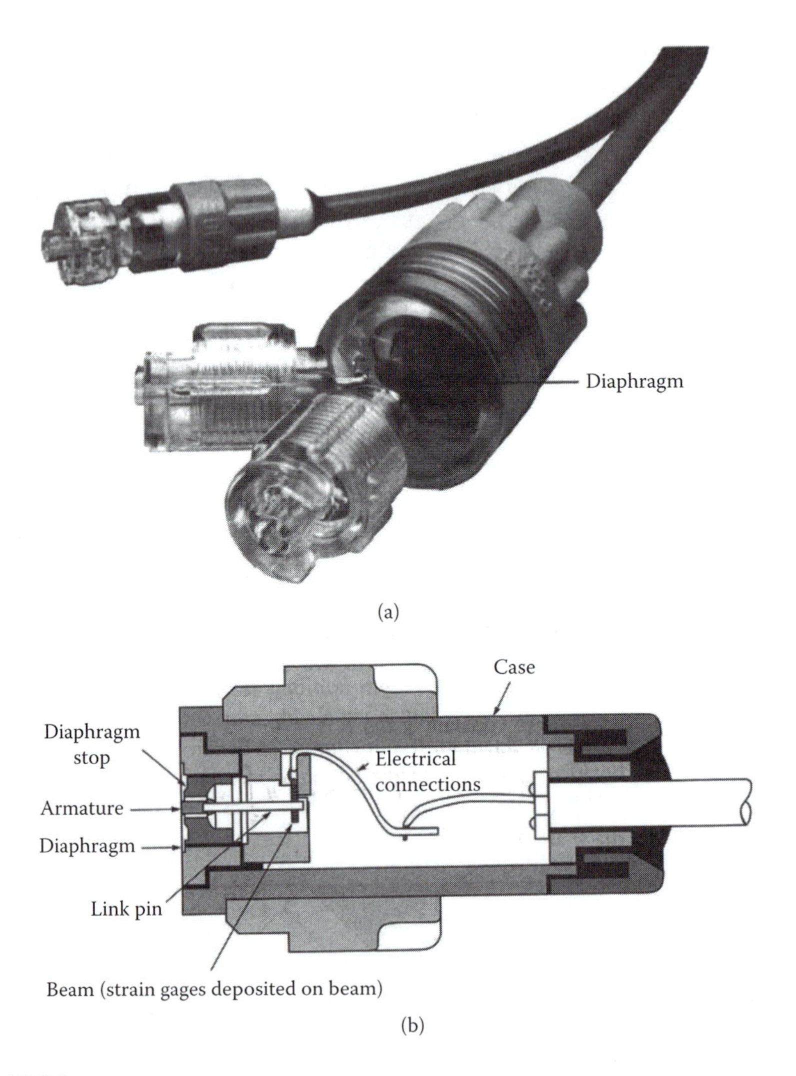

FIGURE 9.6
(a) Photographs and (b) schematic of strain-gauge pressure transducers used for biological flows. (From Munson, B. R., Young, D. F., and Okiishi, T. H., *Fundamentals of Fluid Mechanics*, Wiley, 1998. With permission.)

9.4 Forces on Submerged Surfaces

When we design devices and objects that are submerged within a body of fluid (e.g., dams, ships, holding tanks, bridge supports, artificial reefs), we must consider the magnitudes and locations of forces acting on both plane and curved surfaces due to the fluid. If the fluid is at rest, we know that this force will be perpendicular to the surface (normal stress) since there are no shearing stresses present. If the fluid is also incompressible, we know that the pressure will vary linearly with depth.

For a horizontal surface, such as the bottom of a tank, the force due to fluid pressure is easily calculated. The resultant force is just $F = pA$, where p is the uniform pressure and $= \rho gh$ and A is the area of the surface. Since the pressure is constant and uniformly distributed over the bottom, the resultant force acts through the centroid of the area.

In the case of a vertical surface, the pressure is not constant but varies linearly with depth along the submerged surface. This is sketched in Figure 9.7b and reminds us very much of the distributed loading we have seen acting on beams, as in Figure 9.7a. In our analysis of such beams, we found that the equivalent concentrated force (by which we replaced the distributed load to calculate reactions and internal forces) acted through the centroid of the area between the force profile and the beam surface. For example, for a linearly increasing load as in Figure 9.7a, the shape created is a triangle, and the resultant concentrated load acts at $h/3$ from the right end, the centroid of that triangle. The same is true for submerged surfaces. In Figure 9.7b, a hydrostatic pressure is drawn as a distributed load on a vertical wall. This creates a triangular shape, known as a "pressure prism," whose centroid is at $h/3$ from the deepest point. This is the "center of pressure" (the point at which the resultant force acts, y_R) for this load. The resultant force F_R is simply the pressure integrated over this vertical surface,

$$F_R = \int_A dF = \int_A p dA = \int_A \rho g y dA = \rho g h_c A, \tag{9.18}$$

where h_c is the depth of the surface's centroid, $h/2$, and A is the area of the vertical surface.

We would like to move beyond the idealizations of horizontal and vertical surfaces to more realistic geometries. It is useful to formulate a method for calculating the force due to pressure on an *inclined* surface, at an angle θ to

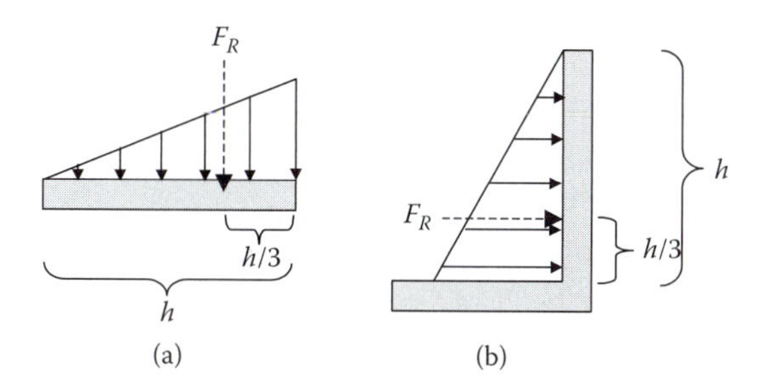

FIGURE 9.7
Pressure prisms for (a) distributed beam loading and (b) hydrostatic pressure.

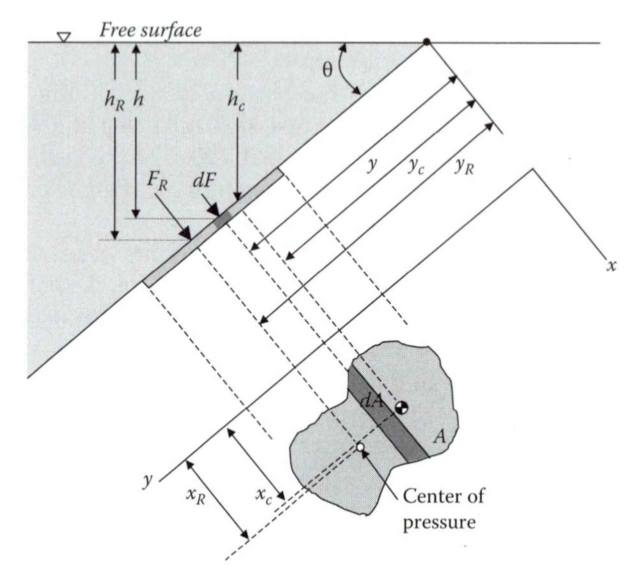

FIGURE 9.8
Hydrostatic force on inclined plane surface of arbitrary shape.

the horizontal fluid surface as shown in Figure 9.8. This general formulation can then be applied to a wide range of problems.

Essentially, we are once again considering the effect of a distributed force, and to deal with the equivalent concentrated load we must find the center of pressure at which this equivalent load acts. We choose coordinates, as shown in Figure 9.8, that are convenient for the surface in question, and we must find the point (x_R, y_R): the center of pressure, at which the resultant force acts.

The total force exerted on the plane surface by the fluid is simply the integral of the fluid pressure over the surface's entire area,

$$F_R = \int_A dF = \int_A p dA, \tag{9.19}$$

where p is gage pressure. For a fluid at rest, the pressure distribution is hydrostatic, and $dF = \rho g h = \rho g y \sin\theta$. For constant ρg and θ, we thus have

$$F_R = \rho\, g \sin\theta \int_A y dA, \tag{9.20}$$

and we recognize that the integral $\int y dA = y_c A$, where y_c is the position of the centroid of the entire submerged surface. So, the resultant force is simply

$$F_R = \rho\, gAy_c \sin\theta = \rho\, gh_c A, \tag{9.21}$$

where h_c is the vertical distance from the fluid surface to the centroid of the area. We notice that this force's magnitude is independent of the angle θ and depends only on the fluid's specific weight, the total area, and the depth of the centroid.

Although we might suspect that this resultant force passes through the centroid of the surface, if we remember that pressure is increasing with increasing depth, we realize that the center of pressure must actually be below the centroid. We can find the coordinates of the center of pressure by summing moments around the x axis, forcing the moment of the resultant force to balance the moment of the distributed force due to pressure, so that

$$y_R = \frac{\int_A y^2 dA}{y_c A} = \frac{I_x}{y_c A}. \tag{9.22}$$

Please note that the xy axes are now playing the same role for our submerged surface that the zy axes did for beam cross sections, so that I_z in the context of beams is the same as I_x in this new context. Since I_{xc} (about the centroid) is typically the easiest second moment of area to calculate, and the one tabulated in handy places,[1] we use the parallel axis theorem to make sure we are considering the second moment of area with respect to our x axis as drawn in Figure 9.8,

$$I_x = I_{xc} + Ay_c^2, \tag{9.23}$$

so that

$$y_R = \frac{I_{xc}}{y_c A} + y_c. \tag{9.24}$$

This expression demonstrates that the resultant force acts on a point *below* the centroid, since $I_{xc}/y_c A > 0$. In a similar way we determine the x coordinate of the center of pressure:

$$x_R = \frac{I_{xyc}}{x_c A} + x_c. \tag{9.25}$$

The second moment of area that appears in this expression, I_{xy}, is the second moment of area with respect to the x and y axes, = $\int xydA$. For symmetric shapes, I_{xyc} is zero, and the resultant force acts at x_c.

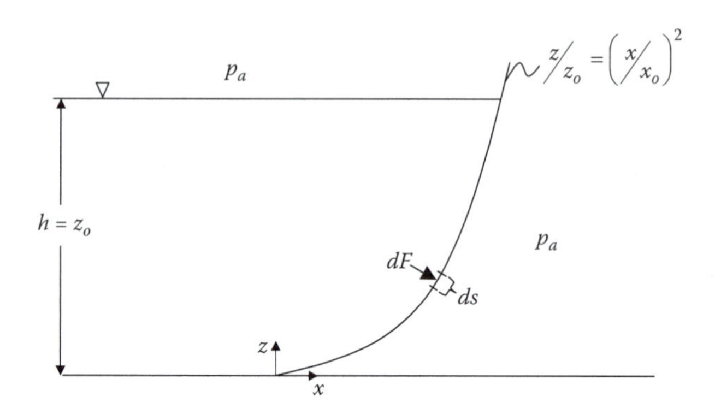

FIGURE 9.9
Curved surface of parabolic dam.

When the submerged surface in question is *curved*, our work is somewhat more complicated. The resultant force due to pressure acts *normal* to the surface, which did not affect our integration over the surface area for a plane surface as the surface had only one outward normal vector. However, for a curved surface, the outward normal changes continuously all along the surface, making our integration less easily simplified.

Rather than accounting for this variation in the outward normal, most analyses simply separate the resultant force on the surface, F_R, into its horizontal (F_H) and vertical (F_V) components. Each of these has a straightforward physical interpretation that becomes clear when it is calculated. As an example, let's consider a parabolic dam as shown in Figure 9.9. The shape of the curved dam surface is described by $z/z_o = (x/x_o)^2$.

The gage pressure at any height z is given by

$$p = \rho g(h - z). \tag{9.26}$$

So, the infinitesimal force at any height z, acting on an infinitesimal area element dA, is

$$dF = \rho g(h - z)\, dA, \tag{9.27}$$

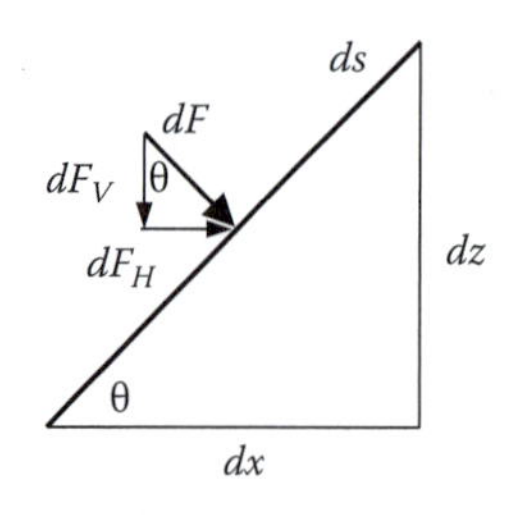

FIGURE 9.10
Infinitesimal segment of curved surface.

and the area dA at any z is simply the width of the dam into the page, w, times the infinitesimal length ds along the curved surface, as shown in Figure 9.10.

We want to find the horizontal and vertical components of this vector dF, so we begin with the horizontal force:

$$dF_H = dF \sin \theta, \tag{9.28}$$

$$= \rho g(h - z)w\, ds \sin \theta. \tag{9.29}$$

From Figure 9.10 we see that $ds \sin \theta$ is just dz, so we have

$$dF_H = \rho g(h - z)w\, dz, \tag{9.30}$$

and we proceed by integrating in z to find the resultant horizontal force on the surface:

$$F_H = \rho g \int_0^h (h - z)w dz \tag{9.31a}$$

$$= \rho gw \left[hz - \frac{z^2}{2} \right]_0^h \tag{9.31b}$$

$$= \frac{1}{2}\rho gwh^2 \tag{9.31c}$$

We can rearrange this horizontal force as

$$F_H = \left(\rho g \frac{h}{2} \right) wh \tag{9.32}$$

and recognize that wh would be the area of a vertical projection of our parabolic curved surface and that $\rho g(h/2)$ would be the resultant force on this vertical projection. We can thus physically interpret the horizontal component of force on a curved submerged surface as the resultant force that would act on a vertical projection (same depth into page, same height) of the curved surface.

Next, we look for the vertical component F_V, integrating $dF_V = dF\cos \theta$ over the surface:

$$dF_V = dF \cos \theta, \tag{9.33}$$

$$= \rho g(h - z)wds \cos \theta. \tag{9.34}$$

In Figure 9.9 we see that $ds \cos \theta$ is just dx, so we have

$$dF_V = \rho g(h - z)w dx. \tag{9.35}$$

To integrate this expression in x we need to express z as a function of x, using the equation of the curved surface:

$$\begin{aligned} F_V &= \rho g \int_0^{x_o} (h - \frac{h}{x_o^2} x^2) w dx \\ &= \rho ghw \left[x - \frac{x^3}{3x_o^2} \right]_0^{x_o} \\ &= \frac{2}{3} \rho ghwx_o. \end{aligned} \tag{9.36}$$

We recognize that

$$\frac{2}{3} x_o z_o$$

is the area contained by a parabolic section with maximum height z_o and maximum width x_o. (In our case, $z_o = h$.) Since w is the width of this section into the page, the vertical force component is the fluid density ρ times the acceleration of gravity g times the volume of fluid; that is, this force is equal to the weight of fluid above the curved surface.

To find the center of pressure at which the resultant of these components acts, we first consider where each component force must act. For the inclined plane surface, these components must induce the same *moment* about a reference point as does the distributed force. Because F_H is the force that would act on a vertical projection of the curved surface, it acts where the equivalent force due to pressure would act on that vertical projection: at $h/3$ up from the base of the surface, the centroid of the pressure prism. Because F_V is the weight of the fluid supported by the surface, it acts at the x coordinate of the centroid of that volume of fluid. These coordinates come naturally out of the moment calculation. Taking our reference point as the origin of the x, z axes, we require that

$$\underbrace{F_H z_H}_{\substack{\text{moment due to concentrated} \\ \text{horizontal force component}}} = \underbrace{\rho gw \int_0^h z(h - z) dz}_{\substack{\text{sum of all moments due to all} \\ \text{infinitesimal forces } dF_H \text{, each with} \\ \text{moment arm } z.}} . \tag{9.37}$$

For any curved surface (since the shape of the curve $x(z)$ does not enter into the integral), this moment equivalence requires that

$$z_H = \frac{1}{3}h\,.$$

We must also have

$$\underbrace{F_V x_V}_{\substack{\text{moment due to concentrated}\\ \text{vertical force component}}} = \underbrace{\rho g w \int_0^h x(h-z)dx}_{\substack{\text{sum of all moments due to all}\\ \text{infinitesimal forces } dF_V\text{, each with}\\ \text{moment arm } x.}} \,. \tag{9.38}$$

For our parabolic surface, we plug in $z = (h/x_o^2)x^2$, and we get

$$x_V = \frac{3}{8}x_o\,.$$

This is in fact the x coordinate of the centroid of a parabolic section. The line of action of the resultant force

$$F_R = \sqrt{(F_H{}^2 + F_V{}^2)}$$

passes through the point (x_V, z_H) with slope = $\tan^{-1}(F_V/F_H)$.

9.5 Buoyancy

An object that is submersed in fluid is subjected to hydrostatic pressure over its entire surface area. In the previous section, we limited ourselves to the consideration of simple surfaces: walls, gates, and dams. However, if we recognize that the hydrostatic pressure acts on *all* surfaces, we see clearly how a resultant *buoyancy* force arises. The sketch in Figure 9.11 illustrates this.

Archimedes (287–212 BC) was a Greek mathematician who invented the lever, fine-tuned the definition of *pi*, and "discovered" buoyancy. Though some details of this story have taken on the distinct patina of apocrypha, it is still a cracking-good yarn. Archimedes's close friend, King Hiero of Syracuse, suspected that the gold crown he had recently received from the goldsmith did not include all of the gold he'd supplied. He shared his suspicions with Archimedes, who (it is said) went home to ruminate in the

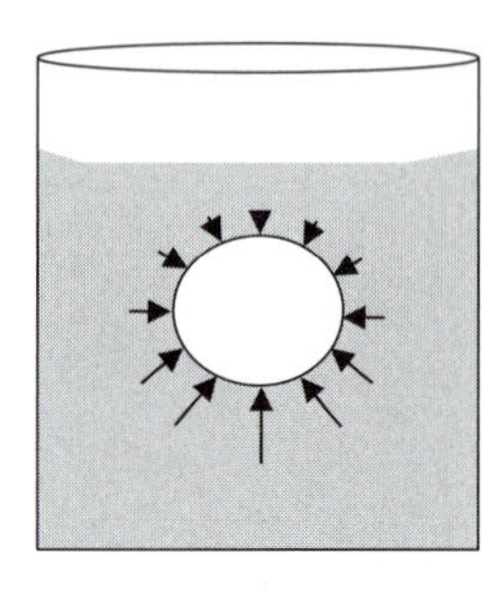

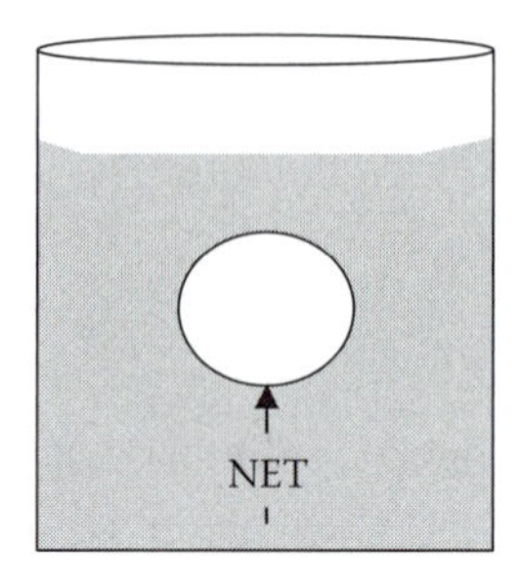

FIGURE 9.11
Distributed force due to hydrostatic pressure on submerged object. Left, distributed; right, net resultant upward vertical buoyancy force.

bathtub. Archimedes, ever observant, noticed that his body displaced the bathwater—when he got into the tub, the water level rose. He quickly calculated that the weight of displaced water balanced his own weight and celebrated this discovery by running through the streets shouting, "*Eureka* (I have found it)," so intoxicated by hydrostatics that he neglected to dry off or wear a bathrobe. The next day, so the story goes, Archimedes dunked his friend's crown, as well as a lump of gold equal to what he'd provided to the goldsmith, and found that they did not displace equal amounts of water. The crown did, in fact, contain less gold than the king had specified. The goldsmith, unable to produce the remainder of the gold, was beheaded posthaste.

Archimedes' Principle states that the buoyant force on an object equals the weight of the volume of fluid the object displaces. The force on a submerged object due to the fluid's hydrostatic pressure tends to be an upward vertical force, as the pressure in the fluid increases with depth and the resultant force is upward. Refer to Figure 9.11 to visualize this. If this buoyancy force exactly balances the weight of the object, the object is said to be neutrally buoyant.

The line of action of the buoyancy force acts through the centroid of the displaced fluid volume. The stability of an object designed to float on or maneuver in a fluid depends on the moments due to the buoyancy and weight forces on the object and on whether the resultant moment tends to right or to capsize the craft. For submerged vessels that operate at a range of depths, mechanisms that allow active control of these forces are necessary. Tanks that can be flooded or filled with air to adjust the vessel's weight mimic the swim bladder in fish to allow vessels to maintain the proper force balance.

9.6 Examples

Example 9.1

Determine the pressure difference between the benzene at *A* and the air at *B* in Figure 9.12.

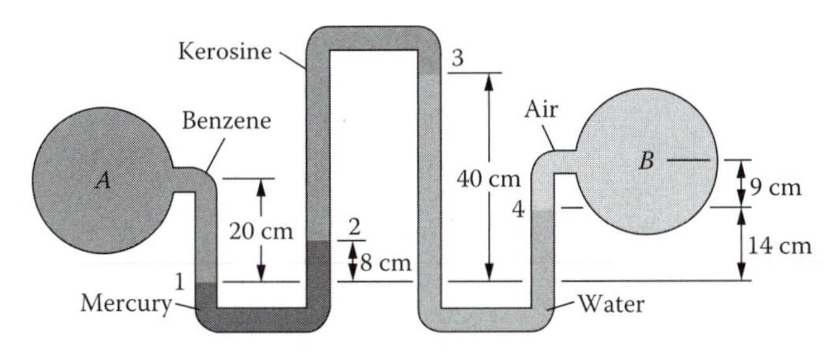

FIGURE 9.12

Given: Manometer geometry and gage fluids; heights of fluid columns.

Find: Pressure difference between A and B.

Assume: No relative motion of fluid elements (hydrostatics); fluids have constant, uniform density, and it is appropriate to evaluate densities at 20°C.

Solution

We look up the properties of the fluids used in our manometer at 20°C and find the following:

Fluid	Density ρ (kg/m^3)
Water	998.0
Mercury	13,550.0
Air	1.2
Benzene	881.0
Kerosene	804.0

We know that in a fluid at rest, the pressure depends only on the elevation in the fluid. Thus, in any continuous length of the same fluid, two points at the same elevation must be at the same pressure. Manometers are based on this principle. We find the requested pressure difference by starting at point *A* and working our way through the manometer, noting that the pressure increases when the fluid level drops and that pressure decreases when the fluid level rises:

- Point 1: $P_1 = P_A + \rho_B g h_1$
- Point 2: $P_2 = P_1 - \rho_M g h_2$
- Point 3: $P_3 = P_2 - \rho_K g h_3$
- Point 4: $P_4 = P_3 + \rho_W g h_4$
- At B: $P_B = P_4 - \rho_A g h_5$

So,

$$P_B = P_A + \rho_B g h_1 - \rho_M g h_2 - \rho_K g h_3 + \rho_W g h_4 + \rho_A g h_5.$$

$$P_B - P_A = (881\ \text{kg/m}^3)(9.81\ \text{m/s}^2)(0.2\ \text{m}) - (13{,}550\ \text{kg/m}^3)(9.81\ \text{m/s}^2)(0.08\ \text{m})$$

$$-(804\ \text{kg/m}^3)(9.81\ \text{m/s}^2)(0.32\ \text{m}) + (998\ \text{kg/m}^3)(9.81\ \text{m/s}^2)(0.26\ \text{m})$$

$$-(1.2\ \text{kg/m}^3)(9.81\ \text{m/s}^2)(0.09\ \text{m})$$

$$= -8885\ \text{Pa}$$

or

$$P_A - P_B = 8.9\ \text{kPa}.$$

Example 9.2

Panel ABC in the slanted side of a water tank is an isosceles triangle with the vertex at A and the base $BC = 2$ m, as shown in Figure 9.13. Find the force on the panel due to water pressure and on this force's line of action.

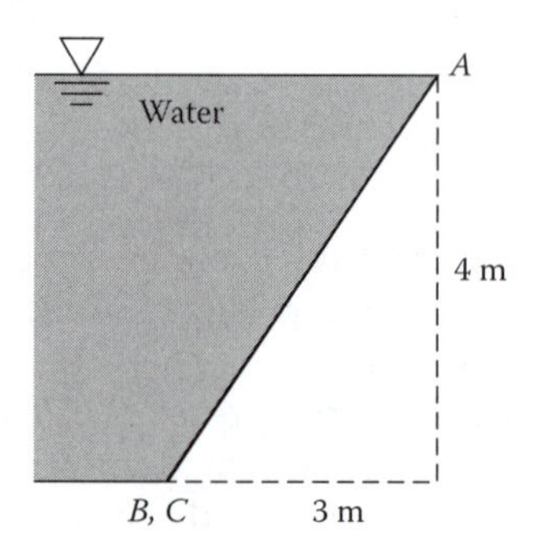

FIGURE 9.13

Given: Dimensions of panel in water tank.

Find: Resultant force on panel; location of center of pressure.

Assume: No relative motion of fluid elements (hydrostatics); water has constant, uniform density, equal to its tabulated value at 20°C (998 kg/m^3).

Solution

We first want to understand the geometry of the triangular panel. We are given a side view of the tank, and the height of the triangle *ABC*. In a head-on view, we would see the panel as sketched in Figure 9.14 at left.

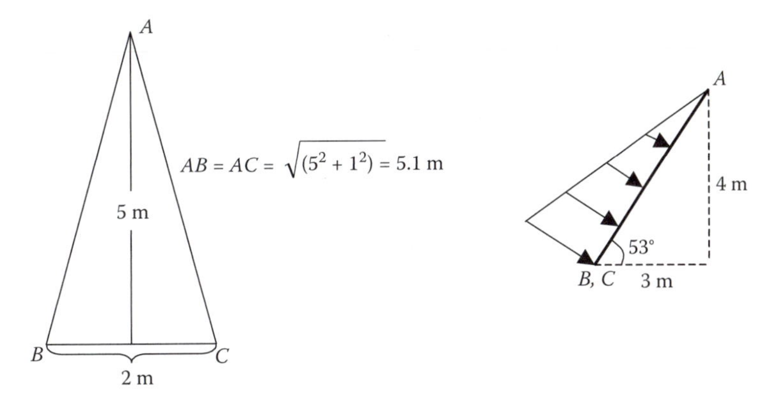

FIGURE 9.14

The water pressure has a hydrostatic distribution, as sketched in Figure 9.14 at right, and the resultant force is found by integrating this pressure over the panel area. This is equivalent to the formula $F_R = \rho g h_c A$, where h_c is the depth of the centroid of the submerged surface, the triangular panel ABC.

The depth of point *A* is zero; the depth of points *B* and *C* is 4 m. The depth of the centroid of the triangular gate *ABC* is 2/3 of the way down. (*Note*: This is because the submerged surface is a triangle, *not* because the pressure has a triangular pressure prism.):

$$h_c = \frac{2}{3}\left(4 \text{ m}\right) = 2.67 \text{ m},$$

$$A = \frac{1}{2}bh = \frac{1}{2}(2 \text{ m})(5 \text{ m}) = 5 \text{ m}^2,$$

so

$$F_R = \rho\, g h_c A = (998 \text{ kg/m}^3)(9.81 \text{ m/s}^2)(2.67 \text{ m})(5 \text{ m}^2)\ F_R = 131{,}000 \text{ N} = 131 \text{ kN}$$

This force acts at the center of pressure of the submerged panel ABC. Due to the symmetry of the panel, this is on the centerline ($x_R = 0$), and we are only required to calculate the y coordinate y_R (Figure 9.15).

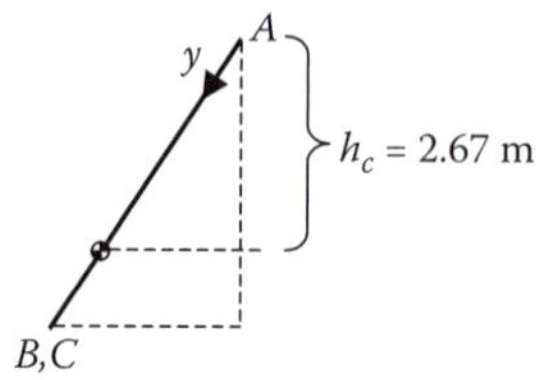

y, A, $h_c = 2.67$ m, B,C

Similar triangles:

$$\frac{4}{5} = \frac{2.67}{y_c}$$

$$y_c = 3.33 \text{ m}$$

FIGURE 9.15

$$y_R = y_c + \frac{I_{xc}}{y_c A} = y_c + \frac{I_{xc}\sin\theta}{h_c A}$$

$$= y_c + \frac{\frac{1}{36} bh^3 \sin\theta}{h_c A}$$

$$= 3.33 \text{ m} + \frac{\frac{1}{36}(2 \text{ m})(5 \text{ m})^3 \sin(53^\circ)}{(2.67 \text{ m})(5 \text{ m}^2)}$$

$$= 3.33 \text{ m} + 0.417 \text{ m}$$

$$= 3.75 \text{ m}.$$

Note that this y_R is measured down from A, along the panel surface, as shown in the sketch in Figure 9.16.

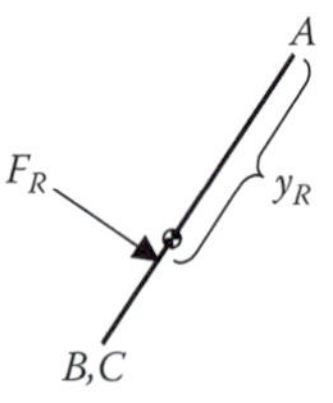

FIGURE 9.16

Example 9.3

Gate AB has an evenly distributed mass of 180 kg and is 1.2 m wide "into the page" as illustrated in Figure 9.17. The gate is hinged at A and rests on a smooth tank floor at B. For what water depth h will the force at point B be zero?

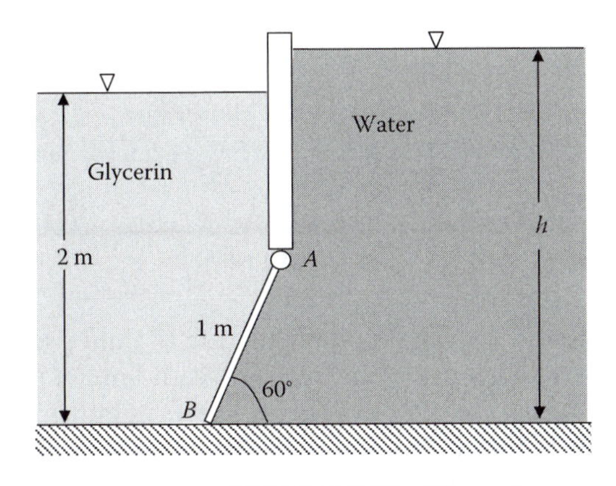

FIGURE 9.17

Given: Gate dimensions and mass; fluids on either side.

Find: Water depth h required for $R_B = 0$.

Assume: No relative motion of fluid elements (hydrostatics); fluids have constant, uniform density, and it is appropriate to evaluate densities at 20°C. Looking up these values, we find that water's density is 998 kg/m^3 and that glycerin's is 1260 kg/m^3.

Solution

We start with a free-body diagram (FBD) of gate AB (Figure 9.18).

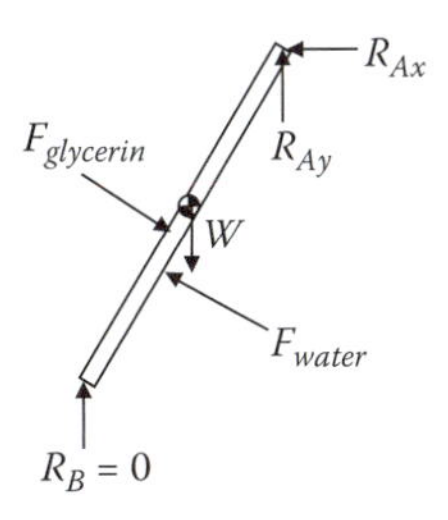

FIGURE 9.18

We intend to apply the equations of equilibrium to the gate to ensure the proper relationships between forces and to meet the constraint that R_B must equal zero. To do this, we first need to evaluate all the forces in the FBD. The weight of the gate is simply mg, or W = (180 kg)(9.81 m/s^2) = 1766 N, and this force acts at the centroid of the gate (Figure 9.19).

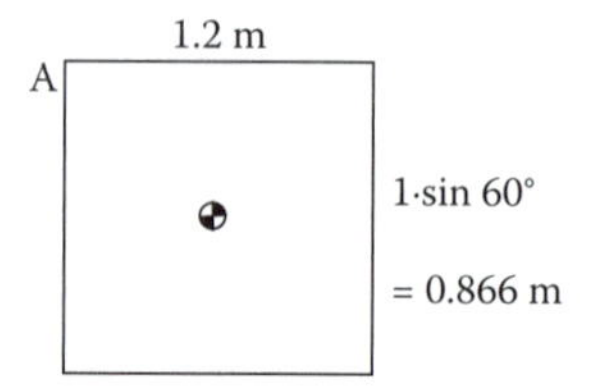

FIGURE 9.19

Next, we must find the forces on the gate due to fluid pressure and where they act. We begin with the glycerin. Since we would prefer not to have to find the reaction forces at A, we plan to sum moments about this point. We need to know the depth of the centroid of the gate, h_c, measured from the surface of the glycerin. The gate is a rectangle, as shown in Figure 9.19, and due to its symmetry its centroid is simply 0.433 m down from the hinge at A. The depth of the centroid is thus 2 m – 0.433 m = 1.567 m below the glycerin surface:

$$F_{glycerin} = \rho_{glycerin} g h_c A$$

$$= (1260\ \tfrac{\text{kg}}{\text{m}^3})(9.81\ \tfrac{\text{m}}{\text{s}^2})(1.567\ \text{m})(1.2\ \text{m}^2)$$

$$= 23.2\ \text{kN}.$$

This force acts at

$$y_R = y_c + \frac{I_{xc}}{y_c A} = y_c + \frac{I_{xc}\sin\theta}{h_c A},$$

and since we intend to sum moments about A, we would like to know the moment arm from $F_{glycerin}$ to point A. Thus, we are most concerned with how much deeper y_R is than y_c, as y_c is clearly 0.5 m from point A:

$$y_R - y_c = \frac{I_{xc}\sin\theta}{h_c A} = \frac{\frac{1}{12}bh^3\sin\theta}{h_c A} = \frac{\frac{1}{12}(1.2\ \text{m})(1\ \text{m})^3\sin 60^\circ}{(1.567\ \text{m})(1.2\ \text{m}^2)} = 0.0461\ \text{m}.$$

We now know that $F_{glycerin}$ = 23.2 kN has a moment arm of 0.5461 m relative to point A.

What remains to be found is the force due to the water on the other side of the gate. Both the magnitude of this force and its moment arm (where it acts) depend on the depth of water, h. For the moment, we leave both these values in terms of the depth of the centroid of the gate, h_c, measured from the water surface – the depth $h = h_c + 0.433$ m:

$$F_{water} = \rho_{water} g h_c A$$

$$= (998\ \tfrac{\text{kg}}{\text{m}^3})(9.81\ \tfrac{\text{m}}{\text{s}^2}) h_c (1.2\ \text{m}^2)$$

$$= (11.75 h_c)\ \text{kN}.$$

And this force acts at y_R, where

$$y_R - y_c = \frac{I_{xc}\sin\theta}{h_c A} = \frac{\frac{1}{12} b h^3 \sin\theta}{h_c A} = \frac{\frac{1}{12}(1.2\ \text{m})(1\ \text{m})^3 \sin 60^\circ}{h_c (1.2\ \text{m}^2)} = \frac{0.0722}{h_c}.$$

So $F_{water} = 11.75 h_c$ kN has a moment arm of $(0.5 + \dfrac{0.0722}{h_c})$m relative to point A (Figure 9.20).

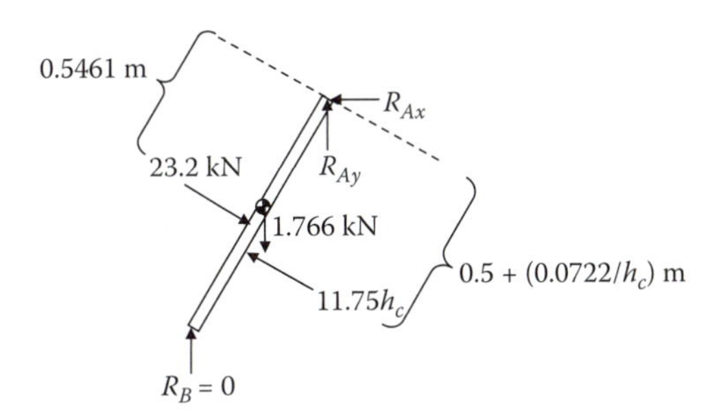

FIGURE 9.20

We choose to sum moments about point A to avoid having to solve for the hinge reaction forces and require that the gate be in equilibrium:

$$\Sigma M_A = 0 = (23{,}200\ \text{N})(0.5461\ \text{m}) + (1766\ \text{N})(0.5\cos 60^\circ) - (11{,}750\, h_c)(0.5 + \frac{0.0722}{h_c}).$$

Solving this expression for h_c, we find that $h_c = 2.09$ m. The depth of the water is then $h = h_c + 0.433$ m $= 2.52$ m.

Example 9.4

The bottle of champagne shown in Figure 9.21 is under pressure, as indicated by the mercury-manometer reading. Compute the net vertical force on the 2-in.-radius hemispherical endcap at the bottom of the bottle.

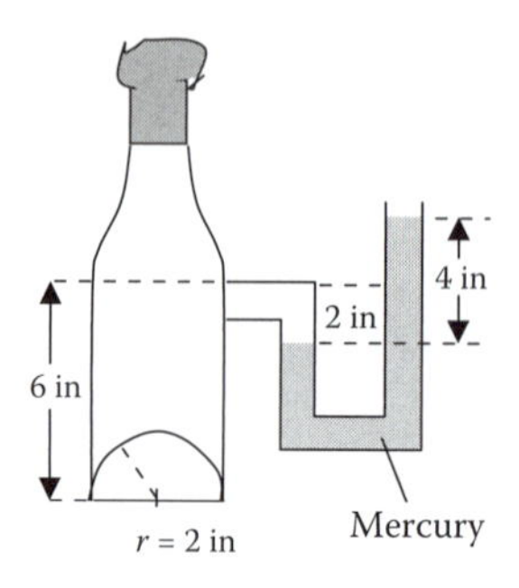

FIGURE 9.21

Given: Pressure measurement; bottle geometry.

Find: Net vertical force on hemispherical surface.

Assume: No relative motion of fluid elements (hydrostatics); fluids have constant, uniform density, and it is appropriate to evaluate densities at 68°F. We look up values for champagne and mercury at this temperature and find that $(\rho g)_C = 59.9\ \text{lbf/ft}^3$ and that $(\rho g)_M = 847\ \text{lbf/ft}^3$.

Solution

We have a manometer that gives us the champagne pressure at a height of 6 in. (We denote values at this position by the subscript "*"), if we work through the U-tube as we did in Example 9.1:

$$p_* + (\rho g)_C \left(\frac{2}{12}\ \text{ft} \right) - (\rho g)_M \left(\frac{4}{12}\ \text{ft} \right) = p_{atm} = 0 \quad \text{(gage)},$$

so

$$p_* = (\rho g)_M (0.333\ \text{ft}) - (\rho g)_C\ (0.167\ \text{ft})$$

$$= (847\ \text{lbf/ft}^3)(0.333\ \text{ft}) - (59.9\ \text{lbf/ft}^3)(0.167\ \text{ft})$$

$$= 272\ \text{lbf/ft}^2 = 272\ \text{psf},$$

This pressure p_* acts on the circular cross-sectional area of the champagne bottle at a height of 6 in., imparting a resultant force of

$$p_* A = (272\ \text{psf})\left(\tfrac{\pi}{4} (\tfrac{4}{12}\ \text{ft})^2 \right) = 23.74\ \text{lbf}$$

on the champagne below it. In addition, the champagne below this 6-in. height imparts its own force on the hemispherical surface. The vertical component of this force, which we are looking for, can also be interpreted as the weight of this champagne above the surface. The net vertical force on the endcap will thus be the p_*A force already calculated, *plus* the weight of the fluid below * and above the hemispherical surface. Since we know the specific weight of the champagne, we need only to find the volume between * and the endcap (Figure 9.22).

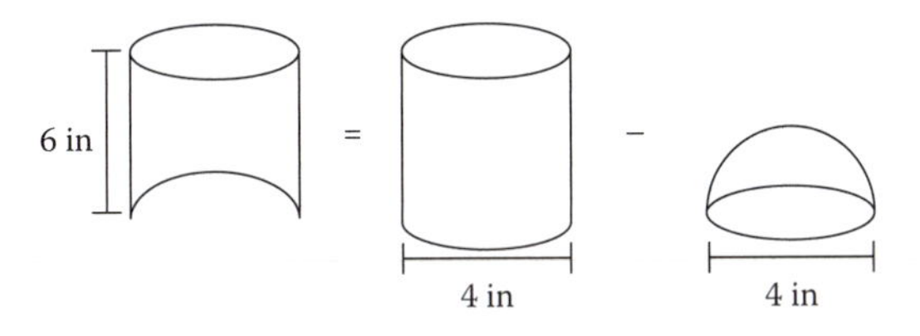

FIGURE 9.22

$$F_V = p_*A + \text{weight of champagne}$$

$$= p_*A + (\rho g)_C\ [\pi(0.167)^2(0.5) - \tfrac{2\pi}{3}(0.167)^3]$$

$$= 23.74 \text{ lbf} + [2.61 - 0.58] \text{ lbf}$$

$$= 25.8 \text{ lbf}.$$

Example 9.5

A parabolic dam's shape is given by $z/z_o = (x/x_o)^2$, where $x_o = 10$ ft and $z_o = 24$ ft. The dam is 50 ft wide (into the page). Find the resultant force on the dam due to the water pressure and its line of action (Figure 9.23).

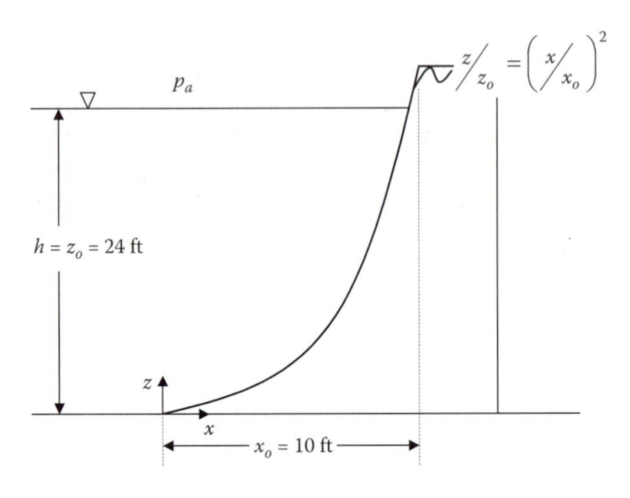

FIGURE 9.23

Given: Geometry of dam; water depth.

Find: Resultant force due to pressure; line of action.

Assume: No relative motion of fluid elements (hydrostatics); fluids have constant, uniform density, and it is appropriate to evaluate densities at 68°F.

Solution

We look up the density of water at this temperature and find that $(\rho g) = 62.4$ lbf/ft^3. We find separately the horizontal and vertical components of the resultant force on the dam surface.

To find the horizontal component F_H, we consider the vertical projection of the curved dam surface, a rectangle that is 24 ft high and 50 ft wide (into the page). This projected surface has an area $A = (24)(50) = 1200$ ft^2, and its centroid is halfway down, at a depth of $h_c = 12$ ft. So,

$$F_H = \rho g h_c A$$

$$= (62.4 \text{ lbf/ft}^3)(12 \text{ ft})(1200 \text{ ft}^2)$$

$$= 899{,}000 \text{ lbf}$$

$$= 899 \text{ kips.}$$

This force acts at the centroid of the pressure prism on the projected vertical surface, at $z_H = h/3 = 8$ ft from the bottom.

The vertical component F_V can be interpreted as the weight of fluid above the curved surface. We consider the properties of a parabolic section, as shown in the sketch in Figure 9.24, to find this value.

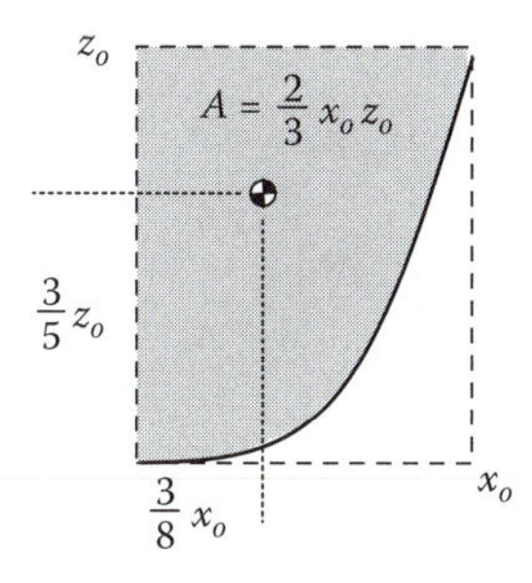

FIGURE 9.24

$$F_V = \rho g V$$

$$= \rho g \left(\frac{2}{3} x_o z_o \right) (50 \text{ ft})$$

$$= (62.4 \text{ lbf/ft}^3)[\tfrac{2}{3}(24 \text{ ft})(10 \text{ ft})](50 \text{ ft})$$

$$= 499{,}000 \text{ lbf}$$

$$= 499 \text{ kips.}$$

This force acts at the x coordinate of the centroid of the volume of fluid, V. From our sketch, we see that this is $3x_o/8 = 3.75$ ft from the origin indicated in Figure 9.23.

The resultant normal force on the surface of the parabolic dam is

$$\begin{aligned} F_R &= \sqrt{F_H^2 + F_V^2} \\ &= \sqrt{(899 \text{ k})^2 + (499 \text{ k})^2} \\ &= 1028 \text{ kips.} \end{aligned}$$

The line of action of this force (Figure 9.25) passes through the point $(x_V, z_H) = (3.75 \text{ ft}, 8 \text{ ft})$ and has slope equal to

$$\tan^{-1}(F_V/F_H) = \tan^{-1}(499/899) = 29°.$$

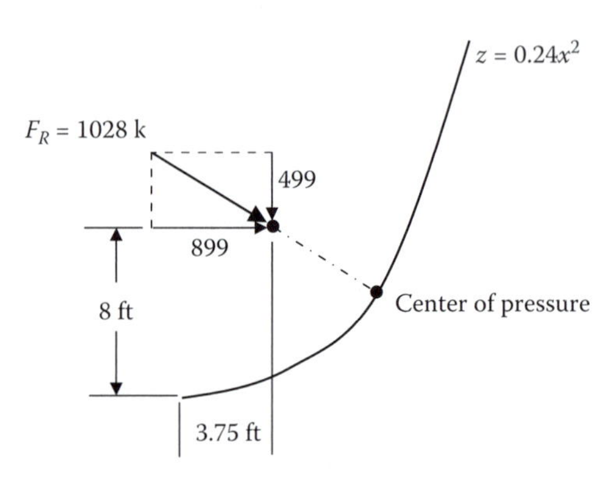

FIGURE 9.25

9.7 Problems

9.1 In 1646 the French scientist Blaise Pascal put a long vertical pipe in the top of a barrel filled with water and poured water in the pipe. He found that he could burst the barrel (not just make it leak a bit as Figure 9.26 shows) even though the weight of the water added in the pipe was only a small fraction of the force required to break the barrel. Briefly explain his finding.

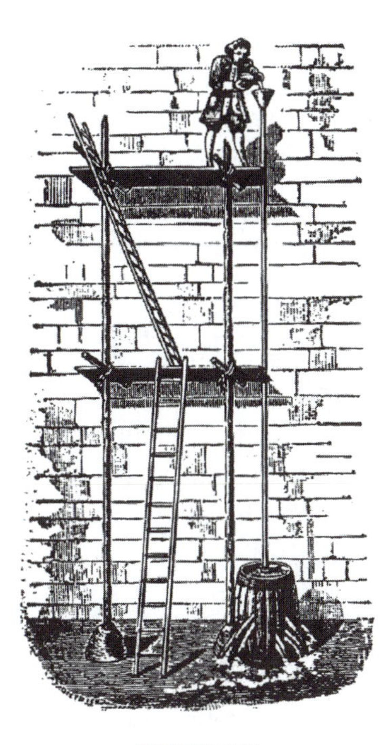

FIGURE 9.26

9.2 In each of the gates shown in Figure 9.27, the top of the gate is supported by a frictionless hinge. Each has a rectangular overhang that sticks out or in a distance a from the gate (a is small relative to the depth h). A stop at the bottom of each wall prevents it from opening in a counterclockwise direction. When the water is the same depth h in both cases, what is the ratio of the horizontal force exerted on stop (a) to the horizontal force exerted on stop (b)? Please draw appropriate free-body diagrams.

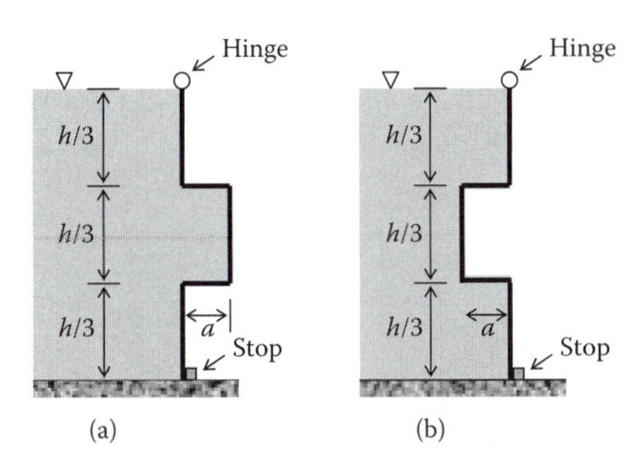

FIGURE 9.27

9.3 The wood (σ_{ys} = 8 ksi, E = 1.5 × 10^6 psi) forms for a concrete wall that is to be 8 ft high are sunk into the ground at the bottom (fixed support) and held in place at the top (10 ft from the ground) by 0.5-in.-diameter tie rods that permit negligible horizontal deflection at that height (Figure 9.28). Each plank in the form may be modeled as an individual beam.

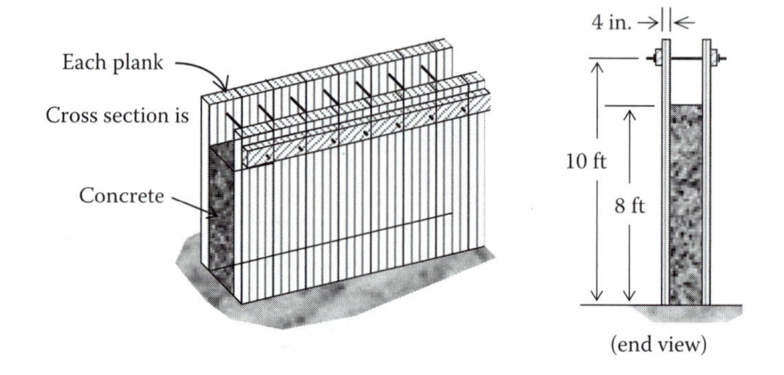

FIGURE 9.28

Assuming that concrete behaves as a liquid (specific gravity = 2.5) just after it is poured, determine (a) the resultant force on a plank due to the concrete and its corresponding center of pressure (height as measured from the ground); (b) the normal stress in a tie rod using the *actual distributed load;* and (c) the maximum normal stress due to bending in a plank using the *actual distributed load.*

9.4 Two very large tanks of water have smoothly contoured openings of equal cross-sectional area (Figure 9.29). A jet of water flows from the left tank. Assume the flow is uniform and unaffected by friction/viscous effects. The jet impinges on a flat plate (and departs the plate in a flow parallel to the surface of the plate) that also covers the opening of the right tank. In terms of the height H, determine the minimum value for the height h to keep the plate in place over the opening of the right tank.

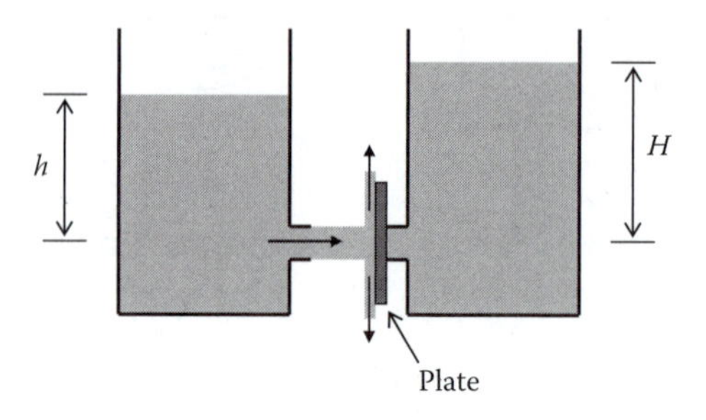

FIGURE 9.29

9.5 You've been contracted to build a cylindrical brick chimney of height H, weighing a = 850 lb/ft of height and fixed securely in a concrete foundation. The inner and outer diameters are d_1 = 3 ft and d_2 = 4 ft, respectively. The chimney must be designed to withstand a distributed load due to a 60 mph (88 ft/sec) wind that we assume is constant at any height from the ground. (a) Find the static pressure at a stagnation point on the chimney. Assume that this pressure acts over the whole projected area (a rectangle $H \times d_2$) for calculation of the force on the chimney by the wind, but this is only an approximation. Find the force/ft of height on the chimney due to the wind. (b) You have a strict design requirement: Considering the weight of the chimney and the wind load together, there is to be no tensile normal stress in the brickwork (because it is brittle and a poor carrier of tensile stress). What is the maximum allowable height H?

9.6 For wall A in Figure 9.30, what is the magnitude and line of action of the horizontal component of the hydrostatic force of the water on the wall (an arc of a circle)? If you were to compare the maximum normal stress due to bending in walls A and B (induced by the hydrostatic loading), would B's be lower, the same, or greater than A's? (Explain briefly in a complete sentence.)

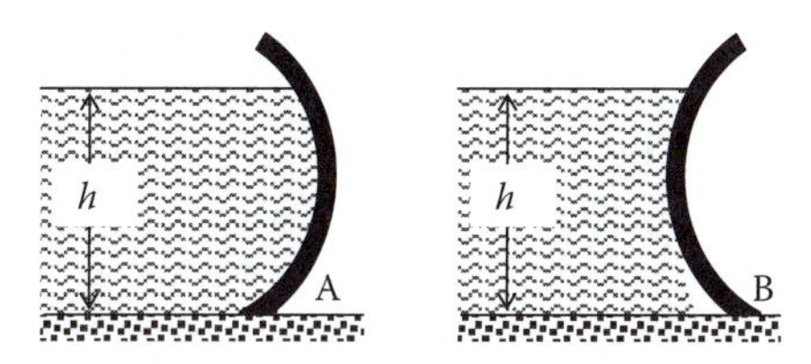

FIGURE 9.30

9.7 A tall standpipe with an open top, as shown in Figure 9.31, has diameter d = 2 m and wall thickness t = 5 mm. (a) If a circumferential stress of 32 MPa is measured in the wall at the bottom of the standpipe, what is the height h of water in the standpipe? (b) What is the axial stress in the standpipe wall due to the water pressure? (c) What is the maximum shear stress induced in the standpipe wall, and where does it occur?

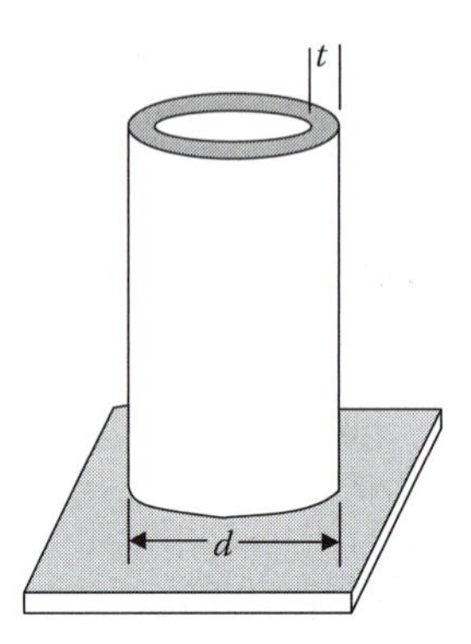

FIGURE 9.31

9.8 A closed tank contains 1.5 m of SAE 30 oil, 1 m of water, 20 cm of mercury, and an air gap on top. The absolute pressure at the bottom of the tank is 60 kPa. What is the pressure in the air?

9.9 Consider a circular cylinder of radius R and length L, and in inverted cone (point/tip up) with base radius R and height L. Both cylinder and cone are filled with water and open to the atmosphere. Write a concise, coherent paragraph that explains the *hydrostatic paradox*: Both containers have the same downward force on the bottom since those bases have the same surface area, even though the cone's volume is only one third of the cylinder's volume.

9.10 The Three Gorges Dam is 2,309 m long, 185 m tall, and 115 m wide at the base. (a) Determine the horizontal component of the hydrostatic force resultant on the dam exerted by water 175 m deep. (b) If all of the people in China (approximately 1.3 billion) were to somehow simultaneously push horizontally against the dam, could they generate enough force to hold it in place with the water at this depth? Support your answer with appropriate calculations.

9.11 What force P is needed to hold the 4-m-wide gate shown in Figure 9.32 closed?

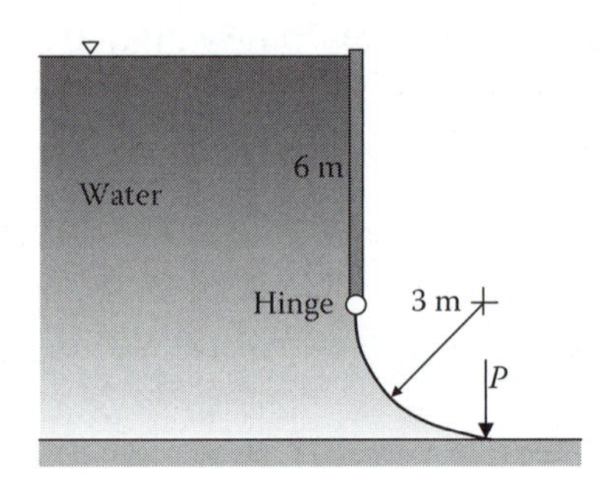

FIGURE 9.32

9.12 Suppose you have three spheres—one of cork, one of aluminum, and one of lead—each with diameter 1.5 cm. You drop all three spheres into a cylinder of water. Explain the different behavior of the three spheres upon their release.

Case Study 6: St. Francis Dam

At three minutes before midnight, March 12, 1928, the St. Francis Dam—built to supply water to the growing city of Los Angeles—collapsed (Figure CS6.1). During the early morning hours of March 13, more than 38,000 acre-ft of water surged down from 1650 ft above sea level. At its highest, the wall of

FIGURE CS6.1
St. Francis Dam before and after the collapse. Photographs courtesy of Santa Clarita Valley Historical Society, Newhall, CA.

water was said to be 78 ft high; by the time it hit Santa Paula, 42 miles south of the dam, the water was 25 ft deep. Almost everything in the water's path was destroyed: livestock, structures, railways, bridges, and orchards. Ultimately, parts of Ventura County lay under 70 ft of mud and debris. Over 500 people were killed, and damage estimates topped $20 million.

William Mulholland (Figure CS6.2), an Irish immigrant who'd risen through the ranks of the city's water department to the position of chief engineer, had proposed, designed, and supervised the construction of the 238-mile Los Angeles Aqueduct, which brought water from the Owens Valley to the city. The St. Francis Dam had been one of the more controversial aspects of his plans. The dam was violently opposed by Owens Valley residents, who sabotaged its construction and often unbuilt portions overnight. The aqueduct itself had been dynamited in 1924. The St. Francis Dam was Mulholland's nineteenth, and final, dam.

The St. Francis was a curved gravity concrete dam, designed to be 62 m high. During construction, the height was increased by 7 m to allow more water to be stored in the reservoir. No change was made to the other dimensions of the dam. In the days before the dam collapsed, the water level in the reservoir was only inches below the top of the dam.

FIGURE CS6.2
Mulholland and H. Van Norman, inspecting wreckage at the St. Francis Dam, March 15, 1928. Photograph courtesy of Santa Clarita Valley Historical Society, Newhall CA.

At the subsequent inquest, it was demonstrated that the dam was leaking as late as the day before the collapse, and it was brought into evidence that the Department of Water and Power (DWP)—and more importantly, Mulholland himself—knew it. Mulholland testified that he'd been at the dam the day before the break but said that he hadn't noticed anything unusual. Leaks, he pointed out, were not particularly unusual in dams, especially dams as large as the St. Francis.

Although the assignation of cause and culpability is still a contentious subject among modern analysts, the 1928 jury ruled that the disaster was caused by the failure of a geological fault and rock formations on which the dam was built. Even so, the public held the DWP, and particularly William Mulholland, responsible. Although no criminal charges were brought against him, he retired from the DWP soon after the jury's verdict and lived in self-imposed exile until he died in 1935 at 79 years old.

Problems

CS6.1 For a dam of height $H = 62$ m, thickness b, and width into the page $w = 75$ m as shown in Figure CS6.3, made of concrete with density 2300 kg/m^3, retaining a body of water that is 60 m deep, find the net moment about point A and the minimum thickness of the dam that will prevent this moment from overturning the dam.

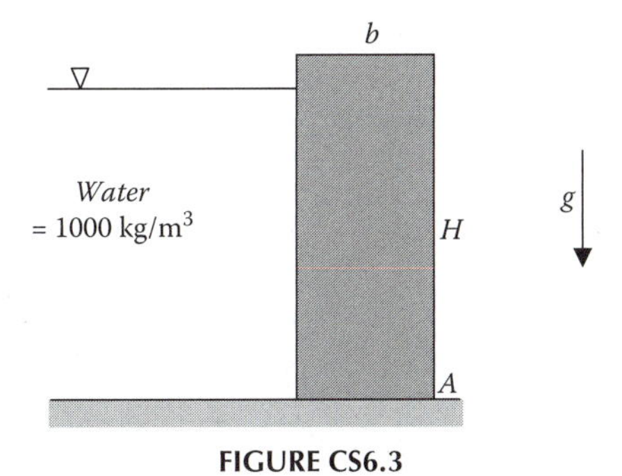

FIGURE CS6.3

CS6.2 If the dam's height is increased to 70 m, and the water depth rises to 68 m, what thickness b is required to prevent tipping?

CS6.3 The St. Francis dam (with dimensions as in CS6.2) was observed to be leaking muddy water at its base, indicating that water was seeping under and around its supports. If water is allowed to penetrate freely under our model dam to point A, what thickness b is necessary to prevent the dam from tipping?

CS6.4 If we refined our model to more accurately represent the geometry of the St. Francis Dam, including the curvature of the surface, would you expect the required thickness b to increase or decrease? Why?

Note

1. For example, in Appendix A of this book.

10

Fluid Dynamics: Governing Equations

In the previous chapter, we considered cases in which there was no relative motion of fluid particles—no velocity gradients and, thus, no shear stress. Now we consider the somewhat more interesting flows in which velocity gradients and accelerations do appear.

10.1 Description of Fluid Motion

You have probably seen the car companies' ads featuring this year's models in wind tunnels, with smoke tracing the flow of air over the new car's streamlined curves. There is a mathematical way to define the equations of these smoke traces, and a physical interpretation of them, that we find quite useful in our discussion of fluid dynamics.

The velocity field, of course, tells the instantaneous speed and direction of the motion of all points in the flow. A *streamline* is everywhere tangent to the velocity field and so reflects the character of the flow field. Streamlines are instantaneous, being based on the velocity field at one given time. The smoke traces just mentioned and illustrated in Figure 10.1 are streaklines, which include all the fluid particles that once passed through a certain point. We may also describe a flow field with pathlines, which represent the trajectory traced out by a given fluid particle over time. When the flow is steady, or independent of time t, the streamlines, streaklines, and pathlines coincide.

For a two-dimensional flow field, we can find the equations of streamlines by applying their definition. Since streamlines are tangent to velocity, the slope of a streamline must equal the tangent of the angle that the velocity vector makes with the horizontal, as shown in Figure 10.2. In mathematical language, this is

$$\frac{dy}{dx} = \frac{v}{u}, \text{ or } \frac{dx}{u} = \frac{dy}{v}, \tag{10.1}$$

so that if the velocity field is known as a function of x and y (and t, if the flow is unsteady), we simply integrate equation (10.1) to find the equation of the streamline. For a three-dimensional flow, we write

FIGURE 10.1
Smoke-traced streaklines of flow around a car. (Source: DaimlerChrysler.)

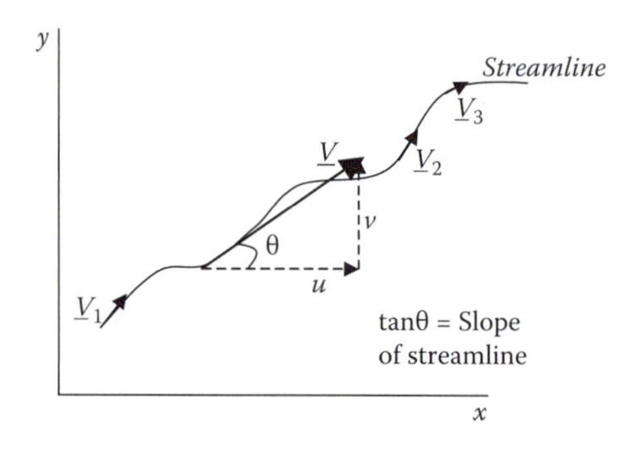

FIGURE 10.2
Streamlines are tangent to the velocity field.

$$\frac{dx}{u} = \frac{dy}{v} = \frac{dz}{w} = ds, \tag{10.2}$$

and we can integrate these expressions with respect to s, holding time constant and using the initial condition (x_o, y_o, z_o, t_o) at $s = 0$, and then can eliminate s to find the equation of the streamline.

For the two-dimensional flow of an incompressible[1] fluid, an even lovelier method of finding streamlines exists. We need only remember that for incompressible flow, $\nabla \cdot \underline{V} = 0$, or

$$\frac{\partial u}{\partial x} + \frac{\partial v}{\partial y} = 0 . \tag{10.3}$$

We can then define a *stream function* ψ by the following:

$$u = \frac{\partial \psi}{\partial y} \text{ and } v = -\frac{\partial \psi}{\partial x} . \tag{10.4a,b}$$

Lines of constant ψ are the streamlines for the flow with $V = (u, v)$.

Streamlines can give us important information about the pattern and relative speed of flow. Closely spaced streamlines reflect faster flow than widely spaced streamlines. There is no flow *across* or *through* streamlines, since the velocity field is purely tangential to these lines. Streamlines can hence be thought of as boundaries for the flow. Fluid particles on one side of a streamline never cross it. Solid boundaries of resting solids are always streamlines, since flow does not penetrate them.

10.2 Equations of Fluid Motion

Although the physics of fluid mechanics are certainly familiar (neither mass conservation nor $F = ma$ is a new concept), fluid particles are much harder to keep track of than the solid bodies we have considered previously. It's very difficult to follow a prescribed amount of fluid mass around. We therefore need some new tools with which to apply the same old physics.

Rather than following the flow of a fixed fluid mass, which would be quite challenging, we keep track of a prescribed *volume* through which fluid may flow. This volume may be thought of as an imaginary "cage" for fluid, though it is more commonly known as a *control volume* (CV). We may choose to have a cage of finite size, in which case we use the integral governing equations of Section 10.3, or of infinitesimally small size, in which case our equations are the partial differential equations of Section 10.4.

10.3 Integral Equations of Motion

We first apply our old physics to a control volume or cage of finite size. This approach is particularly useful when we are interested in the large-scale behavior of the flow field and the effect of a flow on devices such as nozzles, turbine blades, or heart valves.

In our study of solid mechanics we often used a free-body diagram (FBD), in which we isolated an object from its surroundings, replaced these surroundings by the actions they had on our object, and then applied Newton's laws of motion. The fluid mechanics equivalent of this free body or object would be a fluid element, a specific quantity of matter composed of many fluid particles—called the *system approach*. However, fluids move and deform in such a way that it is difficult to keep track of a specific quantity of matter. It's easy to follow a branch moving on the surface of a river, but it's hard to follow a particular portion of water in the river. This is why we consider the flow *through* set boundaries (in this section, finite control volumes) instead of the mass once contained in these boundaries.

10.3.1 Mass Conservation

When a fluid is in motion, it moves in such a way that mass is conserved. This principle, known as mass conservation, places restrictions on the fluid's velocity field. To see this, we consider the steady (i.e., not changing in time) flow of fluid through a duct. A relevant control volume is shown in Figure 10.3.

In this simple example, both inflow and outflow are one-dimensional, so that the velocity V_i and density ρ_i distributions are constant over the inlet and outlet areas. Applying conservation of mass to this control volume means that whatever mass flows into the CV must flow out. In some time interval Δt, a volume of $A_1V_1\Delta t$ flows into the CV, and a volume of $A_2V_2\Delta t$ flows out. These volumes are shaded in Figure 10.3. We multiply these volumes by the fluid density at each control surface (CS) so that we'll have the amount of mass flowing in and out ($\rho_1 A_1V_1\Delta t$ and $\rho_2 A_2V_2\Delta t$). We then balance mass in with mass out, canceling the Δt that appears in both terms, and have

$$\rho_1 A_1V_1 = \rho_2 A_2V_2. \tag{10.5}$$

These terms are now mass flow *rates* (i.e., mass fluxes), as we have divided by Δt. This mathematically states that mass flow rate entering CV is balanced by mass flow rate leaving it. For a more general CV with N inlets and outlets, we would write

$$\sum_{i=1}^{N} \rho_i A_i V_i, \tag{10.6}$$

with inflows negative and outflows positive.

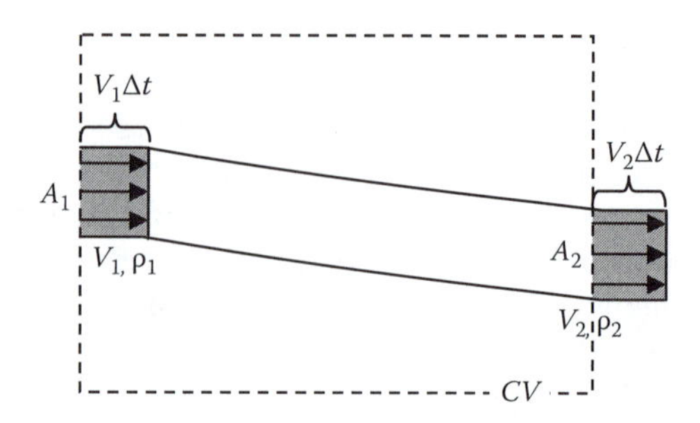

FIGURE 10.3
CV for steady, one-dimensional fluid through a duct.

We now move beyond this simple case of one-dimensional, steady flow. To conserve mass, we require *no net change of mass* in our volume. In the previous example, the amount of mass in our CV changed only due to flow in and flow out. The amount of mass may also change in time due to the flow's unsteadiness. In a more general equation, we must account for both:

total rate of change of property	=	time rate of change in property	+	flux of property across CV surfaces

In our earlier example, this was easy to do: Because it was a steady flow, there was no time rate of change, and because the flow was one dimensional and uniform, the flux was easily computed. We need a more general mathematical statement of mass conservation.

The time rate of change of mass in our control volume is the time derivative of the total product of fluid density in the CV and the volume:

$$\text{time rate of change} = \frac{\partial}{\partial t}\int_{CV} \rho \, d\mathcal{V}. \tag{10.7}$$

The flux of mass across a control surface is the amount of mass per unit time that is transported across the surface's area dA with outward unit normal vector n. Hence,

$$\text{flux across surfaces} = \int_{CS} \rho \, \underline{V} \cdot \underline{n} dA. \tag{10.8}$$

The dot product $(V{\cdot}n)_i$ is simply the normal component of velocity across the ith control surface, and since n is an outward normal vector, product $(V{\cdot}n)_i$ is negative for flows into the CV, and positive for flows out of it. We sum, or integrate, the fluxes across all the surfaces of the CV, as shown in Figure 10.4.

Writing mass conservation requires us to state that the total change in mass in the CV, which must equal the sum of the time rate of change of mass in the CV plus the flux of mass across all control surfaces, is equal to zero:

$$0 = \frac{\partial}{\partial t}\int_{CV} \rho \, d\mathcal{V} + \int_{CS} \rho \, \underline{V} \cdot d\underline{A}. \tag{10.9}$$

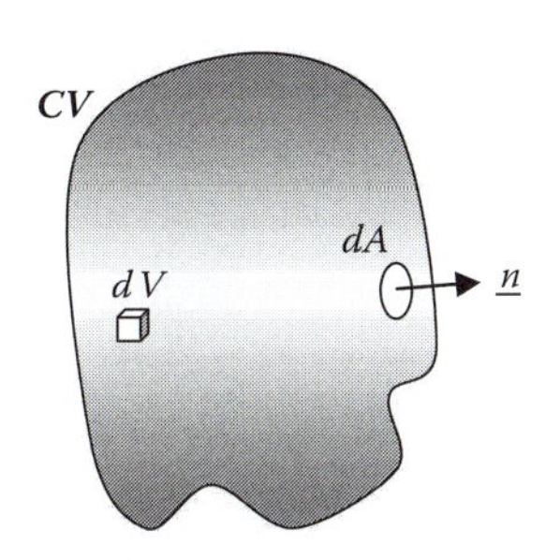

FIGURE 10.4
Standard potato-shaped control volume (CV).

This is how we write mass conservation for a finite-sized control volume.

10.3.2 $F = ma$, or Momentum Conservation

To write down $F = ma$ for a finite control volume, we must consider (1) all forces on the fluid in the CV that may cause an acceleration, and (2) how to express the fluid's *ma*, or the total rate of change of its linear momentum.

Forces on a fluid, just like those acting on a solid, may be either surface (acting through direct contact, on control surfaces) or body forces ("field" forces, acting on the entire control volume without contact). We begin with surface forces. We are already familiar with the notion that a difference in pressure imparts a force. Indeed, a pressure gradient can cause a fluid to move toward the lower pressure. The force on a fluid due to pressure variation must be included in $F = ma$. When there is relative motion of fluid particles, a frictional force (i.e., a viscous stress) is developed and acts on the fluid. Because of the complex form of the constitutive law for fluids, this can be the hardest term to construct; fortunately, it is often possible to *neglect* viscous effects relative to pressure gradients, inertia, and other forces. We add only a rather vague F_{visc} to the equation at this time. The primary body force acting on fluids is gravity. We may also include external reaction forces (from ducts or other surfaces) that act on the fluid.

The total change in linear momentum of fluid in the control volume, like the total change in mass, must be written as its time rate of change plus the flux of it across control surfaces. We again start with a fairly simplistic example: steady, one-directional flow through a duct, as in Figure 10.5.

Because $V_2 \neq V_1$, we know the fluid is accelerating between the inlet and outlet surfaces—even though the velocity is not changing in *time*, there is acceleration due to *spatial* variation in velocity. This fluid acceleration must equal the net force on the fluid in the control volume. The resultant force on the fluid, as we just discussed, must consist of forces: (1) due to pressure variation, (2) due to viscous stresses, (3) due to gravity, and (4) exerted by the duct on the fluid (i.e., reaction forces). To preserve the simplicity of the example, we neglect both viscous and gravitational effects. The resultant force on the fluid in the CV is thus

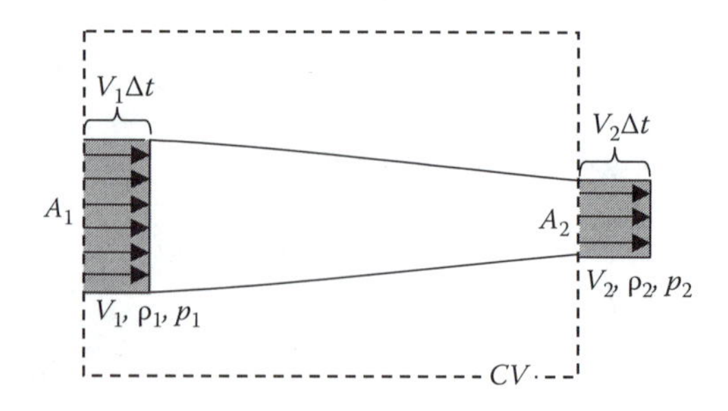

FIGURE 10.5
CV for steady, one-dimensional fluid through a duct.

$$\underbrace{p_1A_1}_{\substack{\text{pressure force}\\ \text{on left CS, in}\\ +x \text{ direction}}} - \underbrace{p_2A}_{\substack{\text{pressure force}\\ \text{on right CS, in}\\ +y \text{ direction}}} + \underbrace{R_x}_{\substack{\text{reaction}\\ \text{on fluid}\\ \text{from duct}}} = F_x \qquad (10.10)$$

and is in the x direction (as reflected by its subscript). It is good practice to draw an FBD of your control volume, indicating the relevant forces and their orientations, as we do in Figure 10.6.

Having written down the resultant force F_x on the fluid, we must now write an expression for the total rate of change in the fluid's x momentum, to balance F_x. Because the flow is steady, the x momentum does not change in time; we simply need to account for the flow of x momentum into and out of the control volume. The difference between inflow and outflow, or the *net change in x momentum*, will balance F_x. In some time interval Δt, an amount of x momentum of $(\rho_1 A_1 V_1 \Delta t)V_1$ flows into the CV, and a volume of $(\rho_2 A_2 V_2 \Delta t)V_2$ flows out. (Note that the x momentum is simply the mass flux, already found in Section 10.3.1, times the x component of velocity at the given CS.) Again, outflow is positive, so the net rate of change of momentum is the difference between these terms, divided by the time Δt: $\rho_2 A_2 V_2^2 - \rho_1 A_1 V_1^2$. We can write the x component of $F = ma$:

$$p_1A_1 - p_2A_2 + R_x = \rho_2 A_2 V_2^2 - \rho_1 A_1 V_1^2. \qquad (10.11)$$

We want to generalize this, so we can apply our physics to less simplistic problems. The forces are easy to generalize: The net pressure force can be written as the pressure p acting on a given control surface times the area of that control surface: $p\underline{n}dA$. We must sum over all the control surfaces of our CV, and we write this as an integral,

$$-\int_{CS} p\underline{n}dA,$$

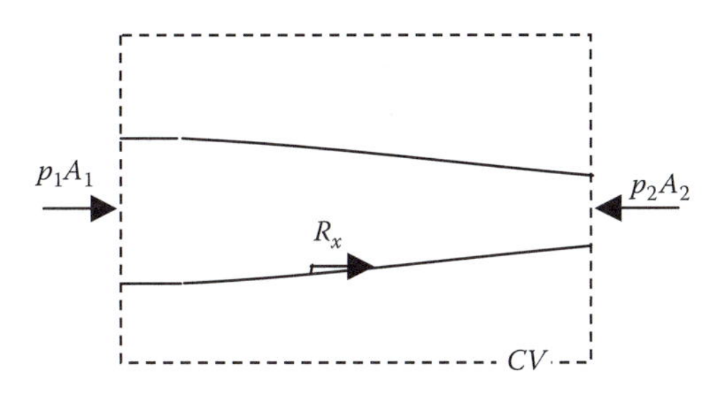

FIGURE 10.6
FBD of fluid in CV.

where the negative sign reflects the fact that a positive p compresses our fluid. The force due to gravity is also easy to write as

$$\int_{CV} \rho\, \underline{g} dV,$$

since it acts on the whole control volume. Next we write a more general expression for the total rate of change of fluid linear momentum. Again, momentum is just mass times velocity, so this extends naturally from our expressions for mass conservation:

$$\text{time rate of change} = \frac{\partial}{\partial t}\int_{CV} \rho\, \underline{V}\, d\mathcal{V}. \tag{10.12}$$

$$\text{flux across surfaces} = \int_{CS} \rho\, \underline{V}\, \underline{V}\cdot \underline{n} dA\,. \tag{10.13}$$

Note that V is a vector [$V = (u, v, w)$], and, hence, there are three components of each of these expressions. We can now write the vector form of our general equation,

$$\underline{F} = \underline{F}_{visc} + \underline{F}_{external} + \int_{CV} \rho\, \underline{g} d\mathcal{V} - \int_{CS} p\underline{n} dA = \frac{\partial}{\partial t}\int_{CV} \rho\, \underline{V}\, d\mathcal{V} + \int_{CS} \rho\, \underline{V}\, \underline{V}\cdot \underline{n} dA \tag{10.14}$$

as well as the form of its three component equations in Cartesian coordinates:

$$F_x = \left(F_{visc}\right)_x + \left(F_{external}\right)_x + \int_{CV} \rho\, g_x d\mathcal{V} - \int_{CS} p\hat{\underline{i}}\cdot \underline{n} dA = \frac{\partial}{\partial t}\int_{CV} \rho\, u\, d\mathcal{V} + \int_{CS} \rho\, u\, \underline{V}\cdot \underline{n} dA, \tag{10.15a}$$

$$F_y = \left(F_{visc}\right)_y + \left(F_{external}\right)_y + \int_{CV} \rho\, g_y d\mathcal{V} - \int_{CS} p\hat{\underline{j}}\cdot \underline{n} dA = \frac{\partial}{\partial t}\int_{CV} \rho\, v\, d\mathcal{V} + \int_{CS} \rho\, v\, \underline{V}\cdot \underline{n} dA, \tag{10.15b}$$

$$F_z = \left(F_{visc}\right)_z + \left(F_{external}\right)_z + \int_{CV} \rho\, g_z d\mathcal{V} - \int_{CS} p\hat{\underline{k}}\cdot \underline{n} dA = \frac{\partial}{\partial t}\int_{CV} \rho\, w\, d\mathcal{V} + \int_{CS} \rho\, w\, \underline{V}\cdot \underline{n} dA. \tag{10.16}$$

This is how we write $\underline{F} = ma$ for a finite-sized control volume.

10.3.3 Reynolds Transport Theorem

As we've said, the big difference between the forms of these governing equations for fluids and solids is that fluids may flow across the surfaces of a control volume. (For a solid, a control volume is a fixed mass of the solid!) The idea of the *Reynolds Transport Theorem* is that we can relate these two approaches by accounting for the flow across control surfaces.

It is possible to obtain the equations for both mass and momentum conservation using the Reynolds Transport Theorem, which says that the total rate of change of some quantity η (a *specific* quantity, some parameter N per unit mass) is equal to the time rate of change of η for the contents of the control volume plus a contribution due to the flow of η through the control surface. For our purposes, N may be the mass of fluid in our control volume (ρdV) or the fluid momentum ($\underline{V}\rho dV$). The Reynolds Transport Theorem has the general form,

$$\underbrace{\frac{D}{Dt}\int\limits_{\substack{mass\\(system)}} \eta\, dm}_{\substack{\text{time rate of change of } \eta\\ \text{for the coincident system}\\ \text{(if we were to follow a mass)}}} = \underbrace{\frac{\partial}{\partial t}\int\limits_{CV} \rho\,\eta\, dV}_{\substack{\text{time rate of change of } \eta\\ \text{of the contents of the coincident}\\ \text{control volume}}} + \underbrace{\int\limits_{CS} \rho\,\eta\, \underline{V}\cdot\underline{n}dA}_{\substack{\text{net rate of flux of } \eta\\ \text{through control surface}}}. \tag{10.17}$$

Once again, to conserve mass, we must have flow in balancing flow out and the Reynolds Transport Theorem with $N = mass$; thus, $\eta = N/mass = \rho dV/\rho dV = 1$ gives us

$$0 = \frac{\partial}{\partial t}\int\limits_{CV} \rho\, dV + \int\limits_{CS} \rho\, \underline{V}\cdot\underline{n}dA, \tag{10.18}$$

where the left-hand side is zero since, by definition, $D(mass)/Dt$ for a system is zero. For the special case of a steady flow, the first term on the right-hand side drops out, and we must have

$$\int\limits_{CS} \rho\, \underline{V}\cdot\underline{n}dA = 0. \tag{10.19}$$

To obtain the control volume form of momentum conservation, we apply the Reynolds Transport Theorem to $N = m\underline{V} = \rho dV\,\underline{V}$, or $\eta = N/m = \rho\underline{V}dV/\rho d\underline{V} = V$, and get

$$\underline{F} = \frac{\partial}{\partial t}\int_{CV} \rho \underline{V} dV + \int_{CS} (\rho \underline{V})\underline{V} \cdot \underline{n} dA, \tag{10.20}$$

where the left-hand side is F since Newton's second law tells us that F = ma = d(mV)/dt. In words, equation (10.20) tells us that the time rate of change of momentum contained in a fixed volume plus the net flow rate of momentum through the surfaces of this volume are equal to the sum of all the forces acting on the volume. (We developed the form of this sum, resultant force, in the previous section.)

10.4 Differential Equations of Motion

We can also construct useful expressions of mass conservation and $F = ma$ for cages or volumes that are infinitesimally small. This is useful when we want to know detailed information about the flow at very small scales at high resolution.

10.4.1 Continuity, or Mass Conservation

The principle of mass conservation is fundamental to the study of mechanics. It states that mass is neither created nor destroyed; hence, the mass of a system remains constant as the system moves through the flow field.

By considering the flow through an imaginary, very small cage in the flow field, we can derive a useful mathematical expression of this principle. (Though this cage has the same dimensions as our frequently discussed fluid element, with volume $dV = dxdydz$, it is stationary.) The fluid has velocity $V = (u, v, w)$ when it is at the center of the cage. Its mass flow rate per unit area may then be written as $(\rho u, \rho v, \rho w)$.

We want to obtain an expression for the mass flow across the cage faces to see what is flowing in and out of the cage. This is shown for one direction of flow in Figure 10.7. We must multiply the face-specific expressions by the appropriate areas ($dxdz$ in both cases), and we can then combine them to get mass flow rate in the y direction:

$$\dot{m}_y = \left[\rho\, v + \frac{\partial(\rho\, v)}{\partial y}\frac{dy}{2}\right] dxdz - \left[\rho\, v - \frac{\partial(\rho\, v)}{\partial y}\frac{dy}{2}\right] dxdz = \frac{\partial(\rho\, v)}{\partial y} dxdydz\,. \tag{10.21}$$

We repeat this in the x and z directions to have an expression for total mass flow rate through the cage:

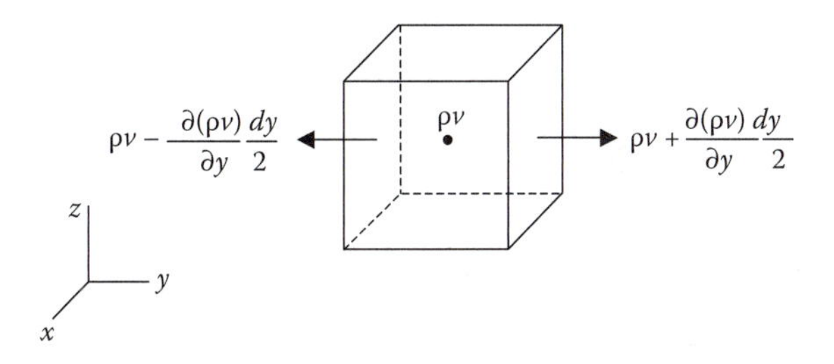

FIGURE 10.7
Mass flow through a cage with volume *dxdydz*. We have used Taylor expansions to express the values of mass flow at all faces, just as we did for the pressure field.

$$\dot{m} = \left[\frac{\partial(\rho\, u)}{\partial x} + \frac{\partial(\rho\, v)}{\partial y} + \frac{\partial(\rho\, w)}{\partial z}\right] dxdydz. \tag{10.22}$$

To satisfy mass conservation, this mass flow rate must balance the rate of mass decrease within the element,

$$-\frac{\partial \rho}{\partial t} dxdydz\ .$$

Balancing these terms and canceling *dxdydz*, we get

$$\frac{\partial \rho}{\partial t} + \frac{\partial(\rho\, u)}{\partial x} + \frac{\partial(\rho\, v)}{\partial y} + \frac{\partial(\rho\, w)}{\partial z} = 0\,. \tag{10.23}$$

This is the mass conservation equation, also known as the continuity equation, for fluids. It is valid for steady or unsteady flow and for incompressible or compressible fluids. We remember that the last three terms, the sum of partials of the vector $\rho\underline{V}$, represent the divergence of this vector, and we can rewrite the equation in vector form as

$$\frac{\partial \rho}{\partial t} + \nabla \cdot (\rho\, \underline{V}) = 0\,. \tag{10.24}$$

In vector form, this equation is independent of coordinate choice and works in cylindrical, spherical, or polar coordinates in addition to our Cartesian *(x,y,z)* friends.

We note that for steady flows, $\frac{\partial \rho}{\partial t} = 0\,.$

For incompressible ($\rho \approx$ constant) fluids, the continuity equation reduces to

$$\nabla \cdot \underline{V} = 0,$$

as we saw in Chapter 8, Section 8.5.1.

10.4.2 $\underline{F} = m\underline{a}$, or Momentum Conservation

Just as for solids, the governing equation for fluid mechanics is Newton's second law. We have inched toward expressing this mathematically for fluids by writing out the forces due to pressure, gravity, and viscosity. All that remains is to formulate the *acceleration* of a fluid element, and since we know the effects of these forces, we can write $\underline{F} = m\underline{a}$.

Although the vast majority of problems we've seen have dealt with nonaccelerating solids, we are familiar with the idea that a solid's acceleration is simply the rate of change of its velocity. This is also true for fluids. We must consider the rate of change of the velocity field, remembering that this change may be in time t and also in x, y, and z.

Given a velocity field $\underline{V} = (u, v, w)$, where each component is allowed to vary in space and time, so that $u_i = f(x, y, z, t)$, we know that the acceleration is the change in this velocity field in a time interval dt. The velocity of some fluid particle L is $\underline{V}_L$ at time t, and at some later time $t + dt$, as shown in Figure 10.8, has evolved:

$$\underline{V}_L\Big]_t = \underline{V}(x, y, z, t), \tag{10.25a}$$

$$\underline{V}_L\Big]_{t+dt} = \underline{V}(x + dx, y + dy, z + dz, t + dt), \tag{10.25b}$$

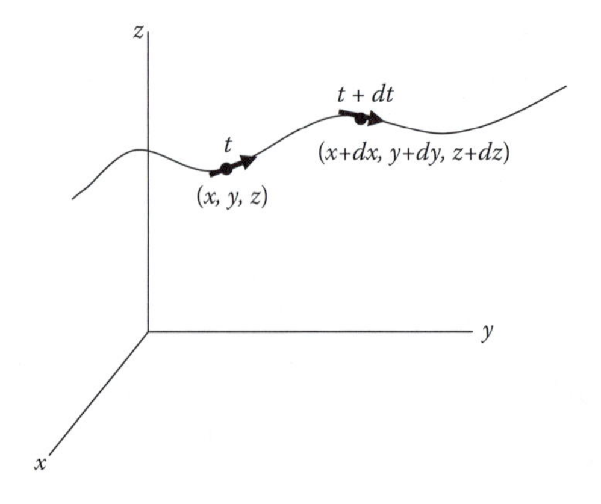

FIGURE 10.8
Evolution of point L and its velocity $\underline{V}_L$.

so the change in V_L may be written as the difference between these two, or

$$d\underline{V}_L = \underline{V}(x+dx, y+dy, z+dz, t+dt) - \underline{V}(x,y,z,t)$$

$$= \frac{\partial \underline{V}}{\partial x}dx + \frac{\partial \underline{V}}{\partial y}dy + \frac{\partial \underline{V}}{\partial z}dz + \frac{\partial \underline{V}}{\partial t}dt \tag{10.26}$$

so that that rate of change in V_L, or dV_L/dt, is

$$\underline{a}_L = \frac{d\underline{V}_L}{dt} = \frac{\partial \underline{V}}{\partial x}\frac{dx}{dt} + \frac{\partial \underline{V}}{\partial y}\frac{dy}{dt} + \frac{\partial \underline{V}}{\partial z}\frac{dz}{dt} + \frac{\partial \underline{V}}{\partial t}, \tag{10.27}$$

or

$$\underline{a} = \underbrace{\frac{\partial \underline{V}}{\partial t}}_{local} + \underbrace{\frac{\partial \underline{V}}{\partial x}u + \frac{\partial \underline{V}}{\partial y}v + \frac{\partial \underline{V}}{\partial z}w}_{convective} \equiv \frac{D\underline{V}}{Dt}. \tag{10.28}$$

This expression is defined as the *material derivative* of the velocity $\underline{V}$. It may also be written in vector form as

$$\underline{a} = \frac{D\underline{V}}{Dt} = \frac{\partial \underline{V}}{\partial t} + (\underline{V}\cdot\nabla)\underline{V}. \tag{10.29}$$

The material derivative is, in a sense, a derivative "following the fluid," as it considers the movement of the fluid particle in question. This way of writing the acceleration takes into account the change in flow velocity in both time and space and equips us to write $\underline{F} = m\underline{a}$ for a fluid element. Notice that it expresses the total change in velocity as the time rate of change, plus a reflection of spatial variation, of velocity.[2]

We have the expression in equation (10.29) for acceleration, and we know that ma is $\rho dxdydz\ a$. We know that both surface and body forces can act on a fluid, and we know how to write the forces due to gravity, pressure, and viscous effects. We also know about surface tension—and because surface tension comes into play only at boundaries between fluids, it affects the boundary conditions but not the governing equations themselves.

If viscous effects are neglected, we can write $\underline{F} = m\underline{a}$ for a fluid element dV:

$$\rho\, dV\frac{D\underline{V}}{Dt} = -\nabla p\, dV + \rho\, \underline{g} dV, \tag{10.30}$$

or, dividing through by dV,

$$\rho \frac{D\underline{V}}{Dt} = -\nabla p + \rho \underline{g}, \tag{10.31}$$

where typically $g = -g\hat{k}$. This equation is Newton's second law for an effectively inviscid fluid, first derived by Leonhard Euler in 1755 and henceforth known as the *Euler equation*. It is also known as the inviscid momentum equation, as it expresses the conservation of linear momentum. Please note that the only assumption made in its derivation is that of *inviscid* behavior—this equation holds for both compressible and incompressible fluids and for steady and unsteady flows. Viscous effects are generally negligible far from flow boundaries, allowing us to rely on Euler's inviscid momentum equation in good faith.

If we include viscous effects, we must write the force on the fluid element due to the viscous stress tensor, which was discussed in Chapter 8, Section 8.5.4. In that discussion, the stress tensor was shown to be composed of a portion due to pressure (already included in the inviscid equation) and another portion that is proportional to the *strain rate tensor,* which itself depends on the velocity gradients in the flow. The force due to stress tensor $\underline{\underline{\tau}}$ on an element dV may be written as $\nabla \cdot \underline{\underline{\tau}}$, which requires the recollection of Gauss's theorem, to change the surface force

$$\int_A \underline{\underline{\tau}}\, d\underline{A}$$

to a force on the entire volume

$$\int_V \nabla \cdot \underline{\underline{\tau}}\, dV.$$

Physically, we could derive this in the same way we determined the force due to pressure on a fluid element dV. For a viscous, incompressible Newtonian fluid,[3] $\underline{F} = m\underline{a}$ is written as

$$\rho \frac{D\underline{V}}{Dt} = -\nabla p + \rho\, \underline{g} + \mu\, \nabla^2 \underline{V}, \tag{10.32}$$

where μ is the fluid's viscosity, and the del-squared operator on V may be written in index notation:

$$\nabla^2 u_i \equiv \frac{\partial^2 u_i}{\partial x_j \partial x_j} = \frac{\partial^2 u_i}{\partial x_1{}^2} + \frac{\partial^2 u_i}{\partial x_2{}^2} + \frac{\partial^2 u_i}{\partial x_3{}^2}. \tag{10.33}$$

Equation (10.32) (as well as its variants with the fuller stress tensor) is known as the *Navier-Stokes equation*. This name is somewhat interesting since Claude Navier got it wrong. He misunderstood viscosity and the dependence of the stress tensor on velocity gradients and published a flawed derivation of $F = ma$ for viscous fluids in 1822. Though his results were correct, his reasoning was flawed. George Stokes later got the derivation of the viscous terms right, so his name was added to the marquee. However, in 1843, two years before Stokes's results were published, a paper appeared by Jean Claude Saint-Venant in which this equation was correctly derived and interpreted. It is a mystery why the equation does not bear his name today. As students of continuum mechanics, we are already grateful to Saint-Venant for his discovery that the details of force application are only relevant in the immediate neighborhood of application, allowing us to use "average" stress relations, and now we have another reason to thank him and to condemn the injustice that leaves his name out of most discussions of the Navier-Stokes equation. Incidentally, Navier was no slouch, despite his errors here—he was a great builder of bridges and did important work in elasticity and solid mechanics.

10.5 Bernoulli Equation

Equation (10.32) for the conservation of momentum is a vector equation, with component equations in each direction of motion. These directions may be (x, y, z), (r, θ), or (r, θ, φ). If we remember that *streamlines* are everywhere tangent to the velocity vector, we can also think of a set of coordinates defined relative to the streamlines—for two-dimensional flow, one coordinate s directed along the streamline, and n defined normal to the streamline. We could then write the component equations of motion in the s and n directions. The resulting equation in the s direction, which states $\underline{F} = m\underline{a}$ along a streamline, may be integrated to yield the following equation:

$$\frac{P}{\rho} + gz + \frac{1}{2}V^2 = \text{constant along a streamline,} \tag{10.34}$$

where we have assumed that gravity acts in the negative z direction, and where V is the velocity in the s direction, simply the magnitude of the velocity vector since $\underline{V} \,||\, s$. This equation is known as the *Bernoulli equation*, and it is true for steady flow of an incompressible fluid under inviscid conditions. For convenience we write the equation together with its restrictions as

$$\frac{P}{\rho} + gz + \frac{1}{2}V^2 = \text{constant:}$$

- On a streamline
- For steady flow
- For incompressible fluid
- If viscous effects neglected

Many problems can be solved using the Bernoulli equation, allowing us to dodge having to solve the full Euler or Navier-Stokes equations. It should not escape our notice that the Bernoulli equation, derived from $F = ma$, looks like an energy conservation equation. This is even easier to see if we multiply through by the (assumed constant) density: Equation (10.34) becomes

$$P + \rho g z + \tfrac{1}{2}\rho V^2 = \text{constant}, \tag{10.35}$$

and we can think of P as a measure of flow work, $\rho g z$ as a gravitational potential energy, and $\tfrac{1}{2}\rho V^2$ as a kinetic energy, all per unit volume of fluid. Daniel Bernoulli actually first arrived at equation (10.34) by performing an energy balance, even though the concept of energy was still a bit fuzzy in 1738.

One of the most useful applications of the Bernoulli equation is a device known as a Pitot[4] tube, as well as its cousin the Pitot-static tube, which is used to measure flow velocities. The tube (Figure 10.9a) contains a column of air. When an oncoming fluid flow impinges on the nose of the Pitot tube, it displaces this air. As we know from hydrostatics, the displacement is proportional to the pressure at the stagnation point on the Pitot tube nose. The difference between this *stagnation pressure* (where the fluid has speed $V = 0$) and the *static pressure* elsewhere in the flow (where the fluid has average speed V_∞), by Bernoulli's equation, is

$$p_{\text{stagnation}} - p_{\text{static}} = \frac{1}{2}\rho V_\infty^2 . \tag{10.36}$$

A Pitot-static tube, as illustrated in Figure 10.9b, contains static pressure ports along the nose to measure the static fluid pressure as well as the stagnation pressure. It's clear from equation (10.36) and Figure 10.9 that the assumptions of steady, incompressible flow should be appropriate when a Pitot tube is used. The flow is also assumed to be inviscid so that there are no boundary layers near tube or other walls to reduce the bulk flow speed from V_∞.

10.6 Examples

Example 10.1

Find the equation of, and sketch, the streamline that passes through (1, –2) for the velocity field given by

$$\underline{V} = xy\hat{i} - 2y^2\hat{j} \text{ m/s}.$$

Given: Velocity vector $\underline{V}$.

Find: Streamline through (x = 1, y = –2).

Assume: No assumptions are necessary.

Solution

By definition, a streamline is everywhere tangent to the velocity field. So, the streamline through (1, –2) is tangent to the velocity V at this point. We state this relationship mathematically as

$$\text{slope } \left.\frac{dy}{dx}\right|_{streamline} = \frac{v}{u}$$

$$\frac{dy}{dx} = \frac{-2y^2}{xy} = \frac{-2y}{x}.$$

Separating variables,

$$\frac{dy}{-2y} = \frac{dx}{x}.$$

Integrating both sides,

$$\int \frac{dy}{-2y} = \int \frac{dx}{x}$$

$$\rightarrow \quad -\frac{1}{2}\ln y = \ln x + C.$$

We have absorbed the constants of integration from both sides into this new constant C:

$$\rightarrow \ln x + \frac{1}{2}\ln y = C.$$

This new C is simply –C from the previous expression. To get rid of the natural logs and find a graphable function *y(x)*, we take the exponent of the entire expression:

$$x\sqrt{y} = C, \text{ or}$$

$$x^2 y = C.$$

This C may no longer bear much resemblance to our initial constant C, but since the product x^2y must equal *a* constant, we might as well use C to represent that constant.

At point (1, –2), $x^2y = (1)^2(-2) = -2$, so the equation of the streamline through (1, –2) is $x^2y = -2$.

We plot this streamline in Figure 10.9.

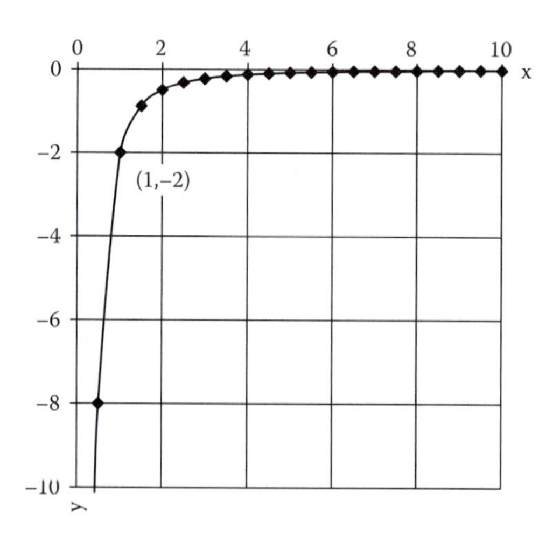

FIGURE 10.9
Pitot tube with differential manometer to measure flow speed: (a) standard arrangement of Pitot tube and static pressure tap; and (b) Pitot-static tube.

Example 10.2

The open tank shown contains water at 20°C and is being filled through section 1 in Figure 10.10. Assume incompressible flow. First derive an analytic expression for the water-level change dh/dt in terms of arbitrary volume flows Q_1, Q_2, Q_3, and tank diameter d. Then, if the water level h is constant, determine the exit velocity V_2 for the given data $V_1 = 3$ m/s and $Q_3 = 0.01$ m^3/s.

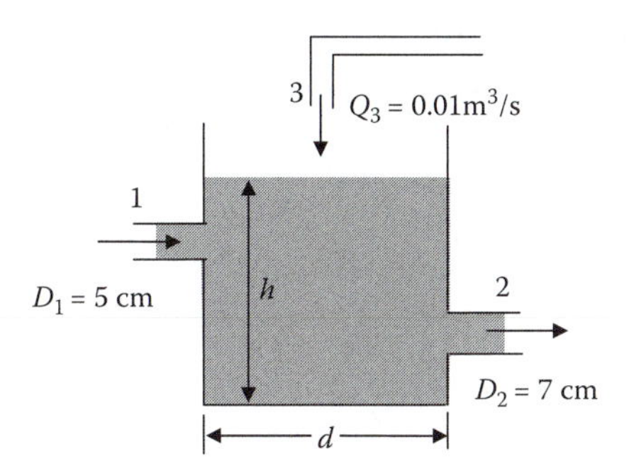

FIGURE 10.10

Given: Tank inlet and outlet information.

Find: dh/dt, unknown exit velocity.

Assume: Inlet and outlet velocity profiles are uniform, one dimensional.
Fluid is incompressible; density is uniform.

Solution

We intend to consider the fluid in the tank shown in Figure 10.11 as the contents of a control volume.

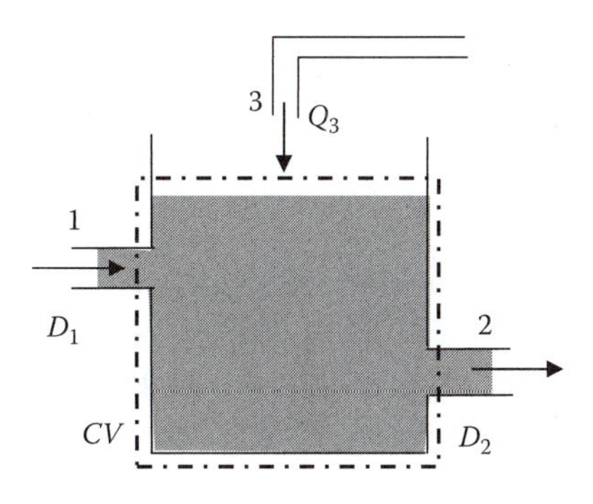

FIGURE 10.11

We must have mass conservation,

$$0 = \frac{\partial}{\partial t}\int_{CV} \rho\, d\mathcal{V} + \int_{CS} \rho\, \underline{V} \cdot d\underline{A},$$

which for this CV can be written as

$$\frac{d}{dt}\left[\rho\frac{\pi\, d^2}{4}h\right]+\rho\left(Q_2-Q_1-Q_3\right)=0,$$

where we have changed the partial derivative to a total one, as time is the only dependence of the quantity in brackets, and where the signs on various flow rates depend on whether they are into or out of the CV: Q_2 is outflow, and thus positive, while Q_1 and Q_3 are both inflow and negative. We further simplify by canceling the common density and by removing the constants from the time derivative. We get

$$\frac{\pi\, d^2}{4}\frac{dh}{dt}+\left(Q_2-Q_1-Q_3\right)=0$$

$$\frac{dh}{dt}=\frac{4\left(Q_1+Q_3-Q_2\right)}{\pi\, d^2}.$$

If h is constant, $dh/dt = 0$ and we must have $Q_1 + Q_3 - Q_2 = 0$. Each $Q_i = V_iA_i$. We can thus solve for the requested value of V_2, which corresponds to $dh/dt = 0$:

$$Q_2=V_2A_2=Q_1+Q_3=0.01\,\tfrac{\text{m}^3}{\text{s}}+\tfrac{\pi}{4}(0.05\text{ m})^2(3\ \tfrac{\text{m}}{\text{s}})=0.0159\,\tfrac{\text{m}^3}{\text{s}}$$

$$V_2=Q_2\,/\,A_2=\frac{0.0159\ \tfrac{\text{m}^3}{\text{s}}}{\tfrac{\pi}{4}(0.07\text{ m})^2}=4.13\ \tfrac{\text{m}}{\text{s}}.$$

Example 10.3

A steady jet of water is redirected by a deflector, as shown in Figure 10.12. The jet has mass flow rate of 32 kg/s, cross-sectional area 2 cm × 40 cm, and speed V_1 when it encounters the deflector. What force per unit width of the deflector (into the page) is needed to hold the deflector in place?

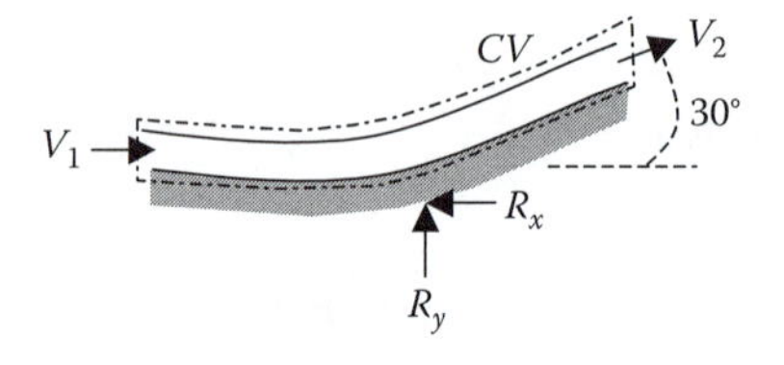

FIGURE 10.12

Given: Geometry of flow deflector.

Find: Reaction forces from deflector support on fluid in CV.

Assume: Jet has constant cross-sectional area, even after being deflected. Flow is steady and incompressible. Density of water is constant, uniformly 1000 kg/m^3. Gravity and viscous effects may be neglected.

Solution

The flow of water imparts a force to the deflector. Reaction forces from the deflector balance these forces and act on the fluid in the CV drawn. We are asked for these reactions, R_x and R_y, if the deflector has width of 1 m into the page.

We begin by finding the inlet velocity V_1. We are given the mass flow rate of the jet, $\dot{m}$, which is $\rho V_1 A_1$. As the cross-sectional area A_1 is also given, we use this to find V_1:

$$V_1 = \frac{\dot{m}}{\rho A_1} = \frac{32 \ {}^{\text{kg}}\!/\!_{\text{s}}}{(1000 \ \frac{\text{kg}}{\text{m}^3})(0.02 \text{ m})(0.40 \text{ m})} = 4 \ {}^{\text{m}}\!/\!_{\text{s}}.$$

If the flow is steady, we must have a constant mass flow rate (what flows into our CV must flow back out again), or $\rho V_1 A_1 = \rho V_2 A_2$. Since we have an incompressible flow and since the jet's cross-sectional area does not change, we must have $V_2 = V_1 = 4$ m/s.

To find the requested reaction forces, we must apply the conservation of linear momentum, or $\underline{F} = m\underline{a}$, in both x and y directions. These equations are

$$F_x = \left(F_{visc}\right)_x + \left(F_{external}\right)_x + \int_{CV} \rho g_x dV - \int_{CS} p\hat{\underline{i}} \cdot \underline{n} dA = \frac{\partial}{\partial t}\int_{CV} \rho u \, dV + \int_{CS} \rho u \, \underline{V} \cdot \underline{n} dA$$

and

$$F_y = \left(F_{visc}\right)_y + \left(F_{external}\right)_y + \int_{CV} \rho g_y dV - \int_{CS} p\hat{\underline{j}} \cdot \underline{n} dA = \frac{\partial}{\partial t}\int_{CV} \rho v \, dV + \int_{CS} \rho v \, \underline{V} \cdot \underline{n} dA.$$

Since our jet is steady, with negligible contributions from gravity, viscosity, and pressure gradients, these equations simplify greatly:

$$-R_x = \int_{CS} \rho u \, \underline{V} \cdot \underline{n} dA .$$

$$R_y = \int_{CS} \rho\, v\, \underline{V} \cdot \underline{n} dA.$$

The x momentum flux therefore balances the reaction force R_x, negative as it is in the negative x direction. Writing out the flux at each of the two control surfaces, we have

$$-R_x = \int_{CS} \rho\, u\, \underline{V} \cdot \underline{n} dA = -\rho\, V_1 V_1 A_1 + \rho\, (V_2 \cos\theta) V_2 A_2.$$

Note that $V_2\cos\theta$ is the x component of velocity at surface A_2 and that V_2 is $\underline{V} \cdot \underline{n}$ at A_2:

$$\begin{aligned} R_x &= \dot{m}(V_1 - V_2 \cos\theta) \\ &= (32\ {}^{kg}\!/\!_{s})(4\ {}^{m}\!/\!_{s})(1 - \cos(30^\circ)) \\ &= 17.2\ \text{N} \end{aligned}$$

in the y direction we have

$$R_y = \int_{CS} \rho\, v\, \underline{V} \cdot \underline{n} dA = 0 + \rho\, (V_2 \sin\theta) V_2 A_2$$

$$\begin{aligned} R_y &= \dot{m} V_2 \sin\theta \\ &= (32\ {}^{kg}\!/\!_{s})(4\ {}^{m}\!/\!_{s}) \sin(30^\circ). \\ &= 64\ \text{N} \end{aligned}$$

Example 10.4

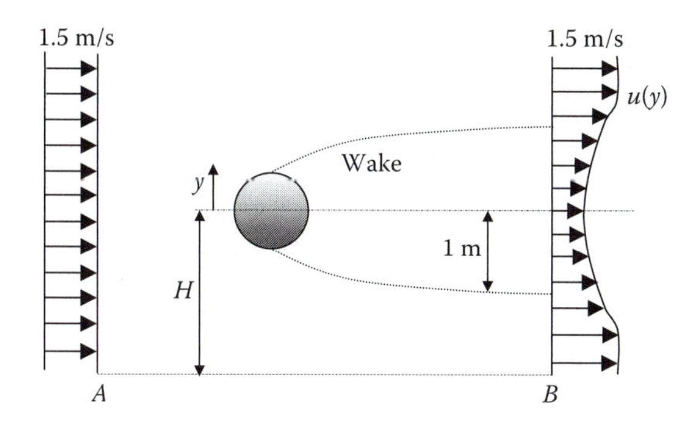

FIGURE 10.13

Uniform air flow with speed U = 1.5 m/s approaches a cylinder as shown in Figure 10.13. The velocity distribution at the location shown downstream in the wake of the cylinder may be approximated by

$$u(y) = 1.25 + \frac{y^2}{4} \qquad -1 < y < 1,$$

where $u(y)$ is in m/s and y is in meters. Determine (a) the mass flux across the surface AB per meter of depth (into the page) and (b) the drag force per meter of length acting on the cylinder.

Given: Flow over cylinder; upstream and downstream velocity profiles.

Find: Mass flux across surface AB, drag force on cylinder.

Assume: Air has constant, uniform density 1.23 kg/m^3. Flow is symmetrical and steady. Pressure differences, gravity, and viscous effects may be neglected.

Solution

We first select a control volume. It is generally wisest to choose control volumes on whose surfaces we have information about the flow. It is also useful to take advantage of symmetry to simplify our calculations.

Our choice of control volume is shown in Figure 10.14.

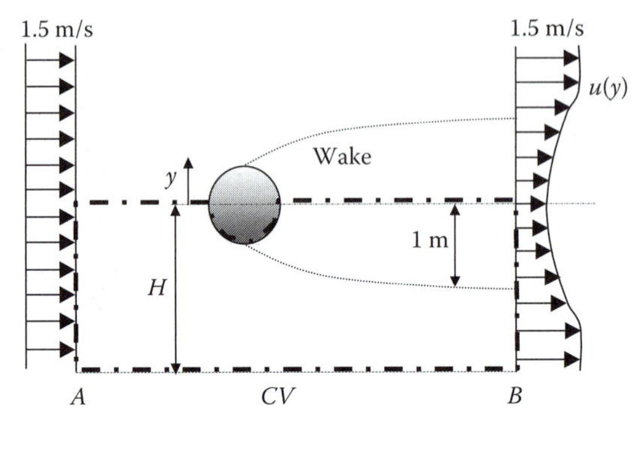

FIGURE 10.14

At its left surface, the normal velocity is U = 1.5 m/s, into the CV. Its top surface is a plane of symmetry for the flow, so there is no mass flux across it. At the right, the wake velocity profile is given by $u(y)$ above, and outside the wake, the velocity is 1.5 m/s, out of the CV. We have a steady flow, so the conservation of mass is written as

$$0 = \int_{CS} \rho \, \underline{V} \cdot \underline{n} dA$$

$$0 = \int_{\substack{left \\ CS}} \rho \, \underline{V} \cdot \underline{n} dA + \int_{AB} \rho \, \underline{V} \cdot \underline{n} dA + \int_{\substack{right \\ CS}} \rho \, \underline{V} \cdot \underline{n} dA$$

$$= -\rho \, UH + \dot{m}_{AB} + \int_0^H \rho \, u(y) dy.$$

Note that since we are not given the length of the cylinder into the page, we must find the mass flux across AB per meter of cylinder length. We account for this by assuming a unit cylinder length. The areas of our control surfaces are thus ldy, with a unit length l. Hence, the area of the left control surface is simply H (m^2).

H, however, is not known. To complete the solution we must investigate the flow field further. Outside the wake region, which is 1 m wide at the control surface, the flow out of the right CS has speed 1.5 m/s. The left control surface has a uniform inflow of 1.5 m/s. Hence, more than 1 m away from the cylinder axis, the flow is unaffected by the cylinder and simply proceeds with constant speed 1.5 m/s. We can therefore assess the amount of mass flux forced across AB by integrating only from 0 to 1, instead of 0 to H:

$$0 = -\rho\, U(1) + \dot{m}_{AB} + \int_0^1 \rho\, u(y)dy$$

$$\dot{m}_{AB} = \rho\, U(1) - \int_0^1 \rho\,(1.25 + \frac{y^2}{4})dy$$

$$\dot{m}_{AB} = (1.23\,\tfrac{\text{kg}}{\text{m}^3})\left\{1.5\ \tfrac{\text{m}^2}{\text{s}} - \left[1.25y + \frac{y^3}{12}\right]_o^1\right\}$$

$$\dot{m}_{AB} = 0.205\ \tfrac{\text{kg}}{\text{s}}\ \text{per meter of cylinder length}\,.$$

To address part (b) of this problem, we conserve linear momentum in the x direction. We may either continue with the same control volume as in part (a), multiplying the fluxes by 2 to obtain the force on the whole cylinder, or we may now use a CV that consists of all the fluid between $-H$ and $+H$, or equivalently –1 and 1. The drag force on the cylinder is in the $+x$ direction; hence, there is an equal and opposite force on the fluid in the $-x$ direction. Conserving x momentum, we have

$$F_x = \left(F_{visc}\right)_x + \left(F_{external}\right)_x + \int_{CV} \rho\, g_x d\mathcal{V} - \int_{CS} p\hat{\underline{i}}\cdot\underline{n}dA = \frac{\partial}{\partial t}\int_{CV} \rho\, u\, d\mathcal{V} + \int_{CS} \rho\, u\, \underline{V}\cdot\underline{n}dA\,.$$

Under the assumptions of steady flow, with negligible contributions from pressure gradients, gravity, and viscous effects, this becomes

$$-F_x = \int_{CS} \rho\, u\, \underline{V}\cdot\underline{n}dA\,.$$

We evaluate the flux at all three control surfaces of the initial CV and multiply each by 2 due to symmetry:

$$-F_x = 2\int_{\substack{\text{left}\\ \text{CS}}} \rho\, u\, \underline{V}\cdot\underline{n}dA + 2\int_{AB} \rho\, u\, \underline{V}\cdot\underline{n}dA + 2\int_{\substack{\text{right}\\ \text{CS}}} \rho\, u\, \underline{V}\cdot\underline{n}dA$$

$$= -2\int_0^1 \rho\, U\cdot U\ 1\cdot dy + 2\int_{AB} \rho\, u\, \underline{V}\cdot\underline{n}dA + 2\int_0^1 \rho\, u(y)\cdot u(y)\ 1\cdot dy.$$

We have again assumed a unit cylinder length into the page so that the area of both right and left control surfaces is $1dy$. We next recognize that the second integral contains the mass flux we just solved for,

$$\dot{m}_{AB} = \int_{AB} \rho \, \underline{V} \cdot \underline{n} dA,$$

and differs from this only by the value of u, the x component of velocity at the surface AB. The surface AB is at a distance of H from the cylinder axis, where, as we have discussed, the cylinder does not influence the x directional flow. The velocity u is therefore $U = 1.5$ m/s on AB. We thus get something even simpler:

$$-F_x = -2\rho\, U^2 (1)^2 + 2U\dot{m}_{AB} + 2\int_0^1 \rho \left[1.25 + \frac{y^2}{4}\right]^2 1 \cdot dy$$

$$-F_x = -2\rho\, U^2 + 2U\dot{m}_{AB} + 2\rho \left[1.25^2 y + \frac{2.5}{12} y^3 + \frac{y^4}{16}\right]_0^1$$

$$-F_x = -2(1.23\, {}^{\text{kg}}\!/\!_{\text{m}^3})\,(1.5\, {}^{\text{m}}\!/\!_{\text{s}})^2 + 2(1.5\, {}^{\text{m}}\!/\!_{\text{s}})(0.205\, {}^{\text{kg}}\!/\!_{\text{s}}) + 2(1.23\, {}^{\text{kg}}\!/\!_{\text{m}^3})\,(1.833\, {}^{\text{m}^2}\!/\!_{\text{s}^2})$$

$$F_x = 4.07 \text{ N per meter of cylinder length.}$$

Example 10.5

For the water siphon in Figure 10.15, find (a) the speed of water leaving as a free jet at point 2 and (b) the water pressure at point A in the flow. State all assumptions. Heights $h_1 = 1$ m, and $h_2 = 8$ m. Drawing is not to scale.

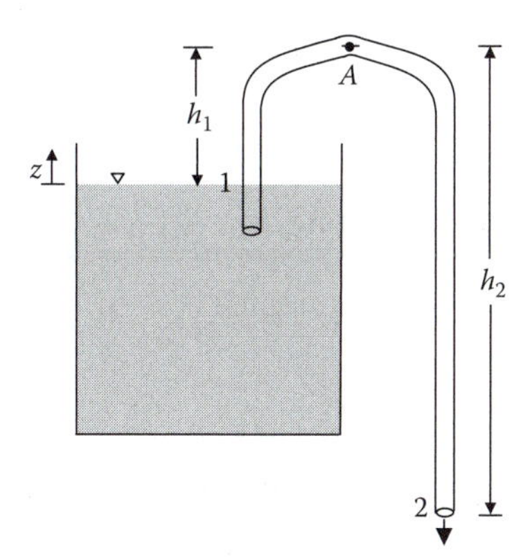

FIGURE 10.15

Given: Length of siphon used to remove water from large tank.

Find: Speed at *2* and pressure at *A*.

Assume: Steady flow (all transient effects associated with flow initiation have died down), incompressible (water has constant, uniform density, equal to 1000 kg/m^3), and negligible viscous effects.

Solution

We would like to use the Bernoulli equation to relate the flow quantities between the labeled points. The conditions necessary for the Bernoulli equation to apply have been reasonably assumed. (We feel least confident in our assumption of negligible viscous losses in the siphon, and in the next chapter we discuss a way to characterize the importance of viscosity in a given flow.)

We must have a streamline on which to apply the Bernoulli equation, so we assume that the one sketched in Figure 10.16 exists. This streamline connects points *1* and *A*, and *A* and *2*.

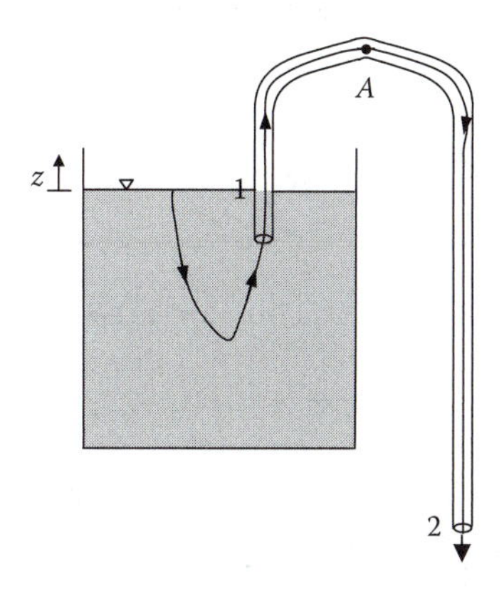

FIGURE 10.16

We make one more assumption to solve this problem: By inspection, the reservoir is much larger than the siphon diameter; that is,

$$A_1 >> A_2.$$

So, if we conserve mass from point *1* to point *2*, we have

$$\rho V_1 A_1 = \rho V_2 A_2.$$

And, with $A_1 >> A_2$, we must have $V_1 << V_2$. We approximate this very small velocity at *1* by saying $V_1 \approx 0$.

We can now apply Bernoulli's equation between points *1* and *2*, to find the unknown V_2:

$$\frac{p_1}{\rho} + \frac{V_1^2}{2} + gz_1 = \frac{p_2}{\rho} + \frac{V_2^2}{2} + gz_2,$$

where, as we have just said, $V_1 \approx 0$, and where $p_1 = p_2 = p_{atm}$. (If we are using gauge pressures, this means $p_1 = p_2 = 0$.) We note from the figure that $z_1 = 0$, and $z_2 = -7$ m, so

$$gz_1 = \frac{V_2^2}{2} + gz_2$$

$$V_2^2 = 2g(z_1 - z_2)$$

$$V_2 = \sqrt{2(9.81 \ \tfrac{\text{m}}{\text{s}^2})(7 \text{ m})} = 11.7 \ \tfrac{\text{m}}{\text{s}}.$$

Our assumed streamline also goes through point A so that the Bernoulli equation is

$$\frac{p_1}{\rho} + \frac{V_1^2}{2} + gz_1 = \frac{p_A}{\rho} + \frac{V_A^2}{2} + gz_A.$$

If we conserve mass within the constant area siphon, we must have

$$\rho V_A A_A = \rho V_2 A_2,$$

or, since $A_A = A_2$, $V_A = V_2 = 11.7$ m/s. We are now equipped to solve the Bernoulli equation for the unknown P_A:

$$\frac{p_1}{\rho} + gz_1 = \frac{p_A}{\rho} + \frac{V_2^{\,2}}{2} + gz_A$$

$$p_A = p_1 + \rho\left(gz_1 - \frac{V_2^{\,2}}{2} - gz_A\right)$$

$$= p_{atm} + \rho\, g(z_1 - z_A) - \rho\frac{V_2^2}{2}$$

$$= p_{atm} + (1000\ \tfrac{\text{kg}}{\text{m}^3})(9.81\ \tfrac{\text{m}}{\text{s}^2})(0 - 1\ \text{m}) - (1000\ \tfrac{\text{kg}}{\text{m}^3})\frac{(11.7\ \frac{\text{m}}{\text{s}})^2}{2}$$

$$p_A = 22.8\ \text{kPa (abs)}$$

$$= -78.5\ \text{kPa (gage)}.$$

We have used standard atmospheric pressure, p_{atm} = 101.325 kPa.

Example 10.6

A person holds her hand out of an open car window while the car drives through still air at 65 mph. Under standard atmospheric conditions, what is the maximum pressure on her hand? What would be the maximum pressure if the car were traveling at 220 mph?

Given: Speed of airflow past hand; standard atmospheric conditions.

Find: Maximum pressure on hand.

Assume: Flow is steady and incompressible, with negligible viscous effects; air has constant, uniform density, equal to its tabulated value at 20°C (1.23 kg/m^3). Standard atmospheric pressure p_{atm} = 101.325 kPa.

Solution

We put ourselves in the frame of the person's hand so that the hand is still and the air moves with speed 65 mph (or 220 mph). We can visualize the airflow as sketched in Figure 10.17.

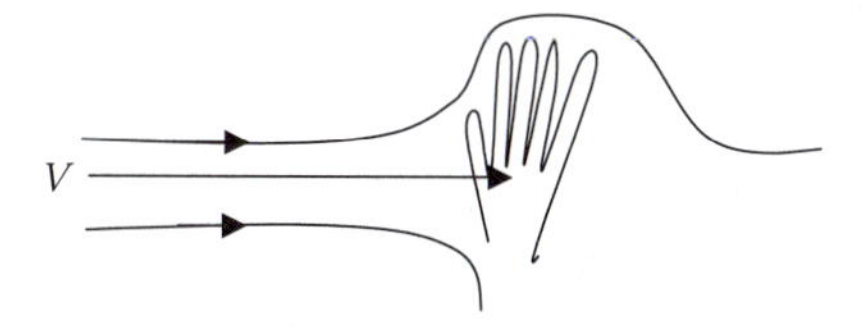

FIGURE 10.17

Note that there is a dividing streamline that impinges on the hand at a stagnation point. (Airflow either goes above this streamline, up and over the hand, or below it.) At this stagnation point, the air will be at its maximum pressure, the stagnation pressure. (Recall that pressure and velocity are inversely proportional.) If we assume that this stagnation streamline is level, so that gravitational effects are easily neglected, we can apply the Bernoulli equation on this streamline to find the stagnation pressure. The Bernoulli equation has the form

$$\left(p+\frac{1}{2}\rho V^2\right)_{upstream}=\left(p+\frac{1}{2}\rho V^2\right)_{SP},$$

or

$$p_{atm}+\frac{1}{2}\rho V^2=p_{\max}.$$

Plugging in the atmospheric pressure, air density, and $V = 65$ mph, we have

$$p_{\max}=101{,}325\text{ Pa}+\frac{1}{2}(1.23\ \tfrac{\text{kg}}{\text{m}^3})\left(65\text{ mph}\frac{0.447\ \frac{\text{m}}{\text{s}}}{1\text{ mph}}\right)^2=101.844\text{ kPa (abs)}$$

$$=520\text{ Pa (gage)}.$$

If the car (and hence the air, in the frame of the hand) moves with speed $V = 220$ mph,

$$p_{\max}=101{,}325\text{ Pa}+\frac{1}{2}(1.23\ \tfrac{\text{kg}}{\text{m}^3})\left(220\text{ mph}\frac{0.447\ \frac{\text{m}}{\text{s}}}{1\text{ mph}}\right)^2=107.28\text{ kPa (abs)}$$

$$=5.95\text{ kPa (gage)}.$$

10.7 Problems

10.1 Five holes are punched in the side of a can of liquid. Which figure shown in Figure 10.18 best illustrates the velocity profile that would result from liquid leaving the five holes?

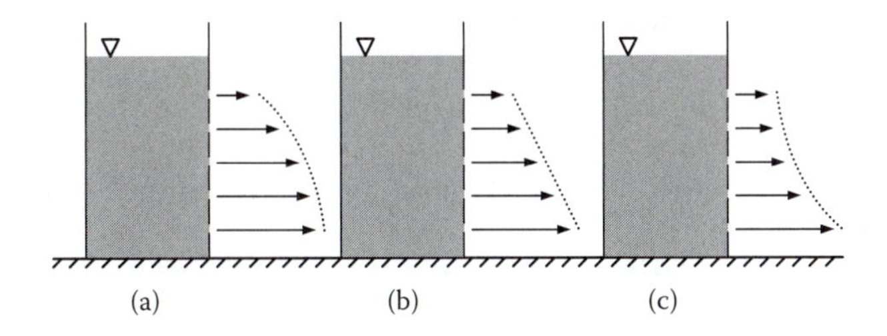

FIGURE 10.18

10.2 A two-dimensional fluid velocity field is given by $u = x(1 + 2t)$, $v = y$. Find the equation of the time-varying streamlines with all pass through the point (x_o, y_o) at some time t. Sketch some of the streamlines at various times t.

10.3 For the three-dimensional, time-varying velocity field $\underline{V} = 3t\hat{\underline{i}} + xz\hat{\underline{j}} + ty2\hat{\underline{k}}$, find the acceleration of a fluid element.

10.4 Consider a two-dimensional velocity field in Cartesian coordinates:

$$(u,v)=\left(\frac{-ky}{x^2+y^2},\frac{kx}{x^2+y^2}\right),$$

where k is a positive constant. Sketch the velocity profiles along the x axis and the line $x = y$. Determine the equation of the streamline passing through $(x, y) = (1, 1)$. What are the velocity and acceleration at this point? Sketch both vectors. Is the flow incompressible?

10.5 The horizontal nozzle shown in Figure 10.19 has $D_1 = 0.3$ m and $D_2 = 0.15$ m, with inlet pressure of the operating fluid (water at 20°C) $p_1 = 262$ kPa (absolute) and $V_2 = 17$ m/s. Compute the normal stress induced in the flange bolts (diameter 1 cm) by keeping the nozzle fixed.

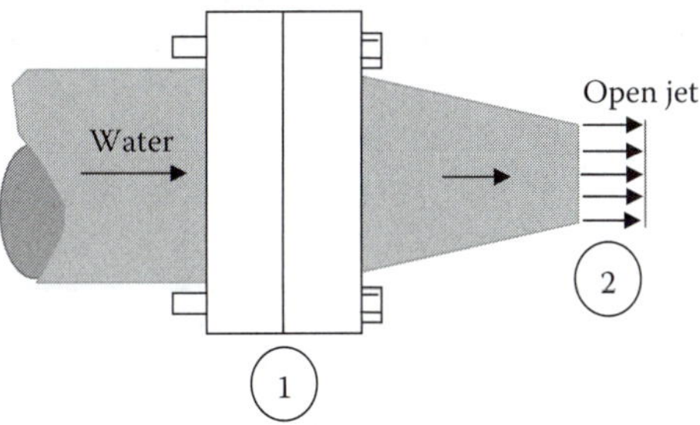

FIGURE 10.19

10.6 Observations show that it is not possible to blow the table tennis ball out of the funnel shown on the left in Figure 10.20. In fact, the ball can be kept in an inverted funnel, like the one on the right, by blowing through it. The harder one blows through the funnel, the harder the ball is held within the funnel. Explain this phenomenon.

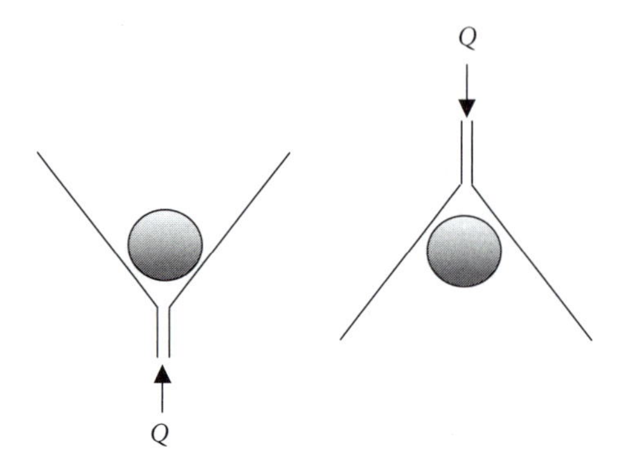

FIGURE 10.20

10.7 An open circuit wind tunnel draws in sea-level standard air and accelerates it through a contraction into a 1 m × 1 m test section. A differential pressure transducer mounted in the test section wall measures a pressure difference of 45 mm of water between the inside and outside. Estimate (a) the test section velocity in mph and (b) the absolute pressure on the front nose of a small model mounted in the test section.

10.8 Blood, an incompressible fluid with density $\rho = 1060\ \text{kg/m}^3$, flows through vessels that often branch. Using the given model (Figure 10.21) for a branching arteriole, and assuming that at the point of interest flow is steady, with negligible contributions from gravity and viscosity, calculate the pressure differences: (a) $P_C - P_A$; and (b) $P_B - P_A$.

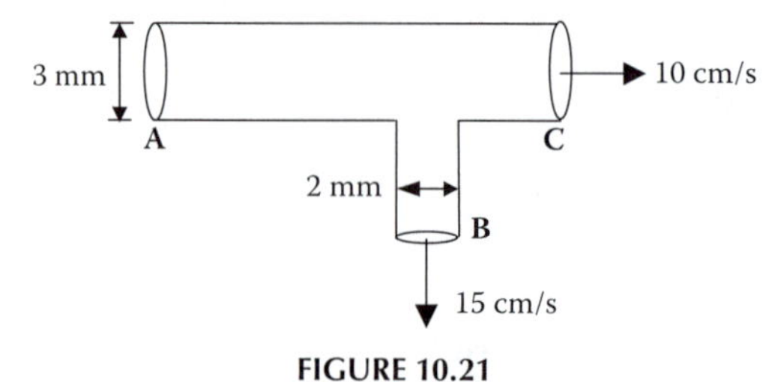

FIGURE 10.21

10.9 Water from a stationary nozzle strikes a flat plate (directed normal to the plate as shown in Figure 10.22). The velocity of the water leaving the nozzle is 15 m/s, and the nozzle area is 0.01 m^2. After the water strikes the plate, subsequent flow is parallel to the plate. (a) Find the horizontal force that must be provided to the plate by the support. (b) Find the maximum longitudinal normal tensile stress the support post if it is a hollow square cross section as shown in the figure. Model the force due to the water as a point load. (c) Find the maximum transverse shearing stress in a cross section of the post.

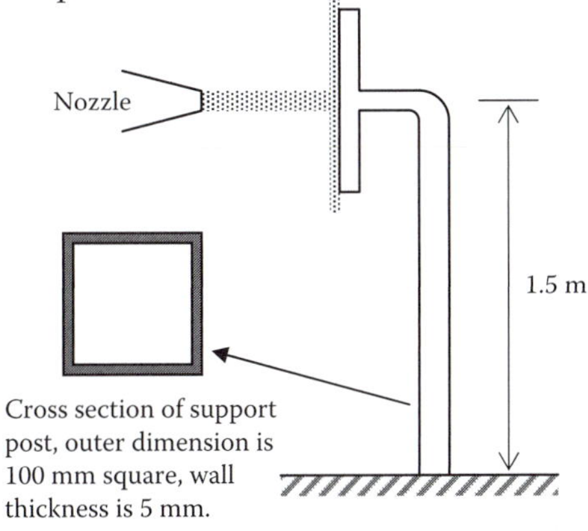

FIGURE 10.22

Notes

1. A similar streamfunction can be derived for compressible flows, though this is outside the scope of this book.
2. Again, we note that for a solid, whose mass could be vigilantly monitored and that would not flow in response to shear, this flux term would not be necessary.
3. The student chafing under the restrictions of incompressibility and Newtonian behavior is encouraged to take further courses in fluid mechanics and to refer posthaste to Kundu's (1990) *Fluid Mechanics*, where he or she will find that the full form of the stress tensor makes things quite a bit more interesting.
4. The Pitot tube is named for Henri Pitot (1695–1771), a French hydraulic engineer who invented it by intuition when measuring the flow in the River Seine in 1732.

11

Fluid Dynamics: Applications

> That we have written an equation does not remove from the flow of fluids its charm or mystery or its surprise.
>
> **Richard Feynman, 1964**

We have found two distinct ways to apply the fundamental concepts of mass conservation and $F = ma$ to fluids. We now want to identify some canonical problems of fluid mechanics, their historical context, and their relevance to us. Both solid and fluid mechanics are enormous fields, with many rich details; in this book, we have been necessarily brief with both of them.

11.1 How Do We Classify Fluid Flows?

The Navier-Stokes equation contains terms corresponding to several possible forces on a fluid element. If we look at it again, we can name each of these forces:

$$\underbrace{\rho \frac{D\underline{V}}{Dt}}_{inertia} = \underbrace{-\nabla p}_{pressure} + \underbrace{\rho\, \underline{g}}_{\substack{gravity \\ (body\ force)}} + \underbrace{\mu \nabla^2 \underline{V}}_{\substack{viscous \\ stress}}. \tag{11.1}$$

We would like to have a way to quantify the relative effects of these forces, and of other factors, on a given flow; this way, when faced with an intriguing fluid mechanics problem we could decisively say whether *viscosity* or *inertia* was the more dominant effect—and how much more dominant. The most useful result would be a dimensionless parameter—that way, it wouldn't matter whether we were dealing with U.S. units or International System of Units (SI); a certain number would represent the same type of flow in either unit system.

It is apparent that by taking the ratio of the inertial and viscous terms of the Navier-Stokes equation, we could obtain this quantification. This ratio clearly goes as ρ/μ. Unfortunately, this ratio ρ/μ has units of time/length^2. To make it dimensionless, we need to multiply it by something with units of length^2/time. The easiest way to construct this "something" is to multiply the velocity V by a characteristic length scale of the problem, say L. It turns out that this is also correct physically, as we can see by a scaling argument:

$$\text{Inertial Force} = mV\frac{dV}{ds} \sim \rho L^3 V \frac{V}{L} = \rho L^2 V^2. \tag{11.2}$$

$$\text{Viscous Force} = \tau A = \mu \frac{du}{dy} A \sim \mu \frac{V}{L} L^2 = \mu V L. \tag{11.3}$$

$$\text{Ratio of Inertial to Viscous Forces} = \frac{\rho L^2 V^2}{\mu V L} = \frac{\rho V L}{\mu}. \tag{11.4}$$

This ratio is known as the Reynolds number, abbreviated *Re*.[1] When *Re* is large, inertial effects dominate the flow, and when *Re* is small, viscous effects dominate. This lets us know what terms we can drop out of the Navier-Stokes equation when we're at the far ends of the *Re* spectrum. As expected, *Re* also tells us something about the character of the flow. Generally, lower *Re* flows are smooth, with parallel streamlines because viscosity tends to diffuse more complex flow patterns. These flows are known as *laminar.* At higher *Re*, viscosity is not strong enough to diffuse eddies and other rotational flow patterns, and the flow tends to be more disorderly. These higher *Re* flows are called *turbulent.* A critical value of Reynolds number, Re_{crit}, is a threshold separating laminar from turbulent flows. The value of Re_{crit} depends on the type of flow being considered, as we'll see.

Other nondimensional parameters serve to measure the relative effects of other forces on a particular flow. For example, the Euler number *Eu* compares pressure drop to inertial forces. The Euler number and a few other relevant parameters are listed in Table 11.1.

The Reynolds, Euler, Froude, and Weber numbers, among others, allow us to quantify the relative importance of different forces on the flow in question; they are also useful in planning experiments. Two flows with the same *Re* have very similar flow patterns and characteristics—inertia and viscosity have the same relationship in both flows. To build an experimental model of a flow whose real dimensions are unwieldy, it is sufficient to match the appropriate nondimensional parameters. It is much more economical to study the influence of *Re* on a given flow than to have to independently vary the density ρ and viscosity μ of a fluid, the size of the model L, and the flow speed V!

TABLE 11.1

Relevant Nondimensional Parameters

Reynolds number	Re	$\frac{\text{inertia}}{\text{viscosity}}$	$\frac{\rho VL}{\mu}$
Euler number	Eu	$\frac{\text{pressure}}{\text{inertia}}$	$\frac{\Delta p}{\rho V^2}$
Froude number	Fr	$\frac{\text{inertia}}{\text{gravity}}$	$\frac{V}{\sqrt{Lg}}$
Weber number	We	$\frac{\text{inertia}}{\text{surface tension}}$	$\frac{V^2 L\rho}{s}$

11.2 What's Going on Inside Pipes?

Pipelines, blood vessels, hallways, and ink jet printers all contain examples of *internal flows*. A fluid's *stiffness*, or viscosity, has a significant effect on the flow of an incompressible fluid through a pipe or between parallel plates. The critical Reynolds number in a pipe is about 2000; for $Re < 2000$, pipe flow is laminar.

Pipe flow is said to be *fully developed* when it does not change in the flow direction; as in Figure 11.1, the velocity profile $u(x, r)$ becomes independent of axial length x. At the pipe inlet, flow is uniform [$u(x, r) = U_o = constant$]; as x increases, a very thin layer near the walls slowly grows outward. Viscous effects dominate these thin *boundary layers*, but viscosity does not yet affect the inner core of the flow. We continue along the pipe length x until viscosity affects the entire cross section. Finally, the thin layers all merge and the flow becomes fully developed.

For a laminar flow, we can determine the entrance length necessary for a pipe flow to become fully developed if we know the Reynolds number:

$$\frac{L_E}{D} = 0.065Re, \tag{11.5}$$

where Re is based on the average velocity ($V = \bar{u}$) and the diameter of the pipe ($L = D$).

Now that we know to expect the flow to become independent of axial length x, we would like to be able to determine the shape of the velocity profile $u(y)$, or $u(r)$ for a circular pipe. We assume steady flow to simplify our

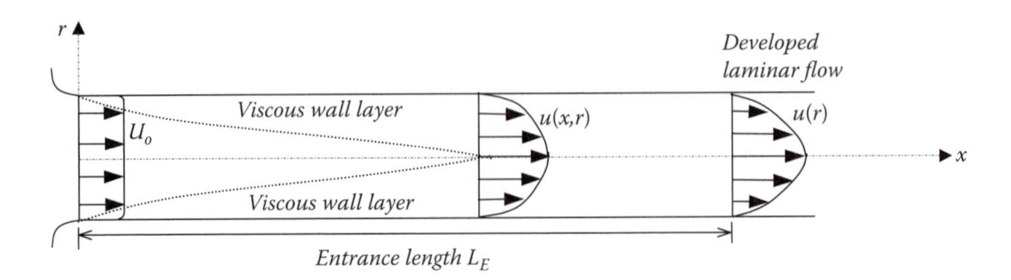

FIGURE 11.1
Laminar flow developing in a pipe or a wide rectangular channel.

lives a bit. Figure 11.2 shows a cylindrical fluid element for which we can now write $F = ma$.

Once the flow is fully developed, it experiences no acceleration. (The local acceleration $\frac{\partial u}{\partial t} = 0$ and the convective acceleration, $u\frac{\partial u}{\partial x} = 0$ since u is a function of r only.) Every part of the fluid moves with constant velocity, although neighboring particles have different velocities and this velocity gradient, as we well know, gives rise to a shear stress.

For this simple analysis, we neglect gravity, assuming that pressure and viscous effects are much more significant. The pressure is constant across vertical cross sections (with no hydrostatic effect due to gravity), though it changes in x. So if pressure is $p = p_1$ at section (1) as shown, it is $p_2 = p_1 - \Delta p$ at section (2). We anticipate that pressure decreases in the direction of flow,

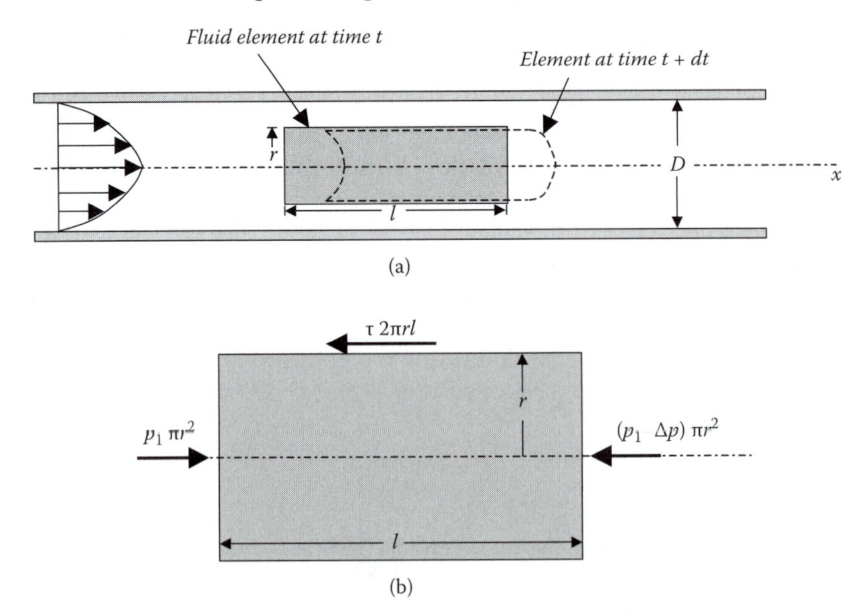

FIGURE 11.2
(a) Motion of a cylindrical fluid element within a pipe flow. (b) Free-body diagram of a cylindrical fluid element.

so that $\Delta p > 0$.[2] A shear stress, τ, acts on the surface of the fluid cylinder. This shear stress is a function of radius r, $\tau = \tau(r)$.

Once again, we need to look at this free-body diagram (FBD) (Figure 11.2b) and write out $(F = ma)_x$ for this cylindrical fluid element. We have $a_x = 0$, and the remaining terms of the force balance are

$$p_1 \pi r^2 - (p_1 - \Delta p)\, \pi r^2 - \tau\, 2\pi r l = 0. \tag{11.6}$$

This expression can be simplified, yielding

$$\frac{\Delta p}{l} = \frac{2\tau}{r}. \tag{11.7}$$

This balance of forces is necessary to drive each fluid particle down the pipe with constant velocity. Since neither Δp nor l depends on radial coordinate r, the right-hand-side term, $2\tau/r$, must not depend on r. That is, $\tau = Cr$, where C is a constant. At $r = 0$, the pipe centerline, there is no shear stress. At r's maximum value of $D/2$, the shear stress has its maximum value, called τ_w, the *wall shear stress*. This boundary condition lets us determine the value of C, which must be $C = 2\tau_w/D$:

$$\tau(r) = \frac{2\tau_w r}{D}. \tag{11.8}$$

From the force balance, then, we must have

$$\Delta p = \frac{4l\tau_w}{D}. \tag{11.9}$$

We see that a small shear stress can produce a large pressure difference if the pipe is relatively long. We also note that if viscosity were zero there would be no shear stress and the pressure would be constant throughout the pipe. To get further with this analysis, we need to know how the shear stress is related to the velocity.

We could proceed by integrating the full Navier-Stokes equation for this steady, incompressible, viscous flow of a Newtonian fluid, or we could simply remember that for a Newtonian fluid, shear stress is proportional to velocity gradient. For our pipe flow, this is

$$\tau = -\mu \frac{du}{dr}, \tag{11.10}$$

where we have included the negative sign to have $\tau > 0$ for $du/dr < 0$, since the velocity decreases from the centerline to the outer wall and shear stress is maximum at the pipe wall; it is more intuitive to keep track of positive τ's. If we combine this equation (the definition of a Newtonian fluid) with the force balance ($F = ma$), and eliminate τ, we get

$$\frac{du}{dr} = -\left(\frac{\Delta p}{2\mu\, l}\right) r, \tag{11.11}$$

which we integrate to find the velocity profile,

$$u(r) = -\left(\frac{\Delta p}{4\mu\, l}\right)^2 r^2 + C_1, \tag{11.12}$$

and use the no-slip boundary condition ($u(r = D/2) = 0$) to find $C_1 = (\Delta p/16\mu l)\, D^2$, so that

$$u(r) = \frac{\Delta\, pD^2}{16\mu l}\left[1 - \left(\frac{2r}{D}\right)^2\right] = V_C\left[1 - \left(\frac{2r}{D}\right)^2\right], \tag{11.13}$$

where *Vc* is the centerline velocity, defined by $\Delta pD^2/16\mu l$. We can also express the velocity profile in terms of the wall shear stress, and in terms of $R = D/2$, as

$$u(r) = \frac{\tau_w D}{4\mu}\left[1 - \left(\frac{r}{R}\right)^2\right]. \tag{11.14}$$

This velocity profile is *parabolic* in the radial coordinate r and has a maximum value, V_c, at the centerline, and minimum values (zero) at the pipe wall. We can next find the volume flowrate Q through the pipe. We integrate over a series of very small rings of radius r and thickness dr to find Q:

$$Q = \int udA = \int_{r=0}^{r=R} u(r) 2\pi\, rdr = 2\pi\, V_c \int_0^R \left[1 - \left(\frac{r}{R}\right)^2\right] rdr\,. \tag{11.15}$$

$$Q = \frac{\pi\, R^2 V_c}{2}. \tag{11.16}$$

The average velocity is defined as the flowrate divided by the cross-sectional area, so for this flow we have

$$V_{avg} = \frac{\pi\, R^2 V_c}{2\pi\, R^2} = \frac{V_c}{2} = \frac{\Delta\, pD^2}{32\mu\, l}, \tag{11.17}$$

and

$$Q = \frac{\pi\, D^4 \Delta\, p}{128\mu\, l}. \tag{11.18}$$

We have found that the average velocity is half the centerline velocity for our laminar parabolic velocity profile. Our results also confirm that the flowrate is (1) directly proportional to the pressure drop; (2) inversely proportional to the viscosity; (3) inversely proportional to pipe length, and (4) proportional to the diameter to the fourth power.

Equation (11.18) for Q is commonly known as Poiseuille's law, so named for a French physician who performed the first analysis of laminar pipe flow with the goal of learning about blood flow.[3] Fully developed laminar pipe flow, with its parabolic velocity profile, is generally known as *Poiseuille flow*.

11.3 Why Can an Airplane Fly?

A body, such as a wing or an airfoil, experiences a resultant force due to the interaction between the body and the moving fluid surrounding it. Figure 11.3 shows a two-dimensional airfoil and the forces on it due to the surrounding fluid: (a) pressure force, (b) viscous force, and (c) resultant force (lift and drag).

You are probably already familiar with the idea that the pressure distribution is responsible for lift. The basic idea is that pressure is lower on the upper surface of a wing, so a net upward force keeps the wing aloft. We could show this using the Bernoulli equation: The flow over the smooth upper surface is much faster (therefore exerts lower pressure) than that past the lower surface.

Knowing that drag and lift are the x and y resultants of the pressure and viscous stress forces, we could obtain expressions for drag and lift by integrating these pressure and viscous forces over the body's surface:

$$D=\int dF_x=\int p\cos\theta dA+\int \tau_w \sin\theta dA, \tag{11.19}$$

$$L=\int dF_y=-\int p\sin\theta dA+\int \tau_w \cos\theta dA, \tag{11.20}$$

where θ is the degree of inclination (with respect to horizontal) of the outward normal at any point along the body surface. To carry out this integration, we must know the body shape, including θ as a function of position along the body and the distribution of τ_w and p. This is quite difficult to do for realistic geometries. As we have seen when finding the resultant pressure forces on submerged curved surfaces, there are sometimes ways to get around messy integrations involving changing orientations. This is also the case for lift and drag.

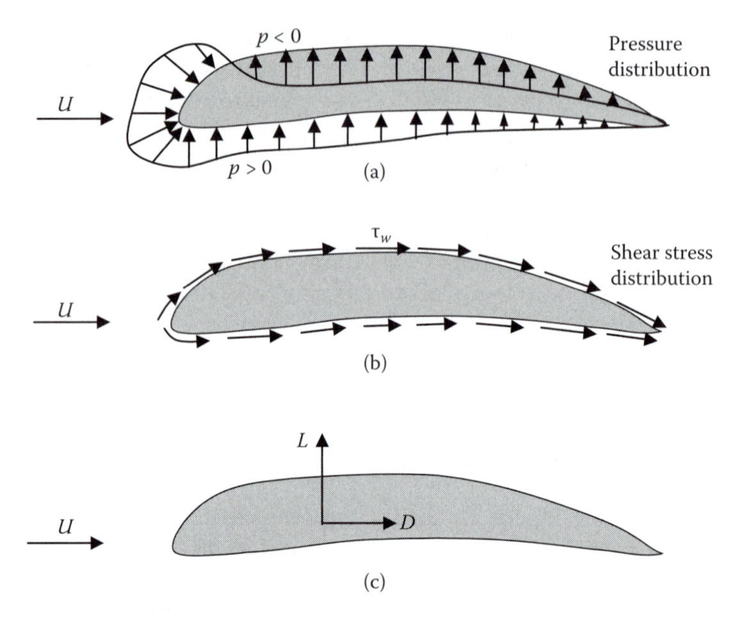

FIGURE 11.3
Forces on two-dimensional airfoil: (a) Pressure force. (b) Viscous force. (c) Resultant lift and drag forces.

In the simplest force analysis of an airplane, the four important forces are lift, drag, thrust (forward propulsion provided by engines), and weight of the plane. Lift must exceed weight and thrust must exceed drag for flight to be possible. We can calculate the lift and drag forces for a certain shape in a certain flow using the following formulas:

$$L = C_L \tfrac{1}{2} \rho\, U^2 A, \tag{11.21}$$

$$D = C_D \tfrac{1}{2} \rho\, U^2 A, \tag{11.22}$$

where A is a characteristic area of the object, typically taken to be the *frontal area*, the projected area that would be seen by an observer riding along with the onrushing flow, parallel to the upstream velocity U. It is important to specify which A one is using in a calculation, and why, when citing lift and drag results. The coefficients C_L and C_D for most common shapes have been determined from experimental data and are tabulated as functions of Reynolds number, as shown in Figure 11.4 for a sphere and a circular cylinder.

We notice in Figure 11.4 that the drag coefficient decreases sharply at a Reynolds number of about 5×10^5. This corresponds to the value of Re_{crit} at which flow transitions from laminar to turbulent. Turbulent flow is characterized by higher fluid momentum, thinner boundary layers, and higher viscous stresses at solid surfaces than laminar flow. For flows over cylinders and spheres, the fluid's higher momentum causes it to more readily follow the body surface without "separating" into a wake region. Turbulent wakes behind cylinders and spheres are therefore generally smaller than laminar wakes. This reduction in the pressure drag on the object overwhelms any increase in viscous drag, and therefore we see a sharp *drag drop* corresponding to the transition to turbulent flow. This phenomenon is sometimes exploited: For example, vortex generators on airplane wings serve to trip flow into turbulent behavior at lower Re than Re_{crit}.

11.4 Why Does a Curveball Curve?

Baseballs and other objects moving through fluids leave *wakes* behind them. These wakes can be either laminar (relatively smooth flow, viscosity damping out disorderly structures) or turbulent (much more disordered, lots of whorls and eddies with no viscosity to wipe them out). Even in laminar flow,

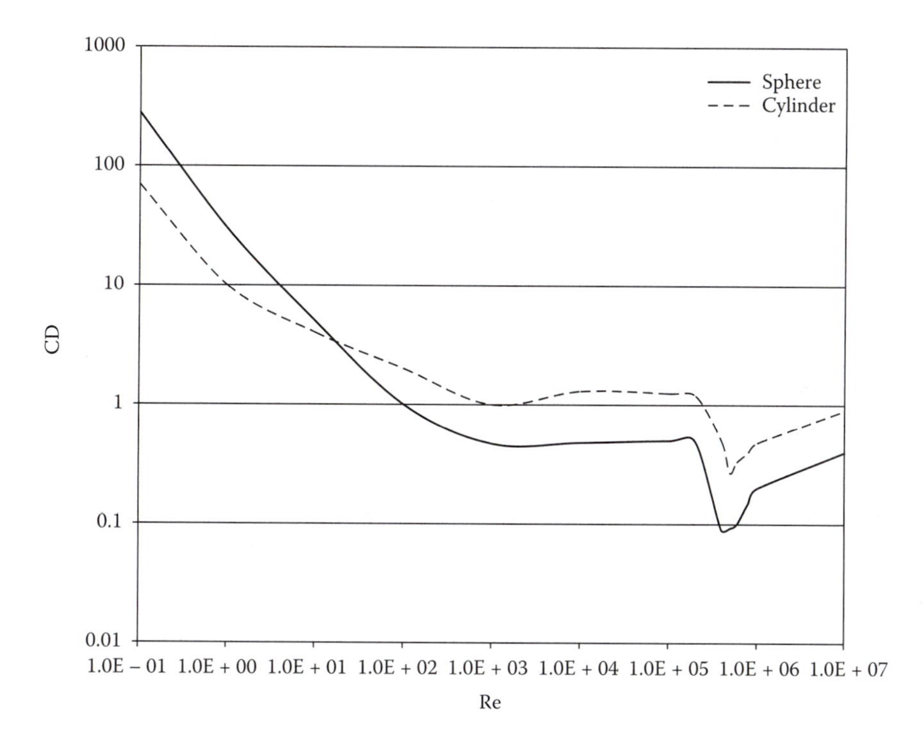

FIGURE 11.4
Drag coefficients for smooth cylinder and sphere, as functions of Re.

obstructions and protrusions such as wings and flaps on airplanes and rocks in streams can cause some rotational flow behind them. Zones of rotational flow are called *vortices*. Figure 11.5 shows the vortices behind several spheres for a range of Reynolds numbers.

There is much that could be said about these flow patterns—a semester's worth—but for now, we are interested in baseballs. A typical pitch has a speed of 75 mph to 90 mph. A regulation Major League Baseball (MLB) ball must have a circumference between 9 and 9.25 in., or a diameter of about 2.9 in. Using the properties of still air at standard atmospheric conditions, we can calculate a typical Reynolds number:

$$\mathrm{Re} = \frac{\rho VD}{\mu} = \frac{(1.2\frac{kg}{m^3})(37\frac{m}{s})(0.0737\ m)}{1.83\times 10^{-5}\ \frac{N\cdot s}{m^2}} = 179000 = 1.79\times 10^5.$$

For a smooth sphere, the transition to turbulence begins at a critical Reynolds number of about $Re_{crit} \sim 5 \times 10^5$. The main difference between a baseball and the smooth spheres in Figure 11.5 is the raised stitching. This unevenness on the ball surface makes the transition to turbulence happen at lower *Re*. This is actually a favorable condition for sports balls—as we saw in Figure 11.4, turbulent drag coefficients are lower than laminar ones. So catalyzing the transition to turbulence can decrease the drag on a ball. This, in fact, is why golf balls are dimpled. The dependence of a baseball's drag coefficient on its speed is shown in Figure 11.6.

If a baseball is thrown without any backspin or topspin imparted by the pitcher, the orientation of the seam causes an asymmetry in the wake, which in turn causes an irregular trajectory. This delivery is commonly known as a *knuckleball*. If, on the other hand, the pitcher *does* impart some spin to the ball as he or she hurls it, the right amount of spin stabilizes this irregularity and helps the trajectory follow a predictable path. This is shown in Figure 11.7, as we see the streamlines over a spinning baseball. The streamlines are crowded near the bottom of the ball (representing faster flow), and the wake is deflected upward by the spin. This deflection is linked to a net downward force on the ball, which is why a pitch thrown in this way will drop or sink as it approaches the batter.

Other types (different in magnitude and direction) of spin can alter the baseball's path in different ways. This effect, known as the Magnus effect, has motivated considerable research into the aerodynamics of baseball. The types of spin imparted for a range of pitch deliveries are sketched in Figure 11.8.

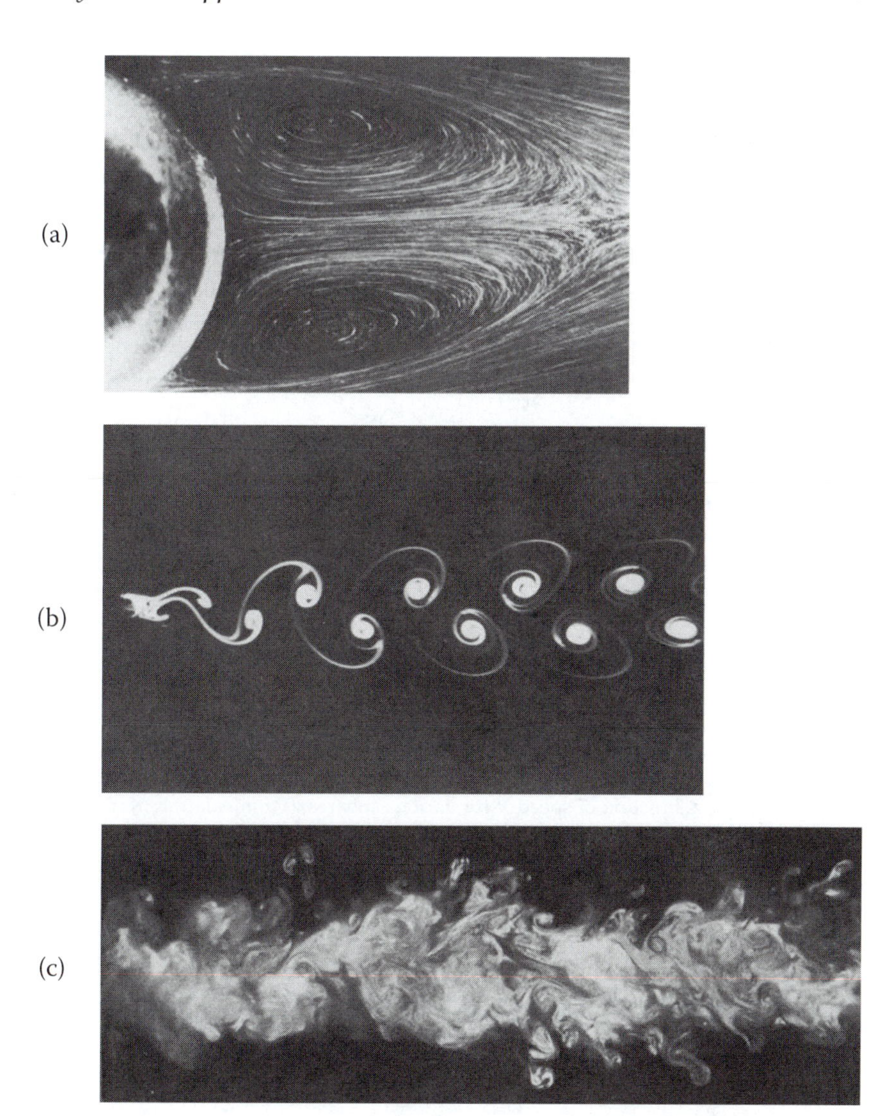

FIGURE 11.5
Wakes behind smooth bodies. Note dependence on Reynolds number. (a) Sphere at Re = 118. Recirculating regions behind sphere still attached. (b) Cylinder at Re = 200. Wake develops into two parallel rows of staggered vortices. (c) Re = 1770. Turbulent wake behind cylinder. Instantaneous flow patterns shown by oil fog. (From Van Dyke, M., *An Album of Fluid Motion*, Parabolic Press, Stanford, CA, 1982. With permission.)

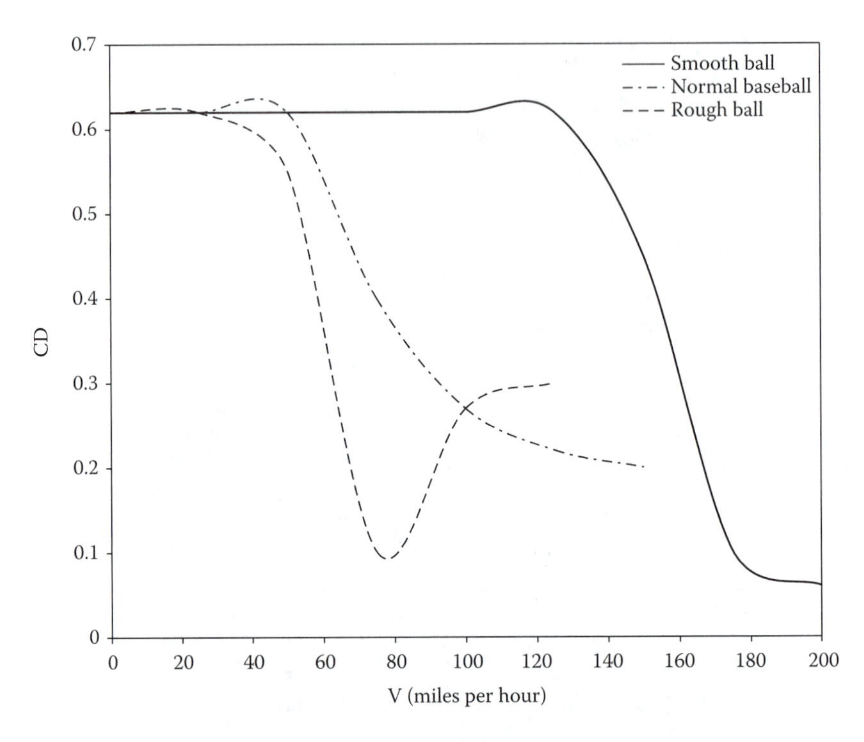

FIGURE 11.6
Drag coefficient as function of speed V for various spheres. (From Adair, R. K., *The Physics of Baseball*, New York, NY: Harper Perennial, 1994. With permission.)

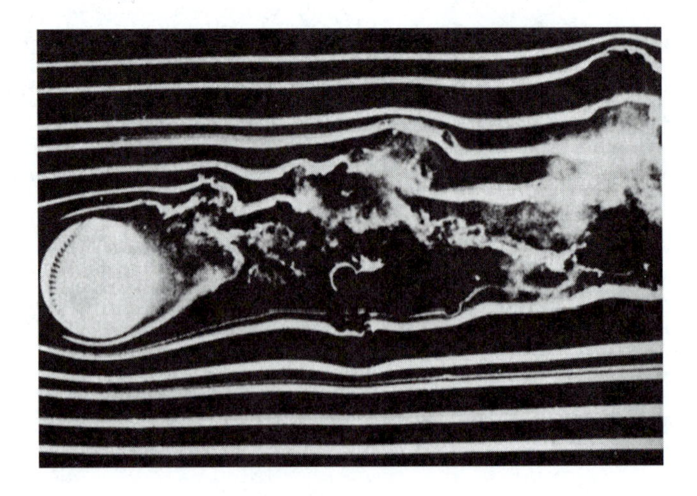

FIGURE 11.7
Smoke photograph of flow around a spinning baseball. Flow is from left to right, flow speed is 21 m/s, and ball is spinning counterclockwise at 15 m/s (= ωr). (Photograph by F. N. M. Brown, Courtesy the University of Notre Dame, South Bend, IN.)

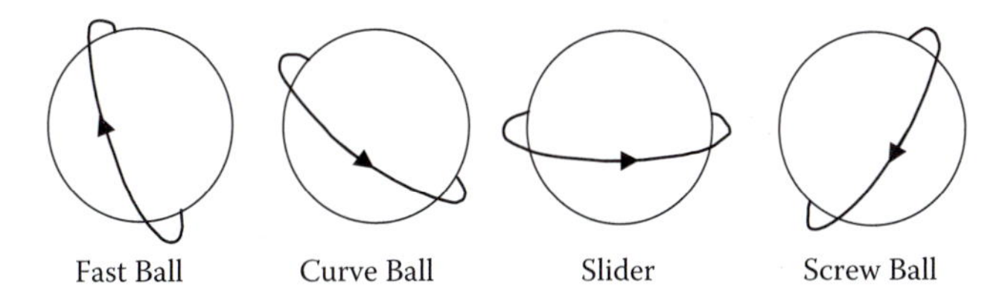

FIGURE 11.8
Ball rotation directions, as seen by the batter, for pitches thrown overhand by a right-handed pitcher. Arrow indicates the direction of rotation. (From Adair, R. K., *The Physics of Baseball*, Harper Perennial, 1994. With permission.)

11.5 Problems

11.1 Typical values of the Reynolds numbers for several animals moving through air or water are listed in the following table. In which cases is the fluid inertia important? In which cases do viscous effects dominate? Do you expect the flow in each case to be laminar or turbulent? Explain.

Animal	Speed	Re
Large whale	10 m/s	300,000,000
Flying duck	20 m/s	300,000
Large dragonfly	7 m/s	30,000
Invertebrate larva	1 mm/s	0.3
Bacterium	0.01 mm/s	0.00003

11.2 The velocity distribution in a fully developed laminar pipe flow is given by

$$\frac{u}{U_{CL}} = 1 - \left(\frac{r}{R}\right)^2,$$

where U_{CL} is the velocity at the centerline and R is the pipe radius. The fluid density is ρ, and its viscosity is μ. (a) Find the average velocity U_{mean} over the cross section. (b) State the Reynolds number for the flow based on average velocity and pipe diameter. At what approximate value of this Reynolds number do you expect the flow to become turbulent? Why is this value only approximate? (c) Assume that the fluid is Newtonian. Find the wall shear stress τ_w in terms of μ, R, and U_{CL}.

11.3 A wing generates a lift L when moving through sea-level air with a velocity U. How fast must the wing move through the air at an altitude of 35,000 ft if it is to generate the same lift? (Assume the lift coefficient is constant.)

11.4 The drag on a 2-m-diameter satellite dish due to an 80 km/hr wind is to be determined through wind tunnel testing on a geometrically similar 0.4-m-diameter model dish. (a) At what air speed should the model test be performed? (b) If the measured drag on the model was determined to be 170 N, what is the predicted drag on the full-scale prototype?

11.5 The viscous, incompressible flow between the parallel plates shown in Figure 11.9 is caused by both the motion of the bottom plate and a pressure gradient, $\partial p/\partial x$. Determine the relationship between U and $\partial p/\partial x$ such that the shear stress on the fixed plate is zero.

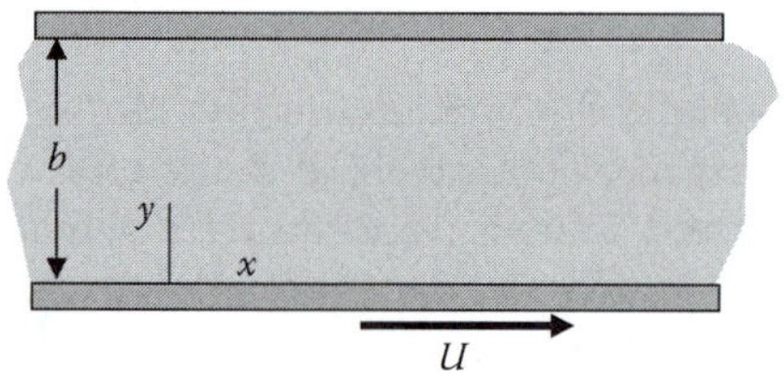

FIGURE 11.9

11.6 Water exits a reservoir at 30-m depth to enter the 150-mm-diameter inlet of a turbine, as shown in Figure 11.10. The turbine outlet is also 150 mm in diameter. The exit flow is then ejected to the atmosphere at 9 m/s through a nozzle with diameter 75 mm. What power is developed by the turbine? What horizontal force is required to anchor the turbine if the inflow and outflow are horizontal?

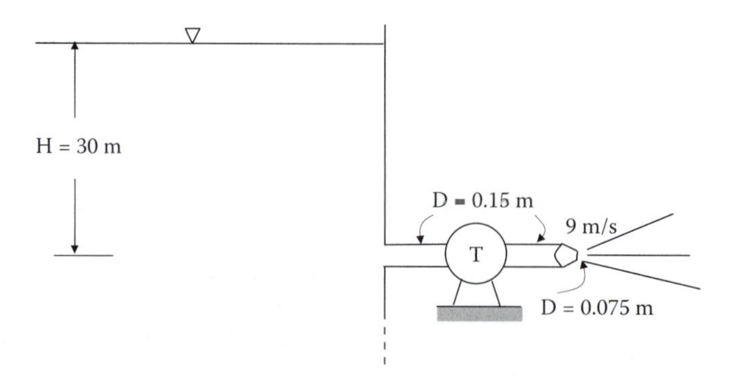

FIGURE 11.10

11.7 Crude oil flows through a level section of the Alaskan pipeline at a rate of 1.6 million barrels per day (1 barrel = 42 gallons). The pipe inside diameter is 120 cm, and its roughness has a characteristic dimension of 1.5 mm. The maximum allowable pressure is 8300 kPa, and the minimum pressure required to keep dissolved gases in solution in the crude oil is 350 kPa. The crude oil has SG = 0.93, and its viscosity at the pumping temperature is $\mu = 0.017$ N·s/m^2. For these conditions, determine the maximum possible spacing between pumping stations.

11.8 Blood is a very interesting fluid: a suspension of red and white blood cells and platelets in a liquid plasma. We would like to be as optimistic as Jean Poiseuille in modeling blood flow, but we know that these cells in the plasma can cause blood's viscosity to be dependent on the shear rate—that is, blood's composition can cause it to behave like a *non-Newtonian* fluid. Especially in regions of very low shear rate, blood's red blood cells have been shown to aggregate and form clumps that cause blood to require a certain *yield stress* to be applied before it flows smoothly again. You are given the data in the following table for an "average" person. This person's cardiac output is 5 L per minute; heart rate is 60 beats per minute; and at a hematocrit of 40%, blood density is 1.06 g/cm^3, and blood viscosity is 3.5 centiPoise (named for Poiseuille and abbreviated cP; 1 cP = 1 mPa•s).

	Internal Diameter (mm)	Wall Thickness (mm)	Percentage of Heart Q	Typical Pressures (mm Hg)
Ascending aorta	20	2	100%	100
Abdominal aorta	12	1.5	50%	90
Femoral artery	8	0.8	10%	80
Random arteriole	0.1	0.02	0.001%	60

Note that the vessels downstream from the heart receive only a portion of its volumetric output due to branching of vessels. The percentages given here are ballpark estimates. Based on these parameters, calculate the following in each of the measured vessels: (a) pressure drop; (b) mean velocity; (c) shear rate at vessel wall; (d) Reynolds number; and (e) percent cross-sectional area change due to pulse pressure, assuming small strain $\varepsilon_{\theta\theta} = \tau_{\theta\theta} / E$.

11.9 Based on the values you calculated in Problem 11.8, answer and explain the following: (a) In which vessels should elasticity of the vessel be considered? (b) In which vessels should the non-Newtonian behavior of blood be considered? (c) Where in the body might turbulence develop? (d) Why does most of the pressure drop in the arterial system occur in the arterioles?

11.10 Wind tunnel testing of the concrete reef balls used in artificial reefs is proposed. A typical reef ball (e.g., Chapter 1, Figure 1.2) has a diameter of 6 ft and is immersed in sea water. A scale model is prepared with diameter of 6 in. (a) At what range of velocities should wind tunnel tests be performed to ensure that the experimental data are relevant to the real reef balls? (b) What effect do you believe that the holes in the reef ball will have on the flow, if any?

Notes

1. The Reynolds number is named for Osborne Reynolds, the son of an Anglican priest who became a noted fluid mechanician (becoming especially active in fluids after 1873). He was particularly influential in the study of pipe flow and the transition from laminar to turbulent flow. He also established the course of study in applied mathematics at the University of Manchester, though sadly, as one biographer reports, "Despite his intense interest in education, he was not a great lecturer. His lectures were difficult to follow, and he frequently wandered among topics with little or no connection" (Anderson 1997).
2. Fluids tend to flow from high pressure toward lower pressure regions, just as mass tends to flow from regions of high to low concentrations.
3. Poiseuille (1799–1869) made the same assumptions we've made: He modeled blood flow as a steady, incompressible flow of a non-Newtonian fluid in rigid circular pipes. Although these are spectacularly inappropriate assumptions for blood flow as it is now understood, Poiseuille flow theory has proven robust in its ability to relate flow rate and fluid mechanical forces for many internal flows. It is even a reasonable ballpark predictor of blood flow, as we see in Problem 11.8.

12

Solid Dynamics: Governing Equations

We have thoroughly considered the equilibrium of solids. By examining the isolated effects of various types of loading and then the methods of combining these effects in more realistic situations, we have come to understand many problems for which $\Sigma \underline{F} = \underline{0}$. However, although external forces often cause a solid to be in equilibrium, it is also possible for them to result in the solid's *motion*. Because both solids and fluids are continua and because they are governed by mass conservation and Newton's second law, we expect their equations of motion to markedly resemble each other. In this chapter we briefly consider the governing equations for the motion of solids and some examples of their solution.

The key concepts of any continuum mechanics problem are continuity, compatibility, and the relevant constitutive law. In the problems we have considered thus far, we have rarely had to check these conditions (and the constitutive law, Hooke's or Newton's, has been a straightforward one); we have been able to implicitly assume they were met. However, as our study of mechanics continues, we encounter more general, less constrained problems of continuum mechanics for which we must apply these three concepts. In this section, we discuss the "next level" of continuum mechanics in the context of these three C's. We begin by briefly defining each of them. As you may recall from Chapter 2 and Chapter 3:

- Continuity: Density must be definable, continuous function.
- Compatibility: Displacements (u, v, w) must be continuous.
- Constitutive law: Deformation (strain) must be related to loading (stress).

12.1 Continuity, or Mass Conservation

If a material is a *continuum* we are able to ignore the fundamentally discrete composition of matter and all those atoms dancing about, and to assume that the substance of material bodies is uniformly distributed. This continuum model allows us to divide matter into smaller and smaller portions, each of which has the physical properties of the original body. So, we can assign quantities such as density and velocity to each point of the space occupied by the body.[1]

Recall that for a continuum we are able to mathematically define a mass density as

$$\rho = \lim_{\Delta V \to 0} \frac{\Delta m}{\Delta V} \tag{12.1}$$

and that this density, like other properties of the continuum, is a continuous function of position and time: $\rho = \rho(x, t)$. We can thus describe the mass of an entire body (of total volume V) by

$$m = \int_V \rho(\underline{x}, t) dV . \tag{12.2}$$

Since mass is neither created nor destroyed, we require that the mass of the body remains invariant under motion. Its total derivative must be zero,

$$\dot{m} = \frac{d}{dt} \int_V \rho(\underline{x}, t) dV = \int_V \left(\frac{\partial \rho}{\partial t} + v_i \frac{\partial \rho}{\partial x_i} + \rho \frac{dv_i}{dx_i} \right) dV = 0, \tag{12.3}$$

where we have, in a sense, used the chain rule to construct a *total* or *material derivative* of the fluid mass $\rho(x, t)V$. Since equation (12.3) must hold for any dV of the body, we must have

$$\frac{\partial \rho}{\partial t} + v_i \frac{\partial \rho}{\partial x_i} + \rho \frac{dv_i}{dx_i} = 0, \tag{12.4}$$

or

$$\frac{\partial \rho}{\partial t} + \frac{\partial}{\partial x_i} \left(\rho \, v_i \right) = 0, \tag{12.5}$$

where the repeated i index, we remember from Chapter 1, section 1.5, represents a summation over i. In vector notation we could write the conservation of mass as

$$\frac{\partial \rho}{\partial t} + \nabla \left(\rho \underline{v} \right) = 0 . \tag{12.6}$$

We note that if the density is constant in x and t, the material is said to be *incompressible*, and in this case our continuity equation requires that

$$\frac{\partial}{\partial x_i} v_i = 0 \quad \text{or} \quad \nabla \cdot \underline{v} = 0 \tag{12.7}$$

for incompressible continua. Each v_i here is the ith component of the vector velocity field, v. This is exactly how we have written mass conservation for a fluid in Chapter 10, Section 10.4.1.

12.2 $\underline{F} = m\underline{a}$, or Momentum Conservation

Newton's second law of motion states that $\underline{F} = m\underline{a}$, the resultant force on an object balances this object's inertia—its mass times its acceleration. This can also be stated as "the resultant force on a body equals the time rate of change of the body's linear momentum." We already understand how to state the resultant force on a body: So far we have been writing $\Sigma \underline{F} = \underline{0}$ for a variety of systems. The stress tensor for a given body reflects its response to all external loads, so by writing the stress tensor we have effectively written the resultant surface force on the body.

We may also consider the effects of a *body force* such as gravity or the force due to an electromagnetic field. A sample tuberous body with resultant surface and body forces is shown in Figure 12.1.

Hence, we understand that the ith component of the total resultant force F on a body is written as

$$F_i = \int_V \rho\, b_i dV + \int_S \tau_{ij} n_j dS. \tag{12.8}$$

All that remains is then to write the change in momentum for the same body, or ma. Again, we write only the ith component of the body's acceleration,

$$\int_V \rho \frac{dv_i}{dt} dV, \tag{12.9}$$

where we have taken the total derivative of the momentum per volume, (ρv), and then have used the conservation of mass to eliminate the derivatives of density. $F = ma$ is simply then the balance of the resultant force and the inertia:

$$\int_V \rho\, b_i dV + \int_S \tau_{ij} n_j dS = \int_V \rho \frac{dv_i}{dt} dV . \tag{12.10}$$

It only remains for us to convert the surface area integral to a volume integral, which we may do by Gauss's Theorem, and to obtain

$$\int_V \rho\, b_i dV + \int_V \frac{\partial}{\partial x_j} \tau_{ij} dV = \int_V \rho \frac{dv_i}{dt} dV . \tag{12.11}$$

As this must be true for any volume, we truly have

$$\rho\, b_i + \frac{\partial}{\partial x_j} \tau_{ij} = \rho \frac{dv_i}{dt}, \tag{12.12}$$

or, in vector form,

$$\nabla \underline{\underline{\tau}} + \rho\, \underline{b} = \rho \frac{d\underline{v}}{dt} . \tag{12.13}$$

For solids in equilibrium, as we have already seen, the resultant forces sum to zero. The x component of the governing equation for such a solid would be

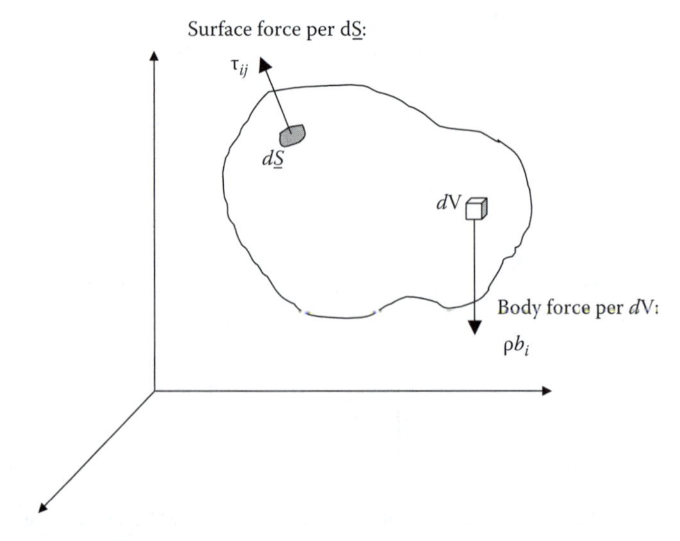

FIGURE 12.1
Forces on a body.

$$\frac{\partial \tau_{xx}}{\partial x}+\frac{\partial \tau_{xy}}{\partial y}+\frac{\partial \tau_{xz}}{\partial z}+\rho\, b_x=0\,. \tag{12.14}$$

Equation (12.13), as expected, looks strikingly like the Navier-Stokes equation developed for fluids in Chapter 10, Section 10.4.2. Here, the viscous force and the pressure force (previously known as F_{visc}, or $\mu\nabla^2\underline{V}$, and $-\nabla p\, dV$) have been combined, as the pressure (i.e., normal stress) and viscous stresses are combined into one stress tensor, $\underline{\underline{\tau}}$. But the form of $F = ma$ looks awfully familiar.

12.3 Constitutive Laws: Elasticity

The behavior of the material in question provides us with our third governing equation. We can then analyze solids in motion by solving these three equations. If a material behaves *elastically,* this means two things to us: (1) The stress is a unique function of the strain; and (2) the material is able to fully recover to its *natural* shape after the removal of applied loads. Although elastic behavior can be either linear or nonlinear, in this textbook we're concerned primarily with *linearly elastic* materials to which Hooke's law applies. The constitutive law for linearly elastic behavior is simply

$$\tau_{ij}=C_{ijkm}\varepsilon_{km}, \qquad \text{or} \qquad \underline{\underline{\tau}}=\underline{\underline{\underline{\underline{C}}}}\underline{\underline{\varepsilon}}, \tag{12.15}$$

where, as we discussed in Chapter 3, Section 3.5, C is a fourth-order tensor whose eighty-one components reduce to thirty-six unique components due to the symmetry of both the stress and strain tensors.

For isotropic materials, we are able to find the exact form of C. If the material is isotropic, then its elastic tensor C must be a fourth-order, isotropic tensor. An isotropic tensor is one whose components are unchanged by any orthogonal transformation from one set of Cartesian axes to another. This requirement guides the form that C must take

$$C_{ijkm} = \lambda\delta_{ij}\delta_{km} + \mu(\delta_{ik}\delta_{jm} + \delta_{im}\delta_{jk}) + \beta(\delta_{ik}\delta_{jm} - \delta_{im}\delta_{jk}), \tag{12.16}$$

where λ, μ, and β are scalars. We remind ourselves that the Kronecker deltas are simple second-order identity tensors [$\delta_{ij} = 1$ if $i = j$, but $\delta_{ij} = 0$ if $i \neq j$]. Due to the symmetry of both the stress and strain tensors, we must have $C_{ijkm} = C_{jikm} = C_{ijmk}$. This requires that $\beta = -\beta$ and thus that $\beta = 0$. Hooke's law (here's the important part) then takes the form

$$\tau_{ij} = [\lambda\delta_{ij}\delta_{km} + \mu(\delta_{ik}\delta_{jm} + \delta_{im}\delta_{jk})]\varepsilon_{km}, \tag{12.17}$$

or, using the Kronecker delta's substitution property,

$$\tau_{ij} = \lambda\delta_{ij}\varepsilon_{kk} + 2\mu\varepsilon_{ij}. \tag{12.18}$$

This is Hooke's law for isotropic elastic behavior. If we rearrange this to make it an expression for strain ε_{ij}, we can obtain the following relations for Young's modulus and Poisson's ratio (the shear modulus $G = \mu$) and, finally, the generalized form of Hooke's law, for linearly elastic materials:

$$E = \frac{\mu\,(3\lambda + 2\mu)}{\lambda + \mu}$$
$$\nu = \frac{\lambda}{2(\lambda + \mu)}. \tag{12.19}$$

$$\varepsilon_{ij} = \frac{1}{E}\left[(1+\nu)\tau_{ij} - \nu\delta_{ij}\tau_{kk}\right]. \tag{12.20}$$

As long as the material in question does not split apart or overlap itself, its displacements must be continuous. This requirement is known as *compatibility*, and it is guaranteed by a displacement field that is single valued and continuous, with continuous derivatives. The strain tensor is composed of the derivatives of the displacement field, as we have seen. So in two dimensions, we may write the compatibility condition in the form

$$\frac{\partial^2\varepsilon_x}{\partial y^2} + \frac{\partial^2\varepsilon_y}{\partial x^2} = \frac{\partial^2\gamma_{xy}}{\partial x\partial y}. \tag{12.21}$$

Alas, in three dimensions we have six strain components to keep track of, and there are five additional compatibility conditions.

Using these governing equations, it is possible to fully describe the equilibrium or motion of a continuum. Often, a constitutive law will be experimentally obtained for a given material, and it is the job of the continuum mechanician to express the governing equations appropriately and to solve them. In most cases it is not possible to analytically solve these equations; generally, it is necessary to solve them numerically.

By integrating the differential equations of equilibrium, we obtain results that agree with our simpler calculations, since our new partial differential equations are simply saying what we've said all semester: For a body in equilibrium, the sum of the forces acting on the body is zero. This is the same statement whether we say it by means of a free-body diagram and average stresses or whether we solve complex partial differential equations.

Note

1. This is the starting-off point for George Mase's *Continuum Mechanics for Engineers* (Mase and Mase 1999), an excellent transitional text to move from the mechanics analyses of this textbook to the level of graduate continuum mechanics.

References

Adair, R.K., *The Physics of Baseball*, New York, NY: Harper Perennial, 1994.

American Society of Mechanical Engineers, *International Boiler and Pressure Vessel Code*, New York, 2001.

Anderson, J.D., *A History of Aerodynamics*, Cambridge, MA: Cambridge University Press, 1997.

Aris, R., *Vectors, Tensors, and the Basic Equations of Fluid Mechanics*, Mineola, NY: Dover, 1962.

Barron's Atlas of Anatomy, Barron's Educational Series, Hauppauge, NY, 1997.

Bassman, L. and Swannell, P., *Engineering Statistics Study Book*, University of Southern Queensland, 1994.

Bedford, A. and Liechti, K.M., *Mechanics of Materials*, Upper Saddle River, NJ: Prentice Hall, 2000.

Beer, F.P. and Johnston, E.R., *Mechanics of Materials*, 2d ed., New York, NY: McGraw-Hill, 1992.

Brinckmann, P., Frobin, W., and Leivseth, G., *Musculoskeletal Biomechanics*, New York, NY: Thieme, 2002.

Chadwick, P., *Continuum Mechanics*, Mineola, NY: Dover, 1976.

"Collapsed Roof Design Defended," *Engineering News-Record*, June 29, 1978.

"Collapsed Space Truss Roof Had a Combination of Flaws," *Engineering News-Record*, June 22, 1978.

Cook, R.D. and Young, W.C., *Advanced Mechanics of Materials*, 2d ed., Upper Saddle River, NJ: Prentice Hall, 1999.

Crowe, M.J., *A History of Vector Calculus*, Mineola, NY: Dover, 1967.

Davis, M., "How the Hero Who Brought Water to L.A. Abruptly Fell From Grace," *Los Angeles Times*, July 25, 1993, p. M3.

Dawson, T.J. and Taylor, C.R., "Energetic Cost of Locomotion in Kangaroos," *Nature* 246, 1973, pp. 313–314.

Debes, J.C. and Fung, Y.C., "Biaxial Mechanics of Excised Canine Pulmonary Arteries," *American Journal of Physiology* 269, 2, 1995, pp. H433–442.

"Design Flaws Collapsed Steel Space Frame Roof," *Engineering News-Record*, April 6, 1978.

Dym, C.L., *Introduction to the Theory of Shells*, New York, NY: Hemisphere Publishing, 1990.

Dym, C.L. and Little, P., *Engineering Design: A Project-Based Introduction*, New York, NY: John Wiley & Sons, 1999.

Enderle, J.D., Blanchard, S.M., and Bronzino, J.D., *Introduction to Biomedical Engineering*, San Diego, CA: Academic Press, 2000.

Evans, K.E.,"Tensile Micro-structures Exhibiting Negative Poisson's Ratios" *Journal of Physics D: Journal of Applied Physics* 22, 1989, pp. 1870–1876.

Feld, J. and Carper, K., *Construction Failure*, New York, NY: Wiley & Sons, 1997.

Frocht, M.M., "Factors of Stress Concentration Photoelastically Determined," *ASME Journal of Applied Mechanics* 2, 1935, pp. A67–A68.

Fung, Y.C., *A First Course in Continuum Mechanics*, Upper Saddle River, NJ: Prentice Hall, 1994.

Fung, Y.C., *Biomechanics: Mechanical Properties of Living Tissues*, New York, NY: Springer, 1993.

Gere, J.M. and Timoshenko, S.P., *Mechanics of Materials*, 4th ed., WS Publishing, 1997.

Gordon, J.E., *Structures: Why Things Don't Fall Down*, New York, NY: DaCapo, 2003.

Gordon, J.E., *The New Science of Strong Materials, or Why You Don't Fall through the Floor*, Princeton, NJ: Princeton University Press, 1988.

Green, K.H., http://www.math.arizona.edu/~vector/.

"Hartford Collapse Blamed on Weld," *Engineering News-Record*, June 24, 1979.

Humphrey, J.D. and Delange, S.L., *Biomechanics*, Springer, New York, 2003.

Isenberg, C., *The Science of Soap Films and Soap Bubbles*, Mineola, NY: Dover, 1992.

Jackson, D.C. and Norris, H., "William Mulholland and the St. Francis Dam Disaster (Privilege and Responsibility)," *California History*, September 2004.

Jacobsen, L.S., "Torsional-Stress Concentrations in Shafts of Circular Cross-Section and Variable Diameter," *Transactions of the American Society of Mechanical Engineering* 47, 1925, pp. 619–638.

Kuethe, A.M. and Chow, C.Y., *Foundations of Aerodynamics*, 4th ed., Wiley, 1986.

Kundu, P.K., *Fluid Mechanics*, Academic Press, 1990.

Lakes, R., "Negative Poisson's Ratio Materials," *Science* 235, 1987, pp. 1038–1040.

Leslie, M., *Rivers in the Desert: William Mulholland and the Inventing of Los Angeles*, New York, NY: HarperCollins, 1993.

Levy, M. and Salvadori, M., *Why Buildings Fall Down*, New York, NY: Norton, 1992.

Macaulay, W.H., "Note on the Deflection of Beams," *Messenger of Mathematics 48*, 1919, pp. 129–130.

Malvern, L.E., *Introduction to the Mechanics of a Continuous Medium*, Upper Saddle River, NJ: Prentice Hall, 1969.

Martin, R. and Delatte, N., "Another Look at the Hartford Civic Center Coliseum Collapse," *Journal of Performance of Constructed Facilities 15*, 1, February 2001, pp. 31–36.

Martin, R.B., Burr, D.B., and Sharkey, N.A., *Skeletal Tissue Mechanics*, New York, NY: Springer-Verlag, 1998.

Mase, G.T. and Mase, G.E., *Continuum Mechanics for Engineers*, 2d ed., Boca Raton, FL: CRC Press, 1999.

Mattson, R., *William Mulholland: A Forgotten Forefather*, Pacific Center for Western Studies, Stockton, CA, 1976.

Munson, B.R., Young, D.F., and Okiishi, T.H., *Fundamentals of Fluid Mechanics*, 3d ed., New York, NY: Wiley, 1998.

"New Theory on Why Hartford Roof Fell," *Engineering News-Record*, June 14, 1979.

Petroski, H., *To Engineer Is Human*, New York, NY: St. Martin's Press, 1982.

Pfrang, E.O., "Collapse of the Kansas City Hyatt Regency Walkways," *Henry M. Shaw Lecture*, North Carolina State University, Raleigh, January 27, 1983.

Pfrang, E.O. and Marshall, R., "Collapse of the Kansas City Hyatt Regency Walkways," *Civil Engineering*, July 1982, pp. 65–68.

Pilkey, W.D., *Peterson's Stress Concentration Factors*, 2nd ed., New York, NY: Wiley-Interscience, 1997.

Potter, M.C. and Wiggert, D.C., *Mechanics of Fluids*, 2d ed., Upper Saddle River, NJ: Prentice Hall, 1997.

Popov, E.P., *Engineering Mechanics of Solids*, 2d ed., Upper Saddle River, NJ: Prentice Hall, 1998.

Reiner, M., *Deformation, Strain, and Flow: An Elementary Introduction to Rheology*, New York: Interscience, 1960.

Ross, S., *Construction Disasters*, New York, NY: McGraw-Hill, 1984.

Rossman, J.C. and Utela, B., Unpublished data.

Sabersky, R.H., Acosta, A.J., Hauptmann, E.G., and Gates, E.M., *Fluid Flow*, 4th ed., Upper Saddle River, NJ: Prentice Hall, 1999.

Schey, H.M., *Div, Grad, Curl, and All That*, W.W. Norton, 1973.

Schlager, N. (Ed.), *When Technology Fails*, Detroit, MI: Gale Research, 1994.

Shepherd, R. and Frost, D. (Eds.), *Failures in Civil Engineering: Structural, Foundation, and Geoenvironmental Case Studies*, New York: The Education Committee of the Technical Council on Forensic Engineering of the American Society of Civil Engineers, 1995.

Smits, A.J., *A Physical Introduction to Fluid Mechanics*, New York, NY: John Wiley & Sons, 2000.

"Space Frame Roofs Collapse Following Heavy Snowfalls," *Engineering News-Record*, January 26, 1978.

Spiegel, L. and Limbrunner, G.F., *Applied Statics and Strength of Materials*, 3d ed., Upper Saddle River, NJ: Prentice Hall, 1999.

Timoshenko, S.P. and Goodier, J.N., *Theory of Elasticity*, New York, NY: McGraw-Hill, 1970.

Timoshenko, S.P. and Woinowsky-Krieger, S., *Theory of Plates and Shells*, New York, NY: McGraw-Hill, 1959.

Van Dyke, M., *An Album of Fluid Motion*, Stanford, CA: Parabolic Press, 1982.

Vogel, S., *Comparative Biomechanics: Life's Physical World*, Princeton, NJ: Princeton University Press, 2003.

Williams, H.E., "On Introducing Engineering Strain," *International Journal of Mechanical Engineering Education* 29, no. 4, 2001, pp. 397–403.

Wylie, C.R. and Barrett, L.C., *Advanced Engineering Mathematics*, New York, NY: McGraw-Hill, 1982.

Young, Y.C., Roark's Formula for Stress and Strain, 6th ed., New York, NY: McGraw-Hill, 1989.

Appendix A

Second Moments of Area

The second moment of area, I, sometimes called the area moment of inertia, is a property of a shape that describes its resistance to deformation by bending. The polar second moment of area, J, often called the polar moment of inertia, describes the resistance of a shape to deformation by torsion. Since the coordinate axes used to obtain the I's and J's listed here run through the centroid of each shape, all moments of area cited here may be thought of as having an additional subscript c denoting that they are taken relative to the centroid.

Remember the following:

$$I_y = \int z^2 dA.$$

$$I_z = \int y^2 dA.$$

$$J = \int r^2 dA.$$

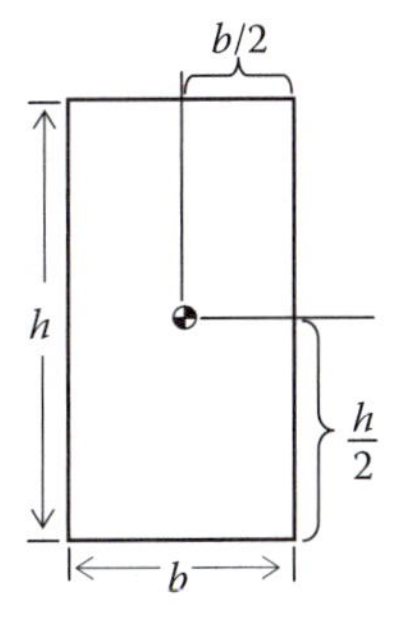

Area (A)	Second Moment of Area (I)	Polar Second Moment of Area (J)
bh	$I_x = bh^3/12$ $I_y = hb^3/12$ $I_{xy} = 0$	$(bh^3/12)\,(h^2+b^2)$

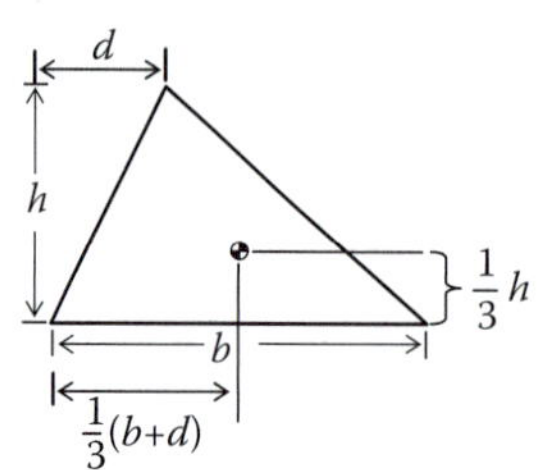

Area (A)	Second Moment of Area (I)	Polar Second Moment of Area (J)
$bh/2$	$I_x = bh^3/36$ $I_{xy} = bh^2(b-2d)/72$	

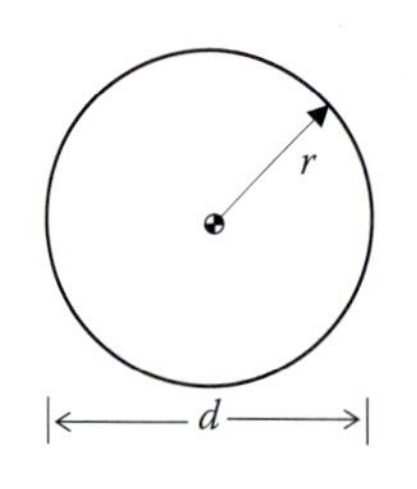

Area (A)	Second Moment of Area (I)	Polar Second Moment of Area (J)
πr^2	$I_x = I_y = \pi r^4/4$ $= \pi d^4/64$ $I_{xy} = 0$	$J = \pi r^4/2$ $= \pi d^4/32$

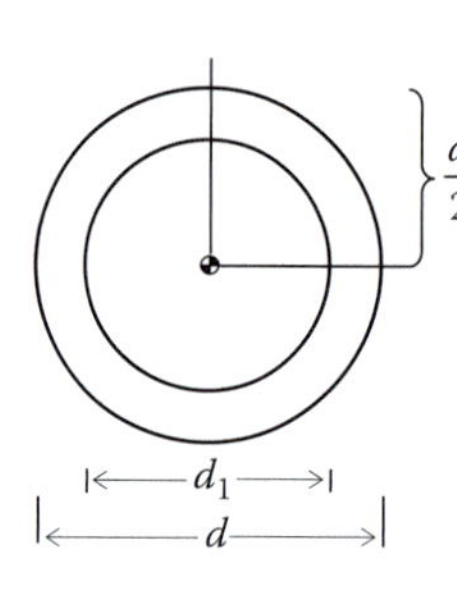

Area (A)	Second Moment of Area (I)	Polar Second Moment of Area (J)
$\pi(d^2-d_1^{\,2})/4$	$I_x = I_y = \pi(d^4-d_1^{\,4})/64$ $I_{xy} = 0$	$\pi(d^4-d_1^{\,4})/32$

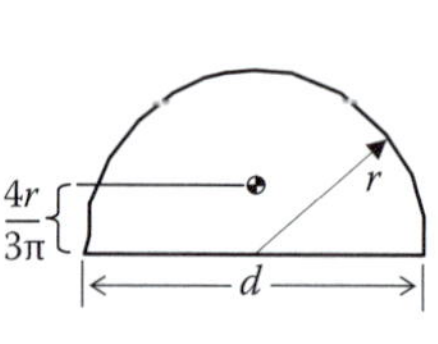

Area (A)	Moment of Inertia (I)	Polar Second Moment of Area (J)
$\pi r^2/2$	$I_x = 0.1098r^4$ $I_y = \pi r^4/8$ $I_{xy} = 0$	$J_{CG} = I_x + I_y$ $J_o = \pi r^4/4$

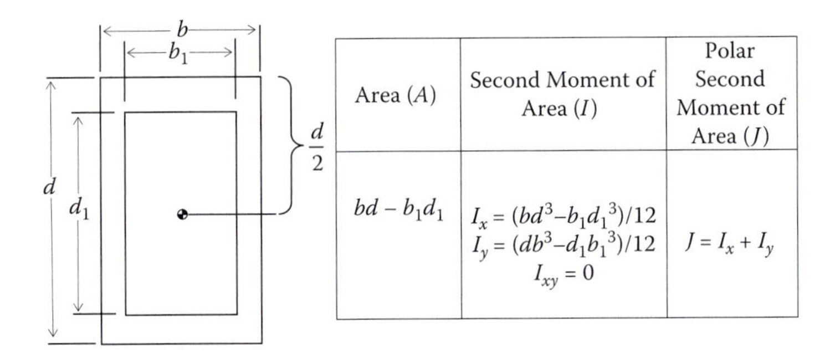

Area (A)	Second Moment of Area (I)	Polar Second Moment of Area (J)
$bd - b_1d_1$	$I_x = (bd^3 - b_1d_1^3)/12$ $I_y = (db^3 - d_1b_1^3)/12$ $I_{xy} = 0$	$J = I_x + I_y$

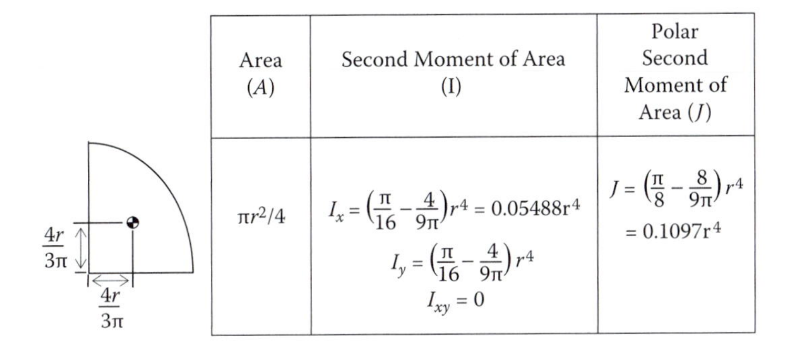

Area (A)	Second Moment of Area (I)	Polar Second Moment of Area (J)
$\pi r^2/4$	$I_x = \left(\frac{\pi}{16} - \frac{4}{9\pi}\right) r^4 = 0.05488r^4$ $I_y = \left(\frac{\pi}{16} - \frac{4}{9\pi}\right) r^4$ $I_{xy} = 0$	$J = \left(\frac{\pi}{8} - \frac{8}{9\pi}\right) r^4$ $= 0.1097r^4$

Appendix B

A Quick Look at the Del Operator

We use the *del operator* to take the gradient of a scalar function, say $f(x, y, z)$:

$$\nabla f = \hat{i}\frac{\partial f}{\partial x} + \hat{j}\frac{\partial f}{\partial y} + \hat{k}\frac{\partial f}{\partial z}.$$

If we "factor out" the function f, the gradient of f looks like

$$\nabla f = \left(\hat{i}\frac{\partial}{\partial x} + \hat{j}\frac{\partial}{\partial y} + \hat{k}\frac{\partial}{\partial z}\right) f.$$

The term in parentheses is called *del* and is written as

$$\nabla = \hat{i}\frac{\partial}{\partial x} + \hat{j}\frac{\partial}{\partial y} + \hat{k}\frac{\partial}{\partial z}.$$

By itself, ∇ has no meaning. It is meaningful only when it acts on a scalar function. The term ∇ *operates* on scalar functions by taking their derivatives and combining them into the gradient. We say that ∇ is a vector operator acting on scalar functions, and we call it the *del operator.*

Since ∇ resembles a vector, we consider all the ways that we can act on vectors and see how the del operator acts in each case.

Vectors		Del	
Operation	**Result**	**Operation**	**Result**
Multiply by a scalar a	Aa	Operate on a scalar f	∇f
Dot product with another vector B	$A \cdot B$	Dot product with vector $F(x, y, z)$	$\nabla \cdot \underline{F}$
Cross product with another vector B	$A \times B$	Cross product with vector $F(x, y, z)$	$\nabla \times \underline{F}$

Divergence

Let's first compute the form of the divergence in regular Cartesian coordinates. If we let a random vector

$$\underline{F} = F_x\hat{i} + F_y\hat{j} + F_z\hat{k},$$

then we define

$$\operatorname{div}\underline{F} = \nabla\cdot\underline{F} = \left(\hat{i}\frac{\partial}{\partial x} + \hat{j}\frac{\partial}{\partial y} + \hat{k}\frac{\partial}{\partial z}\right)\cdot\left(F_x\hat{i} + F_y\hat{j} + F_z\hat{k}\right) = \frac{\partial F_x}{\partial x} + \frac{\partial F_y}{\partial y} + \frac{\partial F_z}{\partial z}.$$

Like any dot product, the divergence is a scalar quantity. Also note that, in general, div *F* is a function and changes in value from point to point.

Physical Interpretation of the Divergence

The divergence quantifies how much a vector field spreads out, or diverges, from a given point *P*. For example, the figure on the left (Figure B.1) has positive divergence at *P*, since the vectors of the vector field are all spreading as they move away from *P*. The figure in the center has zero divergence everywhere since the vectors are not spreading out at all. This is also easy to compute, since the vector field is constant everywhere and the derivative of a constant is zero. The field on the right has negative divergence since the vectors are coming closer together instead of spreading out.

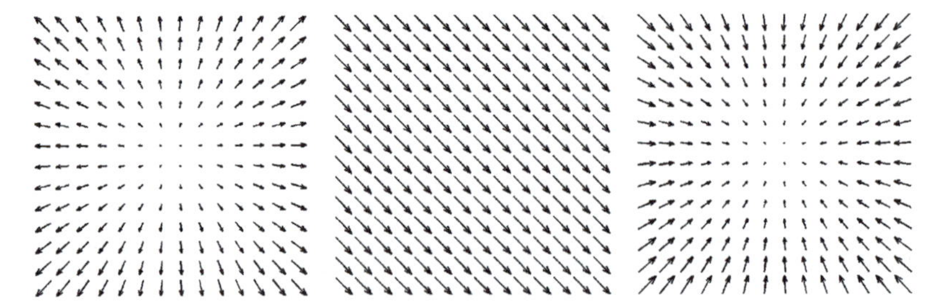

FIGURE B.1
Vector Fields

In the context of continuum mechanics, the divergence has a particularly interesting meaning. For solids, if the vector field of interest is the displacement vector *U*, the divergence of this vector tells us about the overall *change in volume* of the solid. See equation (3.5) and homework problem (3.2), both in Chapter 3, in this textbook. When we have $\nabla\cdot\underline{U} = 0$ we know that the volume of a given solid body remains constant, and we can call the solid *incompressible*. For fluids, we use the velocity vector V to talk about the deformation kinematics. The divergence of the velocity vector tells us about the volumetric strain rate, and when we have $\nabla \cdot \underline{V} = 0$ we say that the flow is incompressible. This generally allows us to neglect changes in fluid density and say that density remains constant (Chapter 8, equation 8.9).

Example

Calculate the divergence of

$$\underline{F} = x\hat{i} + y\hat{j} + z\hat{k}.$$

$$\nabla \cdot \underline{F} = \frac{\partial}{\partial x}(x) + \frac{\partial}{\partial y}(y) + \frac{\partial}{\partial z}(z) = 1 + 1 + 1 = 3.$$

This is the vector field shown on the left om Figure B.1. Its divergence is constant everywhere.

Curl

We can also compute the curl in Cartesian coordinates. Again, let

$$\underline{F} = F_x\hat{i} + F_y\hat{j} + F_z\hat{k},$$

and calculate

$$\text{curl } \underline{F} = \nabla \times \underline{F} = \begin{vmatrix} \hat{i} & \hat{j} & \hat{k} \\ \frac{\partial}{\partial x} & \frac{\partial}{\partial y} & \frac{\partial}{\partial z} \\ F_x & F_y & F_z \end{vmatrix} = \hat{i}\left(\frac{\partial F_z}{\partial y} - \frac{\partial F_y}{\partial z}\right) + \hat{j}\left(\frac{\partial F_x}{\partial z} - \frac{\partial F_z}{\partial x}\right) + \hat{k}\left(\frac{\partial F_y}{\partial x} - \frac{\partial F_x}{\partial y}\right).$$

Not surprisingly, the curl is a vector quantity.

Physical Interpretation of the Curl

The curl of a vector field measures the tendency of the vector field to swirl. Consider the illustrations in Figure B.2. The field on the left, called F, has curl with positive $\hat{k}$ component. To see this, use the right hand rule. Place your right hand at P. Point your fingers toward the tail of one of the vectors of F. Now curl your fingers around in the direction of the tip of the vector. Stick your thumb out. Since it points toward the +z axis (out of the page), the curl has a positive $\hat{k}$ component.

The second vector field G has no visible swirling tendency at all so we would expect $\nabla \times \underline{G} = 0$. The third vector field doesn't look like it swirls either, so it also has zero curl.

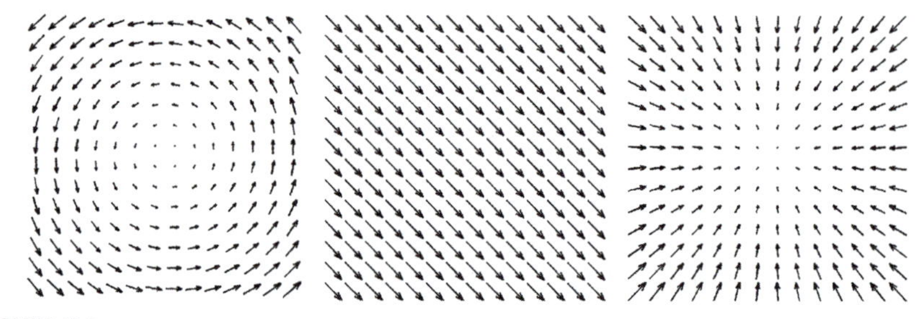

FIGURE B.2
Vector Fields

Examples

Example 1

Compute the curl of $\underline{F} = -y\hat{i} + x\hat{j}$.

$$\nabla \times \underline{F} = \begin{vmatrix} \hat{i} & \hat{j} & \hat{k} \\ \dfrac{\partial}{\partial x} & \dfrac{\partial}{\partial y} & \dfrac{\partial}{\partial z} \\ -y & x & 0 \end{vmatrix} = 2\hat{k}.$$

This is the vector field on the left in Figure B.2. As you can see, the analytical approach demonstrates that the curl is in the positive $\hat{k}$ direction, as expected.

Example 2

Compute the curl of $\underline{H} = x\hat{i} + y\hat{j} + z\hat{k}$, or $\underline{H}(\underline{r}) = \underline{r}$.

$$\nabla \times \underline{H} = \begin{vmatrix} \hat{i} & \hat{j} & \hat{k} \\ \dfrac{\partial}{\partial x} & \dfrac{\partial}{\partial y} & \dfrac{\partial}{\partial z} \\ x & y & z \end{vmatrix} = \underline{0}.$$

This, as you've probably guessed, is the vector field on the far right in Figure B.2.

Laplacian

The divergence of the gradient appears so often that it has been given a special name: the *Laplacian*. It is written as ∇^2 or Δ and, in Cartesian components, has the form

$$\nabla^2 f = \frac{\partial^2 f}{\partial x^2} + \frac{\partial^2 f}{\partial y^2} + \frac{\partial^2 f}{\partial z^2}.$$

It operates on scalar functions and produces a scalar result. When we take the Laplacian of a vector field,

$$\underline{F} = F_x \hat{i} + F_y \hat{j} + F_z \hat{k},$$

we get

$$\nabla^2 \underline{F} = (\nabla^2 F_x)\hat{i} + (\nabla^2 F_y)\hat{j} + (\nabla^2 F_z)\hat{k}.$$

Appendix C

Property Tables

TABLE C1
Typical Properties of Engineering Materials (SI)

		Yield Strength		Moduli			
Material	Density (kg/m³)	Tension (MPa)	Shear (MPa)	E (GPa)	G (GPa)	α (10^{-6}/°C)	Poisson's Ratio ν
Steel							
Structural	7860	250	145	200	77.2	11.7	
Stainless (cold-rolled)	7920	520		190	75	17.3	
Stainless (annealed)	7920	260	150	190	75	17.3	
Gray cast iron	7200			69	28	12.1	0.2–0.3
Malleable cast iron	7300	230		165	65	12.1	0.25
Aluminum							
1100-H14	2710	95	55	70	26	23.6	0.33
6061-T6	2710	240	140	70	26	23.6	0.33
7075-T6	2800	500		72	28	23.6	0.33
Copper							
Annealed	8910	70		120	44	16.9	0.34
Yellow-brass, annealed (65% Cu, 35% Zn)	8470	100	60	105	39	20.9	0.33
Red-brass, annealed (85% Cu, 15% Zn)	8740	70		120	44	18.7	0.33
Tin bronze	8800	145		95		18.0	0.35
Magnesium alloy AZ31	1770	200		45	16	25.2	0.34
Titanium (6% Al, 4% V)	4730	830		115		9.5	0.33
Timber							
Douglas fir	470	56		13	0.7	Varies	
White oak	690	58		12			
Redwood	415			9			
Concrete (high strength)	2320			30		9.9	0.1–0.2
Plastics							
Nylon (molding compound)	1140	45		2.8		144	0.4
Polycarbonate	1200	62		2.4		122	
Polyester, PBT	1340	55		2.4		135	
Polystyrene	1030	55		3.1		135	
Vinyl, rigid PVC	1440	45		3.1		135	
Rubber	910					162	0.44–0.5
Granite (average)	2770			70	4	7.2	
Sandstone (average)	2300			40	2	9.0	
Glass, 98% silica	2190			54	4.1	80	0.2–0.27
Kevlar				130			
Human tendon				2			

TABLE C2

Typical Properties of Engineering Materials (U.S.)

Material	Specific Weight (lbf/in^3)	Yield Strength Tension (ksi)	Yield Strength Shear (ksi)	Moduli E (10^6 psi)	Moduli G (10^6 psi)	α (10^{-6}/°F)	Poisson's Ratio ν
Steel							
Structural	0.284	36	21	29	11.2	6.5	
Stainless (cold-rolled)	0.286	75		28	10.8	6.5	
Stainless (annealed)	0.286	38	22	28	10.8	6.5	
Gray cast iron	0.260			10	4.1	6.7	0.2–0.3
Malleable cast iron	0.264	33		24	9.3	6.7	0.25
Aluminum							
1100-H14	0.098	14	8	10.1	3.7	13.1	0.33
6061-T6	0.098	35	20	10.1	3.7	13.1	0.33
7075-T6	0.101	73		10.4	4	13.1	0.33
Copper							
Annealed	0.322	10		17	6.4	9.4	0.34
Yellow-brass, annealed (65% Cu, 35% Zn)	0.306	15	9	15	5.6	11.6	0.33
Red-brass, annealed (85% Cu, 15% Zn)	0.316	63		17	6.4	10.4	0.33
Tin bronze	0.318	21		14		10	0.35
Magnesium alloy AZ31	0.064	29		6.5	2.4	14	0.34
Titanium (6% Al, 4% V)	0.161	120		16.5		5.3	0.33
Timber							
Douglas fir	0.017	8.1		1.9		Varies	
White oak	0.025	8.4		1.8			
Redwood	0.015			1.3			
Concrete (high strength)	0.084			4.5		5.5	0.1–0.2
Plastics							
Nylon (molding compound)	0.0412	6.5		0.4		80	0.4
Polycarbonate	0.0433	9		0.35		68	

TABLE C2

Typical Properties of Engineering Materials (U.S.)

Material	Specific Weight (lbf/in³)	Yield Strength		Moduli		α (10^{-6}/°F)	Poisson's Ratio ν
		Tension (ksi)	Shear (ksi)	E (10^6 psi)	G (10^6 psi)		
Polyester, PBT	0.0484	8		0.35		75	
Polystyrene	0.0374	8		0.45		70	
Vinyl, rigid PVC	0.0484	6.5		0.45		75	
Rubber	0.033					90	0.44–0.5
Granite (average)	0.100			10	4	4	
Sandstone (average)	0.083			6	2	5	
Glass, 98% silica	0.079			9.6	4.1	44	0.2–0.27
Kevlar				13.8			
Human tendon				0.27			

Note: Exact values of these properties vary widely with changes in composition, heat treatment, and mechanical working. More precise data are available from manufacturers.

TABLE C3

Typical Properties of Common Fluids (SI)

Material	Temperature T(°C)	Density ρ (kg/m³)	Viscosity μ (Ns/m²)	Surface Tension σ (N/m)
Water	0	1000	1.75×10^{-3}	0.0757
	10	1000	1.30×10^{-3}	0.0742
	20	998	1.00×10^{-3}	0.0727
	30	996	7.97×10^{-4}	0.0712
	40	992	6.51×10^{-4}	0.0696
	50	988	5.44×10^{-4}	0.0679
	60	983	4.63×10^{-4}	0.0662
	70	978	4.00×10^{-4}	0.0645
	80	972	3.51×10^{-4}	0.0627
	90	965	3.11×10^{-4}	0.0608
	100	958	2.79×10^{-4}	0.0589
Air	0	1.29	1.72×10^{-5}	
	10	1.25	1.77×10^{-5}	
	20	1.21	1.81×10^{-5}	
	30	1.17	1.86×10^{-5}	
	40	1.13	1.91×10^{-5}	
	50	1.09	1.95×10^{-5}	
	60	1.06	2.00×10^{-5}	
	70	1.03	2.04×10^{-5}	
	80	1.00	2.09×10^{-5}	
	90	0.973	2.13×10^{-5}	
	100	0.947	2.17×10^{-5}	

TABLE C4

Typical Properties of Common Fluids (U.S.)

Material	Temperature T(°F)	Density ρ (slug/ft³)	Viscosity μ (lbf·s/ft²)	Surface Tension σ (lbf/ft)
Water	32	1.94	3.66×10^{-5}	0.00519
	40	1.94	3.19×10^{-5}	0.00514
	50	1.94	2.72×10^{-5}	0.00509
	60	1.94	2.34×10^{-5}	0.00503
	70	1.93	2.04×10^{-5}	0.00498
	80	1.93	1.79×10^{-5}	0.00492
	90	1.93	1.59×10^{-5}	0.00486
	100	1.93	1.42×10^{-5}	0.00480
	212	1.86	5.83×10^{-6}	0.00404
Air	40	0.00247	3.63×10^{-7}	
	50	0.00242	3.69×10^{-7}	
	60	0.00237	3.75×10^{-7}	
	70	0.00233	3.80×10^{-7}	
	80	0.00229	3.86×10^{-7}	
	90	0.00225	3.91×10^{-7}	
	100	0.00221	3.97×10^{-7}	
	200	0.00187	4.48×10^{-7}	

Note: Properties of air are obtained at atmospheric pressure.

Appendix D

All the Equations

	Solids	Fluids
Kinematics What's the vector (u, v, w) everything depends on?	displacement	velocity (i.e.,displacement rate)
Volume change/volume change rate	$\nabla \cdot \underline{U} = \frac{\partial u}{\partial x} + \frac{\partial v}{\partial y} + \frac{\partial w}{\partial z}$	$\nabla \cdot \underline{V} = \frac{\partial u}{\partial x} + \frac{\partial v}{\partial y} + \frac{\partial w}{\partial z}$
Strain/strain rate	$\underline{\underline{\varepsilon}} = \varepsilon_{ij} = \frac{1}{2}\left(\frac{\partial u_i}{\partial x_j} + \frac{\partial u_j}{\partial x_i}\right)$	$\underline{\underline{\varepsilon}} = \varepsilon_{ij} = \frac{1}{2}\left(\frac{\partial u_i}{\partial x_j} + \frac{\partial u_j}{\partial x_i}\right)$
1D Constitutive Law Hookean/Newtonian	$\sigma = E\varepsilon$ $\tau = G\gamma$	$\tau = \mu\gamma$
3D ideal constitutive law	$\tau_{ij} = \lambda\varepsilon_{\alpha\alpha}\delta_{ij} + 2G\varepsilon_{ij}$ $\lambda = fcn(E, \nu)$	$\tau_{ij} = -p\delta_{ij} + 2\mu\varepsilon_{ij}$
General constitutive law	$\tau_{ij} = K_{ijmn}\varepsilon_{mn}$	$\tau_{ij} = K_{ijmn}\varepsilon_{mn}$
Conservation of mass	$\rho = constant$	$\frac{\partial \rho}{\partial t} + \nabla \cdot (\rho \underline{V}) = 0$
Conservation of linear momentum ($\underline{F} = m\underline{a}$)	$\underline{B} + \nabla \cdot \underline{\underline{\tau}} = \rho \underline{a}$	$\underline{B} + \nabla \cdot \underline{\underline{\tau}} = \rho \underline{a}$
B represents the total body force on the element in question, most often represented by ρg. .		
Conservation of angular momentum for an infinitesimal element	$\tau_{ij} = \tau_{ji}$	$\tau_{ij} = \tau_{ji}$

INDEX

D

F

N

T

U